Otoacoustic Emissions:
Principles, Procedures, and Protocols

Core Clinical Concepts in Audiology

Series Editor
Brad Stach, PhD

Basic Audiometry
Editors
Jay Hall & Virginia Ramachandran

Pure-Tone Audiometry and Masking by Maureen Valente, PhD
Basic Audiometry Learning Manual by Mark DeRuiter, PhD and Virginia Ramachandran, AuD

Electrodiagnostic Audiology
Editors
Jay Hall & Virginia Ramachandran

Objective Assessment of Hearing by James W. Hall, III, PhD and De Wet Swanepoel, PhD
Otoacoustic Emissions: Principles, Procedures, and Protocols by Sumitrajit Dhar, PhD
and James W. Hall, III, PhD

Cochlear Implants
Editors
Terry Zwolan & Jace Wolfe

Programming Cochlear Implants by Jace Wolfe, PhD and Erin C. Schafer, PhD

Vestibular Assessment
Editors
Ken Bouchard & Virginia Ramachandran

Vestibular Learning Manual by Bre L. Myers, AuD

Otoacoustic Emissions:
Principles, Procedures, and Protocols

Sumitrajit Dhar, PhD
James W. Hall, III, PhD

PLURAL
PUBLISHING
INC.

SAN DIEGO
OXFORD
BRISBANE

5521 Ruffin Road
San Diego, CA 92123

e-mail: info@pluralpublishing.com
Web site: http://www.pluralpublishing.com

49 Bath Street
Abingdon, Oxfordshire OX14 1EA
United Kingdom

FSC
Mixed Sources
Product group from well-managed
forests and other controlled sources

Cert no. SW-COC-002283
www.fsc.org
© 1996 Forest Stewardship Council

Typeset in 11/13 Palatino by Flanagan's Publishing Services, Inc.
Printed in the United States of America by McNaughton & Gunn

ISBN-13: 978-1-59756-342-0
ISBN-10: 1-59756- 342-0

Library of Congress Cataloging-in-Publication Data:
Dhar, Sumitrajit.
  Otoacoustic emissions : principles, procedures, and protocols / Sumitrajit Dhar and James W. Hall.
  p. ; cm. — (Core clinical concepts in audiology)
  Includes bibliographical references and index.
  ISBN-13: 978-1-59756-342-0 (alk. paper)
  ISBN-10: 1-59756-342-0 (alk. paper)
  1. Otoacoustic emissions. 2. Cochlea. I. Hall, James W. (James Wilbur), 1948- II. Title. III. Series: Core clinical concepts in audiology.
  [DNLM: 1. Cochlea—physiopathology. 2. Otoacoustic Emissions, Spontaneous. 3. Acoustic Stimulation—methods. WV 250]
  RF294.5.O76D43 2010
  617.8'822—dc22
        2011000250

Contents

Foreword

How can we make learning in audiology more effective? This is the question we begin with in designing the Core Clinical Concepts in Audiology series. Our answer is revealed in the construction of the books of the series. Herein we seek to provide palatable and useful information to students and practitioners to develop and refine clinical skills for audiology practice.

By and large, texts available for our field provide exhaustive examination of broad topic areas. Although these texts are useful and necessary for advanced scholarship, we currently lack pedagogic materials that focus on basic clinical methods and knowledge. The books in this series are designed for teaching and learning.

These books are written for the student. The scope of practice for audiology dramatically has expanded since the inception of our field. Today's students must acquire a tremendous arsenal of clinical skills and knowledge in a very short period of time. The books of the CCC series are meant to be clear and comprehensible to students, focusing on the content necessary to achieve knowledge and skills for clinical practice. Furthermore, the books are designed to be economical, both financially and in time spent in learning.

These books are written for the clinician. With expansion of the scope of audiology practice, currently practicing clinicians must acquire new skill sets while continuing to serve their patients—not a small feat. Hardworking practitioners deserve educational materials compatible with the real-world demands of fast-paced and time-limited clinical practice. In response to these needs, the books of the CCC series are designed to be concise. The succinct construction of the series is meant to allow readers efficiently to acquire the essential concepts and skills described in the books.

These books are written for the instructor. Most instructors of audiology courses are familiar with the frustration of searching for materials that cover the topics that reflect the learning outcomes of their courses. Especially lacking are materials designed to promote clinical learning. The books of the CCC series are designed to focus on specific areas of clinical practice. They are targeted toward the learning outcomes commonly found in audiology curricula. Due to the economical nature of the books, instructors can feel comfortable in creatively combining different Core Clinical Concepts in Audiology books to support the unique and diverse learning demands of specific courses.

These books are written for the user. The needs of the reader are our primary concerns. These books aim to help readers learn to be outstanding clinical audiologists. To be sure, these are lofty goals. The authors of the CCC series books have put forth their best effort to accomplish these goals.

Otoacoustic Emissions: Principles, Procedures, and Protocols, by Sumitrajit Dhar and James W. Hall III, is a readable yet comprehensive source of information on otoacoustic emissions (OAEs). It begins with a succinct overview of OAEs and a fascinating historical description of their discovery and emergence as a clinical tool. A chapter is devoted to the anatomic and physiologic underpinnings of OAEs, with an emphasis on cochlear processing important in their generation. Students and clinical audiologists alike will appreciate the way the authors clearly explain and concisely review current research findings on the origins of OAEs and changing perspectives on OAE taxonomy. Another chapter focuses on the important topic of OAE instrumentation, and the often overlooked but critical topic of instrument calibration.

The book includes two chapters offering a detailed yet practical review of the measurement and analysis of transient and distortion product OAEs, with ample reference to screen displays and other features of modern clinical OAE devices. These chapters are followed by two others that provide an up-to-date literature review highlighting all major evidence-based clinical applications of OAEs in children and in adults. Another chapter explores current thinking on the usually neglected efferent auditory pathways, and the role of OAEs in evaluating this important component of the auditory system. The book concludes with an exciting glimpse into the future as the authors introduce the reader to new directions in OAE research and clinical application.

As with the other books of the CCC series, the organization and construction of the book works to provide important and necessary information in a manner consistent with the needs of readers.

James W. Hall III, PhD
Virginia Ramachandran, AuD
Series Editors

Preface

Thirty years ago, beginning in 1980, several groups of hearing scientists independently demonstrated that outer hair cells can elongate and contract. The rather revolutionary discovery of outer hair cell motility suggested an anatomic and physiologic explanation for the generation of otoacoustic emissions, first reported by David Kemp in 1978. Although outer hair cell motility clearly plays an important role in the production of OAEs, ongoing investigations for the past 30 years have yielded a vast amount of information, and even some controversy, as to the precise mechanisms underlying the origins of OAEs. Pursuit of an exact understanding of OAE generation and propagation profoundly will influence their clinical application.

The rather simplistic early classification of OAEs as either spontaneous or evoked has given way to a more complex taxonomy. Importantly, there appear to be major differences in the way TEOAEs and DPOAEs are generated. With the more recently proposed classification system, OAEs are categorized based on their mechanism of generation and grouped into those arising from a nonlinear mechanism and those arising from a linear reflection mechanism. A basic understanding of cochlear physiology is necessary to understand the distinction between these two modes of OAE generation. Unfortunately, the clinical audiologist is hard-pressed to find an up-to-date, straightforward, and clinically focused source of information on the mechanisms of OAE generation. With this in mind, we included in *Otoacoustic Emissions: Principles, Procedures, and Protocols* a clinically oriented review of current explanations for the generation of OAEs, including the latest thinking on the taxonomy of OAEs.

Within the past three decades, over 3,000 articles have appeared in peer-reviewed literature providing evidence in support of dozens of clinical applications of OAEs. For the first time, we summarize this vast amount of information so audiologists and other hearing health professionals can make rational decisions about why and how to use OAEs with children and adults in the clinical setting. Accumulated experience with OAEs has led to some proven procedures and protocols for clinical measurement and analysis, as suggested by the title of our book, *Otoacoustic Emissions: Principles, Procedures, and Protocols*. The vast clinical literature pertaining to OAEs in cochlear pathophysiology in children and adults is also summarized in two chapters of the book.

A book on OAEs would not be complete without a discussion of OAE suppression as a clinical tool. Also, mention must be made of advances in technology that permit the combined and integrated measurement of OAEs and other time-tested clinical techniques. We address each of these important topics in *Otoacoustic Emissions: Principles, Procedures, and Protocols*. Thirty years after their discovery, we are still learning more about the multiple mechanisms responsible for generation of OAEs and, at the same time, witnessing a consistent expansion and refinement of clinical applications. As clinical audiologists and clinical researchers, we gladly convey information and excitement about OAEs to our colleagues in the form of our book, *Otoacoustic Emissions: Principles, Procedures, and Protocols*.

Sumitrajit Dhar, PhD
James W. Hall III, PhD

In memory of

Roger A. Ruth, PhD
September 2, 1950–June 13, 2009
Our audiology colleague, trusted mentor, and beloved friend.

III 1 III

Overview of Otoacoustic Emissions

Introduction

Some delay between the discovery of a new technique for assessing auditory function and its first clinical application is usual and perhaps inevitable. Otoacoustic emissions (OAEs) were no exception. Approximately 5 years separated David Kemp's classic publication describing the first recording of OAEs in 1978 and the publication of peer-reviewed papers describing their use in infant hearing screening (e.g., Johnsen, Bagi, & Elberling, 1983). In the interim, an international collection of scientists and clinicians had initiated independent lines of investigation examining the physiology, biophysics, and clinical applications of OAEs.

Although every new discovery is accompanied by bold predictions of widespread utility in a variety of patient populations, in the long run, few technologies are able to live up to this initial euphoria. Even for techniques that eventually work their way into the clinical test battery, a second delay usually occurs between the first clinical application and widespread clinical acceptance. Again, OAEs were no exception. Another decade passed before OAEs were commonplace in clinical settings. By the mid-1990s, clinical audiologists could purchase an OAE device from a variety of manufacturers and, in the United States, two OAE billing codes were approved. Although David Kemp is certainly responsible for initiating the last phase of this journey to the addition of OAEs in the clinician's arsenal, the voyage began almost two centuries earlier.

Early Contributions: Tartini, Other Musicians, and Psychophysicists

A great wealth of knowledge has now accumulated about OAEs, as evidenced by the information in this textbook. When asked about the origins of this knowledge, most students and practitioners in the field trace the history of OAEs back to David Kemp or farther back to Thomas Gold (their contributions are discussed below). However, the knowledge that our ears generate sounds had existed much before the time of Kemp or Gold, with its first documentation dating back to the 1700s. That story begins with Guiseppe Tartini (Figure 1–1).

Tartini was born on April 8, 1692 in the town of Piran in current-day Slovenia, then a part of the Istrian peninsula in the Republic of Venice. Music historians believe that he received general musical training in his childhood but music was not the focus of his life and career until much later. Tartini studied law at the University of Padua

FIGURE 1–1. Statue of Giuseppe Tartini by Antonio dal Zotto, Tartini Square, Piran, Slovenia. From Wikimedia Commons. Photograph by Stephen Turner. Permission is granted to copy, distribute, and/or modify this document under the terms of the GNU Free Documentation License, Version 1.2 or any later version published by the Free Software Foundation; with no Invariant Sections, no Front-Cover Texts, and no Back-Cover Texts. A copy of the license can be viewed at http://en.wikipedia.org/wiki/GNU_Free_Documenta-

and became an adept fencer during his time at the university. Tartini's early life was far removed from serious music. In fact, we would not even mention Tartini in passing if he had continued on the same path. However, everything changed when he married Elisabetta Premazone in 1710. Elizabetta was a favorite of a powerful cardinal by the name of Georgio Cornaro who promptly charged Tartini with abduction. To escape prosecution, Tartini fled Padua and hid at the monastery of St. Francis in Assisi. It was here that Tartini became a serious student of the violin.

Tartini's skills as a violinist improved so greatly that he was appointed the Maestro di Capella at the Basilica di Sant'Antonio in Padua by 1721. Incidentally, legend has it that Tartini also was the first known owner of a violin made by Antonio Stradivari in 1715. In 1726, Tartini started a violin school, which attracted students from all over Europe, and gradually Tartini became more interested in harmony and the acoustics of music, which led to various treatises authored by him after 1750.

A signature of Tartini's music is the double stop trill where the performer plays two notes simultaneously, often in rapid succession. Considered a difficult skill to master even by modern standards, legend has it that Tartini had six fingers that allowed him to play these trills with relative ease. One of Tartini's better-known compositions is the Violin Sonata in G Minor, also known as the Devil's Trill Sonata, for the frequent use of these trills. While playing these double stop trills, Tartini recognized the presence of audible notes that were not being produced by the violin. He concluded that these notes must be generated within the ear, and he started using them regularly in his compositions. Essentially, he would have the ear "fill in" to create a sensation of more sound than the violin was producing. Other musicians, such as the German organist Sorge and the French composer Romieu, recognized, researched, and used these ear-generated sounds in their music. Among musicians and scientists interested in hearing, these extra sounds became known as the Tartini tones. What were known as Tartini tones in the 1700s are distortion product OAEs today. The reader is directed to Plomp (1976) for a detailed early history of Tartini tones.

Soon after the musicians were aware of "Tartini tones," psychoacousticians became interested in the phenomenon. Perhaps the earliest reports in this area are by Vieth in the early 1800s (Plomp, 1976). Vieth coined the term *combination tones* to describe the sensation of extra tones generated in the ear. The possibility of combination tones being generated in the inner ear was incompatible with the dominant model of auditory physiology of that time—that proposed by Hermann Ludwig Ferdinand von Helmholtz (1821–1894), a German physician and physicist.

Helmholtz postulated that the ear essentially worked as a frequency analyzer, decomposing complex signals into their constituent elements before passing them on to the brain. Because this was a linear model, it could not account for the generation of additional sounds within the ear. In order to accommodate combination tones, Helmholtz introduced an "overloading" type of nonlinear component in the middle ear based on the assumption that the displacement of any elastic body is linearly related to the incident pressure only for infinitesimally small amplitudes. Thus, for larger pressures nonlinearities should be expected in the ear (Plomp, 1976).

Psychoacousticians continued to be interested in combination tones and used them to explore the nonlinearities of the inner ear noninvasively. Between about 1950 and 1980, psychoacousticians developed complex experimental techniques to record combination tones from human subjects and used these data to try to understand the source and characteristics of these tones (see Goldstein, 1970, for a review). It was indeed through psychoacoustic experiments that David Kemp (Figure 1–2) became interested in this area of work and later performed the first physiologic recordings of OAEs.

FIGURE 1–2. David Kemp, a British physicist who discovered otoacoustic emissions.

The Prophet Thomas Gold

New ideas in science are not always right just because they are new. Nor are the old ideas always wrong just because they are old. A critical attitude is clearly required of every scientist. But what is required is to be equally critical to the old ideas as to the new. Whenever the established ideas are accepted uncritically, but conflicting new evidence is brushed aside and not reported because it does not fit, then that particular science is in deep trouble—and it has happened quite often in the historical past. If we look over the history of science, there are very long periods when the uncritical acceptance of the established ideas was a real hindrance to the pursuit of the new. (Gold, 1989, p. 103)

Dr. Thomas Gold (Figure 1–3) wrote these words while chair of the Department of Astronomy at

Cornell University in Ithaca, New York more than 40 years after his innovative investigations in Cambridge England of cochlear physiology.

The theme in this passage, however, accurately reflects the approach taken by Gold in the late 1940s when he challenged predominant theories of passive linear cochlear function espoused by Nobel Prize winner Georg von Békésy and, before him, the 19th century scientific giant Helmholtz. Another quote from a 1948 article by Thomas Gold lays the foundation for the subsequent discovery of OAEs 30 years later by another British physicist Dr. David Kemp.

It is shown that the assumption of a "passive" cochlea, where elements are brought into mechanical oscillation solely by means of the incident sound, is not tenable. The degree of resonance of the elements of the cochlea can be measured, and the results are not compatible with the very heavy damping which must arise from the viscosity of the liquid. For this

FIGURE 1–3. Thomas Gold, a British physicist who in the late 1940s recognized the nonlinear nature of the cochlea.

eries and their less-than-enthusiastic reception by Dr. von Békésy. With a simple Internet search, readers will find a wealth of information about Dr. Gold and his many accomplishments.

David Kemp, Discoverer of OAEs

Dr. David Kemp (see Figure 1–2) alone can be credited with the discovery of OAEs. He became involved in auditory research in a roundabout way, beginning with studies in general physics, electronics, and atmospheric physics, then industrial noise control. In the mid-1970s, Dr. Kemp conducted a series of psychoacoustic and then physiologic investigations on basic cochlear function that confirmed the presence of active mechanisms that could produce energy. In his truly classic paper on OAEs in 1978, Kemp unequivocally showed that, following stimulation with tones or with clicks, additional sound could be measured with a small microphone in the external ear canal of animal models and humans.

Kemp's article is filled with meaty yet characteristically restrained quotations summarizing his breakthrough discovery. For example,

> . . . the response [sic OAEs] appears to have its origin in some nonlinear mechanism probably located in the cochlea, responding mechanically to auditory stimulation, and dependent upon the normal functioning of the cochlea transduction process. (Kemp, 1978, p. 1386)

And also,

> In the absence of a complete understanding of the mode of action of the sensory cells in the cochlea, it is tempting to suggest that one of the functions of the outer hair cell population is generation of this mechanical energy. If a cochlear origin is confirmed by experiments currently in progress, the technique developed in this study will provide a new avenue for investigation of the auditory system, with applications in both research and audiological medicine. (Kemp, 1978, p. 1391)

reason the "regeneration hypothesis" is put forward, and it is suggested that an electromechanical action takes place whereby a supply of electrical energy is employed to counteract the damping. (Gold, 1948, p. 492)

The interposition of a feedback stage . . . makes construction possible where the nerve ending abstracts much energy from a mechanical resonator. (Gold, 1948, p. 498)

Like a good Hollywood movie, the fascinating story of Thomas Gold (see obituary in text box) has a happy ending. In the later years of Dr. Gold's highly varied and productive professional career, he witnessed not only the vindication of his then-heretical ideas about cochlear function, but also the emergence of OAEs as an important clinical tool. Dr. Kemp has priceless tape recordings of telephone conversations during which Thomas Gold relates in his own words his exciting discov-

The rest, as the saying goes, and as described in the remainder of this book, is history.

Thomas Gold (1920–2004)

Thomas Gold died on June 22, 2004 in Ithaca, New York at the age of 84 after a long and varied academic career that included often-controversial research from the inner ear to outer space. Audiologists will remember Dr. Gold for his novel explanation of cochlear physiology in the late 1940s. During WWII, Gold worked for the British Admiralty on top-secret research projects to further develop radar technology. In the late 1940s, as a young graduate student, he conducted innovative studies of cochlear mechanics and physiology at the prestigious Cavendish Laboratory at Cambridge University in England. Gold's claim that the inner ear contained "mechanical resonators" and operated actively in the processing of auditory information and tuning of the auditory system were almost heretical at the time, and certainly at odds with much of the mainstream thinking dominated by the work of the eminent Georg von Békésy. Around 1948, Gold met with von Békésy and proceeded to expound on his novel theories. Some 50 years after the meeting, Gold recounted the event with characteristic British understatement and humor, and von Békésy's less than enthusiastic response to the new ideas on cochlear function. Dr. Gold's observations on active and nonlinear processes in the cochlea prophesied the discovery 30 years later of otoacoustic emissions by another British auditory scientist, Dr. David Kemp.

Dr. Gold was born in Vienna in 1920. After schooling in Switzerland and his tenure at Cambridge University, he accepted a position at Harvard University. Then, in 1959, Gold accepted a faculty position at Cornell University where he had a distinguished career serving as chair of the astronomy department and director of the Center for Radiophysics and Space Research. As an aside, at Cornell the popular astronomer Carl Sagan was a colleague and a friend of Dr. Gold. Gold repeatedly generated controversy with his innovative research and provocative publications. In 1955, for example, he made the then outrageous claim that the surface of the moon was covered with fine powder. Although he was criticized rigorously at the time, his theories were proven in 1969 when the Apollo 11 crew returned to earth with samples of the rock powder. A fascinating obituary of Gold published in *Physics World* by deputy editor Martin Durrani documents numerous other controversies that characterized Gold's illustrious career.

As noted by Louis Friedman, Executive Director of The Planetary Society and a former student of Gold's, "Whether he was ultimately proved right or wrong, his (Gold's) ideas always challenged his colleagues to think deeply about any subject he pursued. His approach exemplified the scientific method at its best, posing hypotheses and testing them to advance our basic understanding of the universe." Clearly, this statement appropriately describes Dr. Gold's contributions to our understanding of the workings of the inner ear. His innovative approaches to research questions and his willingness to face criticism for ideas that challenge accepted thinking and wisdom of the day offer a valuable model for auditory scientists of our day.

Obituary by James W. Hall III, University of Florida, Gainesville, FL. Reprinted with permission from: *Audiology Today*, *16*, (September/October), 2004, p. 42

Evolution of OAEs as a Clinical Tool

The earliest clinical application of OAEs was newborn hearing screening. Certain advantages of OAEs as a hearing-screening tool were recognized almost immediately following the 1978 Kemp publication (Kemp, 1978). Abnormalities in OAE findings typically were associated with common causes of hearing loss in young children, namely, middle ear disorder and outer hair cell dysfunction. Consequently, OAE screening outcome (e.g., Pass versus Fail) quickly, objectively, and rather effectively differentiated children who were likely to have reasonably normal peripheral auditory function and hearing sensitivity within normal limits from those with peripheral auditory dysfunction and perhaps some degree of hearing loss.

In the early years of OAE screening, from about 1985 to 1995, some authors inevitably compared OAEs very favorably to the then-established ABR screening technique. OAEs were praised for both the relative brevity and simplicity of the technique ("no electrodes are required!"). These specific advantages, however, were not always supported by evidence from formal investigations with head-to-head comparisons of the OAE and ABR techniques. Limitations of OAEs as a hearing screening tool also became quite apparent with accumulated clinical experience, particularly problems with unacceptably high failure rates associated with noise in the test environment and vernix caseous in the external ear canal of newborn infants. More recently, the insensitivity of OAEs to "auditory neuropathy" has been added to the list of limitations of the technique in selected patient populations. Still, on balance, the many clinical advantages of OAEs as a screening technique outweigh the drawbacks. OAEs unquestionably remain an attractive option for hearing screening in varied pediatric populations, ranging from newborn infants to school-age children.

The introduction of commercially available OAE devices, and especially instrumentation for recording distortion product OAEs (DPOAEs) and transient evoked OAEs (TEOAEs) from an international list of manufacturers, led naturally to multiple nonscreening applications of the new technique. Beginning in the mid-1990s, clinical researchers enthusiastically explored the potential diagnostic value of OAEs in virtually every imaginable etiology for hearing loss. We can attest personally to the excitement produced by evaluating, for the first time in a specific patient population, auditory function with a new procedure. As soon as the newly purchased OAE device was unpacked and given a test-run on the clinician's ears, and maybe one or two handy coworkers (or offspring!), and certainly before the manual was reviewed, OAEs were somewhat shakily recorded from the first patient to arrive in the clinic. Typically, the new "service" initially was provided as a professional courtesy (i.e., free of charge).

Arbitrary and somewhat random application of OAEs by individual clinicians within a short time gave way to systematic clinical investigations, and the development of evidence-based rationale for OAE measurement. By the year 2000, the diagnostic value of OAEs as a component within an audiologic test battery was reported for a wide spectrum of etiologies in pediatric and adult patient populations. Presentations at scientific meetings, and then peer-reviewed publications, soon appeared describing patterns of OAE findings in disorders from malingering to Ménière's disease. The unique sensitivity and specificity of OAEs to cochlear—specifically outer hair cell—dysfunction logically led to the measurement of OAEs in at risk persons, such as patients exposed to hazardous levels of sound and those with tinnitus.

The etiology now typically known as "auditory neuropathy spectrum disorder (ANSD)" is perhaps the best example of the dramatic impact of OAEs on diagnosis of auditory dysfunction. The term "auditory neuropathy" was actually coined in 1996 just as the application of DPOAEs was rapidly expanding around the world as a clinical procedure. The different types and sites of auditory dysfunction included within the general term auditory neuropathy spectrum disorder were relatively unexplored, and inadequately appreciated prior to the widespread clinical use of OAEs. It is not an exaggeration to claim that the advent of OAEs as a clinical procedure contributed directly to the recognition of the clinical entities known by the term "auditory neuropathy."

OAEs continue to evolve as a clinical tool as does our understanding of the physiology associated with OAEs. Much of this book is devoted to a review of the current understanding of the physiology and biophysics related to OAEs as well as the numerous applications of OAEs in children and adults. There is still plenty of room for development and implementation of more sophisticated and rigorous strategies for OAE measurement and analysis. Without doubt, new clinical applications for OAEs will be discovered and developed in the years to come. Also, technological advances in instrumentation will considerably enhance the clinical value of OAEs and their role in the audiologic test battery. An as example, newly introduced devices combining OAE and ABR or OAE and tympanometry technology (reviewed in Chapter 10) likely will contribute to more efficient and effective identification and diagnosis of auditory dysfunction.

INTRODUCTION TO AUDITORY SYSTEM ANATOMY AND PHYSIOLOGY IN OAE MEASUREMENT

Introduction

A good understanding of the anatomic and physiologic underpinnings of OAE generation and measurement is essential for recording, analyzing, and interpreting findings in the clinical setting. This topic, particularly current knowledge and theories regarding cochlear anatomy, physiology, and mechanisms that play a role in the generation of OAEs, is reviewed in far more detail in the next chapter. As illustrated schematically in Figure 1–4, four general regions of auditory system anatomy are involved in the generation and measurement of OAEs. What follows is an introduction to the role of each of these four regions in the generation of OAEs activity, and the influence

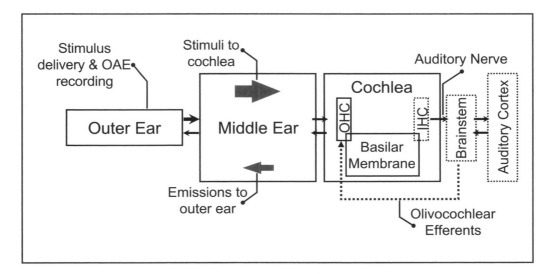

FIGURE 1–4. A simple block schematic of the major regions of the auditory system that influence the measurement of otoacoustic emissions. Note the directional arrows in the middle ear depicting the bidirectional energy transfer through this space. The size of the arrows is representative of the relative magnitudes of the energy traveling into and out of the cochlea. Structures outlined in dashed lines play a secondary role in modulating OAEs.

of abnormal function of each of the regions on OAEs recordings. The reader is encouraged to refer to the next chapter for a more in depth treatment of current theories and controversies related to OAE generation.

External Ear Canal

As the interface between the auditory system and the measurement apparatus, the external ear canal plays a crucial role in stimulus delivery and OAE recording. The typical OAE probe houses one or two signal delivery tubes that are connected to miniature speakers as well as one or more microphones that collectively record the acoustic signal in the ear canal. The probe, fitted with an appropriately sized ear tip, has to fit snugly in the ear canal to create an approximately isolated acoustic chamber between the probe tip and the eardrum. Improper orientation of the probe could lead to a blockage of the signal delivery system if one or more sound delivery tubes are blocked against the ear canal wall. On the other hand, a leaky fit of the probe tip could contaminate the OAE recording by allowing environmental sounds in and biological signals out of the ear canal.

Let us first consider the normal external ear canal. Ear canal acoustics critically influence the stimuli used to record OAEs. Clinical OAE systems attempt to account for the influence of ear canal acoustics by careful calibration of the stimuli prior to each recording. However, adequately controlling for ear canal acoustics is a complex and difficult issue and calibration techniques have continued to evolve three decades after the discovery of OAEs. Given the importance of stimulus calibration, we devoted an entire chapter (Chapter 4, Instrumentation and Calibration) to the topic.

A variety of pathologic and nonpathologic conditions of the external ear have serious ramifications on OAE measurements. These negative effects are manifest either through the compromise of the stimuli needed to generate the OAEs or through interference in the recording of the generated OAE. Clinical audiologists who apply OAEs regularly are able to readily list a variety of external ear canal deviations that negatively influence OAE measurement. These external ear

canal factors, from vernix caseous in infants to excessive cerumen in adults, are noted in discussions of the clinical applications of OAEs later in the book.

Middle Ear

The middle ear (tympanic membrane and ossicles) is a vital, and sometimes weak, link in OAE measurement. As is seen in Figure 1–4, stimuli used to elicit OAEs are transmitted to the cochlea via the middle ear. The OAEs generated in the cochlea then travel back through the middle ear on the way to the external ear canal. In Figure 1–4, the direction of stimulus and OAE propagation (inward and outward, respectively) is indicated by the direction of the arrows, whereas the intensity of each sound is depicted by the thickness or size of the arrows. The rather complex influence of the middle ear in OAE measurement is reviewed more rigorously in Chapter 2 under the heading "How Do OAEs Return to the Ear Canal?"

The fact that both the stimuli and the OAEs have to travel into and from the cochlea through the middle ear suggests that the health of the middle ear has the opportunity to influence OAE recordings twice. Thus, middle ear condition is significantly more important for OAE recordings as compared to other electrophysiologic tests such as the ABR. In the case of tests such as the ABR, the stimuli are influenced by middle ear condition on their inward journey to the cochlea. However, the middle ear does not have the opportunity to influence the biologic response. Reverse propagation of energy through the middle ear is thought to be less efficient than forward propagation, leading to a loss of as much as 30 dB (see Chapter 2 for a detailed discussion). Thus, even under normal circumstances, the OAE signal reaching the ear canal is greatly attenuated. It stands to reason then that even a minor pathology can greatly influence OAEs by first compromising the inward flow of the stimuli and then the propagation of OAE energy from the cochlea to the ear canal.

In patients with middle ear dysfunction, secondary to most etiologies, OAEs are abnormally reduced in amplitude or not detectable in the external ear canal. As a rule, the likelihood of recording OAE activity in patients with middle ear dysfunction is inversely related to the extent of the abnormality. One clinically important implication of this discussion is that middle ear status should always be evaluated formally (e.g., with tympanometry) whenever OAEs are reduced in amplitude or not detectable, particularly within the lower frequency region. Middle ear dysfunction should be ruled out before concluding that OAE abnormalities are secondary to cochlear auditory dysfunction. Because the topic is so important for clinical measurement of OAEs, we review middle ear status as a factor in OAE analysis in detail in Chapter 6.

Cochlea

Since David Kemp cautiously speculated over 30 years ago on the likely role of outer hair cells in the generation of OAEs (highlighted with a quotation earlier in this chapter), hundreds of peer-reviewed journal articles have been published in support from experiments conducted on humans and other laboratory animals. Without any doubt, OAEs reflect outer hair cell activity and functional integrity of outer hair cells is essential for generation of OAEs. Functional integrity of another cochlear structure, the stria vascularis, also appears to be very important for the generation of OAEs. This may not be surprising given the known role of the stria vascularis in maintaining the electrochemical gradients in the cochlea essential for normal outer hair cell function. Inner hair cells do not appear to be a factor in the generation or measurement of OAEs. Among cells in the body, outer hair cells are uniquely capable of motility (movement—change in cell length, in this case).

When the energy from sound stimulation reaches the cochlea and moves the basilar membrane, a remarkable cascade of events occurs within the outer hair cells. The sequence of events begins with rather straightforward deflection of stereocilia on the top of the outer hair cells that leads to the opening of ion channels and depolarization of the cells. Morphologically, during depolarization the outer hair cells become short and fat, whereas during hyperpolarization the cells are tall and thin. This vertical movement caused by the length change of outer hair cells is fed into the basilar membrane, enhancing its motion and ultimately leading to greater hearing sensitivity and frequency selectivity. This, of course, is a gross oversimplification of outer hair cell function. In the next chapter, we explain the role of outer hair cells in OAE generation in greater detail. A firm grasp of this topic is very important for audiologists. An up-to-date understanding of OAE generation, including current theories of generation of different OAE types as well as the role of different parts of the outer hair cells in OAE generation, is essential for the modern audiologist.

Efferent Auditory System

The fourth anatomic region or system that is involved in OAEs is the efferent auditory system (refer again to Figure 1–4). Because the efferent system is not central to the generation or clinical application of OAEs, we provide only a very brief summary of the anatomy and its relation to OAE measurement. The efferent system and its influence on OAE measurement is reviewed more fully in the next chapter. The efferent auditory system descends from the cortex, through the brainstem, to reach the hair cells in the inner ear where it exerts mostly an inhibitory influence on cochlear activity. Although the efferent auditory system is not as well understood as the traditional ascending (afferent) auditory pathways, in recent years, efferent system anatomy and physiology has generated considerable research interest. This interest and excitement is understandable as OAEs offer an objective, noninvasive, clinically feasible tool for measuring efferent effects on cochlear function.

2

Anatomy and Physiology

INTRODUCTION

The normally functioning human ear can detect sounds produced by vibrations as miniscule as a tenth of a nanometer (nm). Expressing this quantity in units that may be a little more familiar, a tenth of an nm would be equal to approximately 0.0000000004 (4.00^{-10}) inches or 0.000000001 (1.00^{-9}) cm. If the ear were any more sensitive, we would hear the constant buzz of the random movement of air molecules around us. At the other end of the continuum, the ear tolerates sounds that are 12 orders of magnitude more intense, resulting in a dynamic range of 120 dB. In comparison, the almost extinct magnetic tape could only encode a dynamic range of 55 dB, and the still popular compact disc has a dynamic range of 96 dB. The ear can perceive sounds over a frequency range spanning 10 octaves and is capable of differentiating sounds only a few Hz apart. Those with unimpaired hearing can pick out a stream of speech even against the background of several other similar streams of speech or other noise. Combining sounds received through the two ears in a complex neural network in the brainstem and cortex, humans can also resolve two sound sources that are a few centimeters apart and sounds in time that are separated by mere tens of microseconds.

Any one of these capabilities would be a tremendous feat, but the fact that our auditory system can achieve all of these is nothing short of incredible. These functions are performed by a combination of the acoustic characteristics of the outer ear, the mechanical properties of the middle ear, the hydrodynamic and electrochemical functions of the inner ear or cochlea, and the neural properties of a few cranial nerves, the brainstem, and the auditory centers of the cortex. A schematic of these structures was shown in Figure 1–4 and their basic functions were explained in Chapter 1. This chapter discusses the anatomic and physiologic details of the parts of the auditory system that are involved in the generation and propagation of otoacoustic emissions (OAEs). A detailed discussion of general auditory anatomy and physiology is beyond the scope of this text. Readers unfamiliar with the auditory system are encouraged to, at the least, revisit the general overview of auditory anatomy and physiology in Chapter 1 before proceeding with the information on the generation of OAEs presented in this chapter. The following discussion is constructed entirely from the point of view of OAEs. Thus, we refrain from discussing in detail important parts of the auditory anatomy that are not involved in the stimulation, generation, and recording of OAEs. Our journey through the auditory system proceeds in an OAE-centric fashion. That is, we begin with the stimulus signals and follow their route into the cochlea. Then we discuss the structures and functions important in the generation and propagation of OAEs.

MAIN STRUCTURES INVOLVED IN OAES

The outer ear and middle ear, along with the cochlea, are the main auditory structures involved in the generation, propagation, and recording of OAEs (Figure 2–1). In addition, there is considerable evidence that the olivocochlear efferents are effective modulators of OAEs. As the olivocochlear efferents form the terminal extension of the descending corticothalamic efferent pathway, their modulation of OAEs opens the interesting possibility of central control of the peripheral auditory system. However, the anatomy and physiology of efferent modulation of OAEs is not reviewed here. Another chapter (9) is dedicated to that important topic. The following discussion focuses on the role of the: (1) outer ear as the junction where OAE instrumentation meets the biology of the ear, (2) middle ear in forward propagation of stimulus signals and backward propagation of the OAEs, and (3) cochlea in the generation of OAEs. These structures along with their constituent elements are illustrated in Figure 2–1. Once our journey reaches the cochlea, much attention will be directed to the outer hair cell at the expense of the inner hair cell and the afferent neural elements. There is little evidence that inner hair cells or the afferent nerve are involved in the generation or measurement of OAEs. Let's begin with the stimulus signals.

The External Ear Canal

Notice that the title of the section does not read "The External Ear." The pinna, a part of the external ear, is extremely important in hearing natural sounds generated in the free field. However, the pinna's role in modifying the incoming acoustics and aiding in sound localization in both horizontal and vertical planes, especially for frequencies

around and above 6 kHz, as well as front-to-back discrimination are certainly not utilized during OAE recordings. The acoustic influences of the pinna essentially are eliminated as the OAE probe is inserted into the ear canal. Clinicians who make OAE recordings know, of course, that the pinna does play a role in the fit of the OAE probe into the ear canal. It should be mentioned here that OAEs have been recorded (from guinea pigs) using "open" systems where the probe is not hermetically sealed into the ear canal (Withnell, Kirk, & Yates, 1998). The goal of this work was to extend the frequency range of recordable OAEs. We do not see such systems forthcoming to the audiology clinic. Research on methods to overcome the limited recording bandwidth of OAEs is discussed in later chapters of the book.

The external ear canal is, on average, 26 mm long in the typical adult. Given its known length, and because it is closed at one end by the tympanic membrane, the ear canal has generally definable and well studied resonance characteristics. The open ear canal has a resonance peak around 2.6 kHz in adults (Shaw, 1974). The resonant frequency of the external ear canal is higher in newborns (more on this later), but it approximates adult-like values after the first 2 to 3 years of life (Kruger, 1987). However, the act of inserting an OAE probe into the ear canal effectively modifies the ear canal. With the probe in place, the resonance characteristics are determined by the cavity that is enclosed between the probe and the eardrum and hence is entirely dependent on the depth of insertion and the quality of the seal.

Visual inspection of a fair sampling of human ear canals inevitably reveals the diversity in their geometry. Some ear canals are practically uniform cylinders. With an otoscope placed at the very entrance of a straight ear canal, it is possible to visualize the entire eardrum. At the other end of the spectrum, there are treacherously curvy ear canals where even considerable manipulation of the pinna to alter the shape of the canal along with a painfully deep insertion of the otoscope speculum reveals only a portion of the eardrum. Thus, the acoustics of an external ear canal artificially created during an OAE recording are unpredictable and difficult to measure. The challenges of estimating ear canal acoustics has turned out

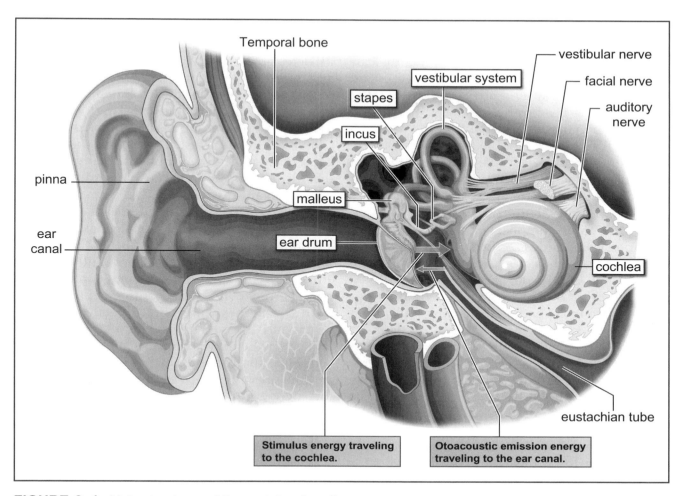

Temporal bone

vestibular nerve

vestibular system

facial nerve

stapes

auditory nerve

incus

pinna

malleus

ear canal

ear drum

cochlea

eustachian tube

Stimulus energy traveling to the cochlea.

Otoacoustic emission energy traveling to the ear canal.

FIGURE 2–1. Main structures of the peripheral auditory system involved in the generation and recording of OAEs. The outer ear is the site of signal delivery for evoked emissions and also for the recording of OAEs. The middle ear serves as the bidirectional passageway for the stimulus inward and the emissions outward to the ear canal. This signal transfer pathway is represented using the two arrows in the middle ear space. Notice the difference in the width of the two arrows. The arrow pointing inward, toward the cochlea, marks the input pathway and is considerably thicker than the arrow marking the outward pathway. This is indeed the case, as the middle ear is more efficient in transferring energy in toward the cochlea. The active biophysical processes that are responsible for the generation of OAEs are housed in the cochlea. The mode of transmission of the OAE energy from the region of generation to the middle and outer ears is an issue of current controversy. The junction between the middle ear and the cochlea also is an important element in the generation and particular characteristics of OAEs.

to be a major hurdle in designing OAEs systems that can deliver signals calibrated at the plane of the eardrum—an issue discussed in detail in Chapter 4.

Delivering Stimulus Signals

The external ear canal, modified in length by the insertion of the OAE probe, is the location where stimulus signals are delivered for their ultimate entry into the cochlea. The signals are then modified by the transformation characteristics of the residual ear canal as they impact the tympanic membrane. It is important to remember that the OAE system can monitor the stimulus signals only through the microphone(s) located in the OAE probe. Thus, the OAE system will fail to register the transformation of the stimulus signals by the resonance characteristics of the residual external ear canal. Without an effective correction for this resonance, the stimulus signals reaching the eardrum only nominally maintain their prescribed characteristics. That is, if a 65 dB SPL signal at 1 kHz is being used to record OAEs, that signal is only nominally at 65 dB SPL at the eardrum because it has been modified by the resonance characteristics of the ear canal en route to the tympanic membrane. In reality, such transformations lead to significant alterations of signals only above 3 to 4 kHz (Siegel & Hirohata, 1994). To make matters a little more complicated, the magnitude of deviation from the desired level also is dependent on the characteristics of the residual ear canal between the OAE probe and the eardrum. As a result, during OAE measurement, stimulus signals vary from one patient to the next, and within a single individual, across repeated insertions.

In addition to the normal variations in geometry that affect external ear canal acoustics, cerumen or other debris in the ear canal impedes accurate recording of OAEs. In the extreme, the debris can block one or more of the sound delivery channels, thereby causing an important deviation from the desired stimulus levels. Most, if not all, OAE systems today have the capability of detecting such gross anomalies, and most instruments are programmed to interrupt progression of the test. However, minor variations in stimulus levels may go unnoticed by both the equipment software and the clinician, leading to results that are subtly different. Such small differences may not be important in screening applications of OAEs, but they may become significant in applications such as ototoxicity monitoring where the agreement between repeated measures is the metric under examination.

The geometry of the external ear canal also may cause one or more of the signal delivery ports to be blocked against the canal wall. Audiologists have experience with errors of this nature from acoustic impedance measurement, when the estimated ear canal volume is impossibly small and the test does not proceed because a change in ear canal pressure cannot be effectively implemented. In the case of OAEs, the recording software also can detect gross errors between the actual signal level and the expected signal level. Provided the OAE recording software presents a graphic display of the stimulus signals, these errors are easily detectable visually by the clinician. However, an interesting problem can arise if the probe was placed without error at the onset of the test, when the stimulus check typically is performed, but then slips into a compromised position during the course of the OAE test. The clinician is encouraged to periodically examine the stimulus level recorded *during* the actual OAE recording, if such data are available.

In summary, signals used to evoke OAEs are delivered to an external ear canal whose acoustics are unpredictably modified by the insertion of the probe. Probe placement can significantly alter the stimulus arriving at the eardrum, sometimes in ways detrimental to the recording of OAEs. Most OAE systems have incorporated methods of identifying such gross errors, and the test process often is halted when such errors occur. The clinician should be familiar with the details of the software and the display of such errors. Minor aberrations that may go unnoticed can significantly impact OAE recording, and the analysis and interpretation of OAE findings.

The external ear canal is not just the location where all the action is initiated by delivering the stimulus signals. It also is the final collection area for the OAE. The salient aspects of this function are discussed next.

Recording OAE Signals

The characteristics of the OAE recorded in the ear canal undoubtedly are associated with the biophysical processes in the cochlea that are responsible for their generation. However, the final OAE recorded using a microphone(s) in the ear canal is greatly influenced by what lies in between the site of OAE generation and their final registration in the recording system. As we have stressed already, the external ear canal plays an important role in the delivery of the stimulus signal to the cochlea. The external ear canal also is a very important segment of the pathway leading from the site of OAE generation and the recording device. Given the miniscule magnitude of the OAE being recorded, it is critical that almost all the acoustic signal is captured in the ear canal and registered by the microphone. Any leaks in the seal between the OAE probe and the external ear canal allow portions of the OAE signal to escape prior to its capture. The frequencies most affected and the degree of signal loss depends on the dimensions of the leak, much like that of a vent in a hearing aid shell or earmold. Given this need for a tight and secure seal for the proper recording of OAEs, the anatomy of the ear canal assumes great importance. In cases where the ear canal is small in diameter or treacherously curvy, care has to be taken in placing the probe such that no acoustical leaks are present and the probe is in a stable location for the duration of the test.

A good seal in the ear canal is also important in obtaining an acceptable noise floor. External noise may enter the ear canal and become mixed in with the OAE recording when the acoustic isolation of the sealed ear canal is compromised. The change in the noise floor due to a small leak between the probe and external ear canal wall is evident in Figure 2–2. In this example, spectra of two spontaneous (S)OAE recordings from the same ear are presented. In the case of SOAEs,

leaks due to a poor fit of the OAE probe within the external ear canal causes an elevation of the noise floor at low frequencies, possibly masking the presence of SOAEs in the affected frequency range. SOAEs are not regularly evaluated in the clinic. However, other OAE types are not exempt from the ill effects of interference from ambient noise.

All OAE recording systems employ intelligent "stopping rules" where averaging is continued until a desired signal-to-noise ratio (SNR) is reached. These systems typically also have a preset maximum duration for signal averaging after which the recording is stopped irrespective of the level of the absolute noise floor or the SNR. A leak caused by a poor match between the ear tip and the external canal anatomy can increase test time because the presence of noise in the ear canal leads to an increase in the number of averages necessary to reach a target SNR. In the extreme, noise leaked into the ear canal can lead to compromised measurement of transient or distortion product OAEs. That is, the desired noise floor or SNR cannot be reached after the maximum averaging time has been expended. Up to this point,

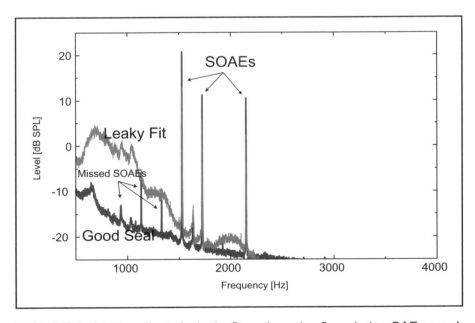

FIGURE 2–2. The effect of a leaky fit on the noise floor during OAE recordings. SOAE spectra from the same ear are presented from two recordings. A slit leak in one of the recordings caused an elevation of the noise floor at low frequencies. Notice that several SOAEs would go undetected in the leaky recording as they are "hidden" by the elevated noise floor.

we have discussed the negative impact of ambient environment noise on OAE measurement. Physiologic sources of noise, arising from patient movement, breathing, and even vascular (blood flow) sources near the ear, also can play a role in OAE recording.

Just as stimulus levels can be compromised by the blockage of the signal delivery ports by cerumen, debris, or a bend in the ear canal, so can the recorded OAE signal. The microphone and the recording system must function adequately throughout the calibration process in the ear canal for the test to proceed. The clinician should be cognizant of the usual noise floor on a given measurement system. If the noise floor or the stimulus levels appear to be unusually low, the explanation might be partial blockage of the microphone port. Under such circumstances, the level of the recorded OAE will be low or the OAE might even be absent. Abnormal or totally absent OAEs due to this measurement condition could easily be misinterpreted to be biological in origin, that is, secondary to cochlear dysfunction.

The acoustic characteristics of the external ear change during the first decade of life, and perhaps over a longer time frame (Okabe, Tanaka, Hamada, Miura, & Funai, 1988). The changing anatomy of the external ear canal plays an important role in the developmental differences observed in OAEs. Keefe and Abdala (2007) recently presented a detailed theoretical analysis of the effect of a growing ear canal on distortion product (DP)OAEs and wideband reflectance data. The authors started with the assumption that the cochlear processes responsible for OAE generation are mature at birth and, hence, all differences in DPOAEs observed between newborns or infants and adults were accounted for by differences in outer and middle ear anatomy and physiology. In their model, Keefe and Abdala (2006, 2007) separated the developmental effects of the outer and middle ears on the forward as well as reverse transfer functions involved in the generation and recording of OAEs. Their acoustic estimates indicated that ear canal area increases from approximately 10 mm^2 in the newborn to approximately 60 mm^2 in the adult. This difference in ear canal area results in a 17-dB difference in the level of the OAE recorded in a newborn versus an adult ear.

As the ear canal grows, the difference is reduced to approximately 5 dB between a 6-month-old and adult ear canal. The difficulty in recording OAE data from newborns and infants dictated that the results reported by these authors were limited in frequency range. However, these differences in the anatomy of the external ear canal between infants and adults significantly alter the OAE recorded and are important to account for before clinical interpretation of data.

To summarize, the anatomy of the external ear canal influences the recorded OAE via the goodness of the fit between the ear canal and the OAE probe, the blockage of the acoustic path to the microphone by debris or geometry of the ear canal, and also by developmental differences that alter the acoustics of the ear canal. The clinician can control for the first two variables by a careful visual examination of the ear canal and proper placement of the probe. Developmental differences in ear canal anatomy must be accounted for in normative ranges and interpretation of test results.

The Middle Ear

Much like the external ear canal, the middle ear serves two purposes in the context of OAEs: stimulus signals must travel inward into the cochlea via the middle ear, much like environmental sounds do in natural hearing; and OAEs must travel outward from the cochlea to be recorded in the ear canal. The mechanical advantages that are available to natural sounds entering the cochlea are also available to stimulus signals as they travel in the "natural" direction. That is, the area ratio between the eardrum and the oval window, the lever advantage due to the articulated ossicles, and the mechanical advantage afforded by the geometry and placement of the eardrum all are utilized by the stimulus signals during their inward journey to the cochlea. To the middle ear, the stimuli for generating OAEs are nothing special. Thus, the stimuli also will evoke the middle ear muscle reflex when the intensity level is greater than the acoustic reflex threshold. The inadvertent elicitation of the acoustic reflex becomes salient when sounds are used to stimulate the olivocochlear efferent system during OAE measurement. These higher intensity signals, used in activating the efferent system, can easily trigger the middle ear muscle reflex. Under these test conditions, OAE recordings are more difficult to interpret as the effects of both the middle ear muscle reflex and the olivocochlear reflex are represented in them (more on this in Chapter 9).

Unfortunately, the mechanical advantages imposed on sounds traveling inward through the middle ear to the cochlea are not available to sounds traveling outward through the middle ear from the cochlea to the outer ear. This inefficiency is directly reflected in the low levels of OAEs that are typically recorded in the ear canal. This inefficiency in reverse transmission through the middle ear is discussed next.

The Middle Ear Does Not Work as Well in Reverse

It makes intuitive sense that reverse sound propagation through the middle ear is not very efficient. The middle ear has evolved as an impedance matching device for the efficient coupling of sound in air in the external ear to that in a fluid medium in the cochlea. Each mechanism that aides in this impedance matching process for forward transmission of sound becomes an impediment for its reverse transmission from the cochlea to the external ear. This intuitive expectation appears to hold in the limited data available in the literature. In the gerbil, forward pressure gain of between 20 and 25 dB and a reverse pressure loss of approximately 35 dB has been observed (Dong & Olson, 2006). On second thought, the fact that the middle ear system is not exactly symmetric (forward gain = reverse loss) is interesting and perhaps reflects complex modes of vibration of the eardrum.

Understandably, such direct measures of forward and reverse sound transmission characteristics are much harder to accomplish in humans. To circumvent this practical limitation, (Keefe, 2002) deduced the forward and reverse transmission characteristics of the human middle ear from distortion product (DP) OAE input/output functions. The basic assumption that allowed this derivation was that cochlear input/output functions are invariant over the length of the cochlea, or at least over finite portions of the cochlea. Thus,

it followed that any frequency-dependent differences observed in DPOAE input/output functions were imposed by the forward and reverse transmission characteristics of the middle ear. The results indicated narrowband transmission characteristics in both the forward and reverse directions that in combination mimicked the band-pass characteristics of the behavioral threshold curve.

One advantage in examining the middle ear, however, is the fact that its transmission characteristics are not critically dependent on metabolic forces. That is, unlike measurements in the cochlea, middle ear function results obtained from cadavers should reasonably represent those in live ears. Puria (2003) took advantage of this observation in conducting detailed measurements from four human cadavers. Simultaneous measurements of sound pressure in the ear canal, stapes velocity, and fluid pressure in the vestibule were obtained. With a sound source in the ear canal, the ratio between the pressure measured in the vestibule and the sound pressure in the ear canal is equivalent to the gain in forward transmission. Conversely, with a sound source in the vestibule, the ratio between the sound pressure in the ear canal and that in the vestibule is a measure of the reverse gain. These results indicate a peak gain of approximately 18 dB in forward transmission around 0.9 kHz. The average gain dropped off at rates of approximately 10 and 7 dB per octave below and above this peak. The reverse pressure gain showed a peak *loss* of −30 dB around 1.4 kHz, with steeply increasing losses above and below the peak frequency.

In the case of OAEs, it is immaterial to think about forward and reverse middle ear transmission separately. The measured OAE is influenced by both forward and reverse transmission and, hence, what matters is the combined effect of transmission in both directions. Puria (2003) computed this net effect on OAEs and estimated it to be a *net loss*. The peak of this combined function was approximately −7 dB around 1.3 kHz, dropping off at approximately 11 dB/octave both above and below this peak frequency. That is, the best the middle ear does in the context of OAEs is a net loss of 7 dB at around 1.3 kHz, with the losses increasing rapidly at frequencies above and below this peak.

A very interesting observation related to this narrowband nature of round-trip middle ear transmission is the relatively invariant levels of evoked OAE observed as a function of frequency. The middle ear is increasingly inefficient at frequencies above 1.3 kHz, yet we do not observe a similar drop off in evoked OAE levels with increasing frequency (Puria, 2003). Puria (2003) argues, therefore, that the cochlea must generate greater OAE levels at these higher frequencies to compensate for the increasingly inefficient middle ear transmission.

Developmental Differences in Middle Ear Transmission

The maturing middle ear significantly influences the evoked OAE levels measured in newborn and infant ears. It is well documented in the literature through the work of various groups, and everyday clinical experiences, that higher levels of evoked OAEs are observed in newborns and infants as compared to adults (see Abdala, Oba, & Ramanathan, 2008, for a recent report). Abdala and Keefe (2006) recently investigated the influence of the outer and middle ears in these observed differences. They measured DPOAE input/output functions, suppression tuning curves, and ear-canal reflectance in infants through 6 months of age and adults. Middle ear and cochlear input/output functions were assumed to be linear and nonlinear, respectively, and a model of signal transmission through the infant middle ear was constructed based on these assumptions. The results of this modeling suggested that the signal levels delivered to the newborn cochlea are attenuated due to immaturities in middle ear signal transmission. Once these immaturities were compensated for, the adult and infant DPOAE suppression tuning curves approximated each other. In a more elaborate theoretical follow-up study, Keefe and Abdala (2007) deduced that the infant middle ear attenuates forward transmission of signals by approximately 16 dB compared to the adult middle ear. However, this loss is more than compensated by the boost that the OAE receives in the infant ear canal due to the seven times smaller ear canal area. These differences in forward and reverse transmission through the

middle ear arguably account for much of the differences in OAE levels seen between infants and adults. More recent work on the details of DPOAE level and phase characteristics, however, suggests that developmental differences in the cochlea also may influence the differences in OAEs observed between adults and infants (Abdala & Dhar, 2010; Dhar & Abdala, 2007).

Negative Middle Ear Pressure

The effects of negative middle ear pressure on OAEs, especially DPOAEs, have been investigated relatively extensively. The interest on this topic is not only theoretical. It has important clinical implications, as negative middle ear pressure is a common condition, especially in young children. Interestingly, negative middle ear pressure appears to affect DPOAE levels in a complex manner (Sun & Shaver, 2009). DPOAE levels are most reduced at frequencies below 1 kHz, with minimal effects at 2 kHz and variable effects above 2 kHz. Sun and Shaver even observed increases in DPOAE level at some (high) frequencies in the presence of negative middle ear pressure. Fortunately, introducing negative air pressure in the ear canal can compensate the ill effects of negative middle ear pressure (Sun & Shaver, 2009). We are beginning to witness instrumentation converging for multiple techniques. That is, OAE measurements and tympanometry can be performed using the same device. New directions in OAE technology are reviewed in Chapter 10. With such instrumentation, it would be logical to always make OAE recordings under compensated pressure conditions, much like we make acoustic reflex measurements today.

In summary, the middle ear plays a vital role in the stimulation and recording of OAEs. The role of the middle ear in OAEs is perhaps more important than in other audiologic procedures such as hearing threshold or brainstem response measurements, where only the stimulus has to travel through the middle ear to the cochlea. In recording OAEs, not only must the stimulus travel inward to the cochlea through the middle ear, but the OAE energy also has to travel outward to the ear canal through the middle ear. Thus, middle ear function has a double impact on OAEs.

The Cochlea

The mammalian inner ear or cochlea, a coiled structure comprising of three fluid-filled chambers and some of the most interesting tissue in our body, is seated within the temporal bone. The cochlea houses the sensory structure for hearing, the organ of Corti, and also is the site of generation of OAEs. A complete review of cochlear anatomy and physiology is well beyond the scope of this text. The important anatomical elements of the cochlea are introduced below with emphasis on the structures specifically responsible for the generation and propagation of OAEs.

The fact that the cochlea is the site of OAE generation was proven in the very first publications on OAEs. Kemp (1978) recorded the responses of a healthy ear to tone bursts of different frequencies. He clearly demonstrated that responses to low-frequency tone bursts arrived at the ear canal microphone with a greater delay than responses to tone bursts of higher frequencies. Kemp argued that this difference in delay was reflective of the tonotopic organization of the cochlea. That is, responses generated at the base of the cochlea arrived at the ear canal with a smaller delay as compared to responses generated at more apical locations. OAEs actually provided support for the notion that "active processes" in the cochlea selectively provided amplification to low-level inputs several years before the discovery of motility in outer hair cells (Brownell, Bader, Bertrand, & de Ribaupierre, 1985).

There is now wide acceptance that OAEs are by-products of the active processes in the cochlea responsible for our acute sensitivity to sounds and for the sizable dynamic range of hearing, as well as our ability to make fine frequency discrimination. More specifically, investigations have unequivocally linked OAEs to the integrity of outer hair cells (OHC). Damage to OHC through experimental exposure to noise or chemical toxins eliminate OAEs (see Kemp, 2002, for a recent review). Although the origin of OAEs in the cochlea, and specifically the OHC, appears to be certain, several important questions about the generation and propagation of OAEs are still being debated. It also is important to note that, although individual OHC may be the source of power behind OAEs, mechanical properties of the entire cochlear chamber play an important role in determining the exact details of the OAE recorded in the ear canal. A brief overview of current thinking about mechanisms of OAE generation is provided in the text box titled, "Classifying OAEs." The next chapter is dedicated to this topic as well. We begin our exploration of the cochlea with an overview of the overall structure and function.

General Structure of the Cochlea

The cochlea in humans (as in other mammals) is a tapering, coiled cavity, separated by membranes into three, fluid-filled chambers. The broader end of the coil is considered to be the base of the cochlea and is proximal to the middle ear. The other end of the cochlea, considered its apex, is narrower than the basal portion. The three chambers or scalae within the cochlea are formed by Reissner's membrane, which separates the scala media (middle chamber) from the scala vestibule on top, and the basilar membrane, which separates the scala media from the scala tympani. The scala media (also called the cochlear duct), is bounded by tissue membranes above and below. As a result, the scala media is elastic and can move in either direction in response to pressure differentials between the chambers above and below.

The cochlea is encased in bone except for two membranous windows, the oval window separating the scala vestibule and the middle ear, and the round window separating the scala tympani and the middle ear. The footplate of the stapes occupies the oval window. Thus, motion of the ossicular chain in the middle ear is communicated to the cochlear fluids through the oval window. Any pressure applied to the oval window is communicated via the cochlear fluids from the scala vestibule to the scala tympani at the helicotrema, a small opening at the apical end of the cochlea where the two outer chambers meet. As a result of these anatomic features, an inward push of the footplate of the stapes results in a complementary expansion of the round window membrane into the middle ear space. This also creates a pressure differential above and below the cochlear duct (scala media) causing it to move, moving all structures contained within it, and setting off a chain of

Classifying OAEs

Immediately after their discovery, OAEs were classified on the basis of the nature of the stimuli used to evoke them. Thus, OAEs recorded without any explicit stimulation were named *spontaneous* OAEs. Those evoked using tonal pairs were named *distortion product* OAEs as they appeared at frequencies consistent with intermodulation distortion of the two stimulus tones. OAEs evoked using very brief stimuli (clicks) were named *Click-evoked* or *transient* OAEs, and those evoked using a single tonal stimulus were named *stimulus frequency* OAEs as the frequency of the emission was identical to that of the stimulus.

Although the above nomenclature is still prevalent and acceptable, especially as it is codified for billing purposes, a parallel classification system based on mechanisms of OAE generation has emerged. Kemp and his collaborators have proposed a dichotomy between *wave-fixed* and *place-fixed* OAEs (e.g., Knight & Kemp, 2000, 2001). OAEs are said to be wave fixed when their generators (assumed to be OHC) move or translate along with the envelope of the traveling waves of the stimuli. On the other hand, OAEs are said to be place fixed when their generators are fixed in place along the basilar membrane.

Shera and Guinan (1999) have developed a more formal model based on generation mechanisms. In their scheme, OAEs are generated by two mechanisms, *reflection* and *distortion*. Reflection emissions can be equated roughly to the place-fixed OAEs, and distortion emissions can be equated roughly to the wave-fixed OAEs. The OAEs measured in the ear canal could also be "mixed" in origin where contributions from each mechanism are present.

electrochemical events that lead to the generation of OAEs. Of course, the same sequence of cochlear events also results in the sensation of hearing.

The scala vestibule and tympani are filled with typical extracellular fluid (i.e., high sodium content) called perilymph. In contrast, the fluid in the scala media, known as endolymph, is low in sodium and high in potassium. This unique difference in chemical composition between the scala media and the other two chambers turns out to be very important in powering the active cochlear processes or the cochlear amplifier, believed to be the driving force behind OAEs.

The Cochlear Partition

The various structures central to the sensory function of the cochlea are collectively called the cochlear partition (Figure 2–3). The basilar membrane provides the foundation on which the cochlear partition rests. Bounded by the tectorial membrane on top, the basilar membrane houses the OHC and IHC along with a variety of supporting cells. The cochlear partition becomes approximately 10 times wider, 3 times heavier, and approximately 100 times less stiff as it extends from the base of the cochlea to the apex. These changes in the anatomy of the cochlear partition form the basis of the tonotopic organization of the cochlea. That is, the base of the cochlea is more sensitive to high frequencies and the apex of the cochlea is more sensitive to the low frequencies. It should be noted, however, that these anatomic changes by themselves cannot account for frequency range of human hearing (Ashmore et al., 2010).

The organ of Corti has two types of hair cells. Approximately 4000 flask-shaped IHC are arranged in one row along the length of the cochlea. Approximately 12,000 rod-shaped OHC are arranged in three rows. The hair cells are so named because of the "hair" or stereocilia found on the top surface of the hair cells. The stereocilia are

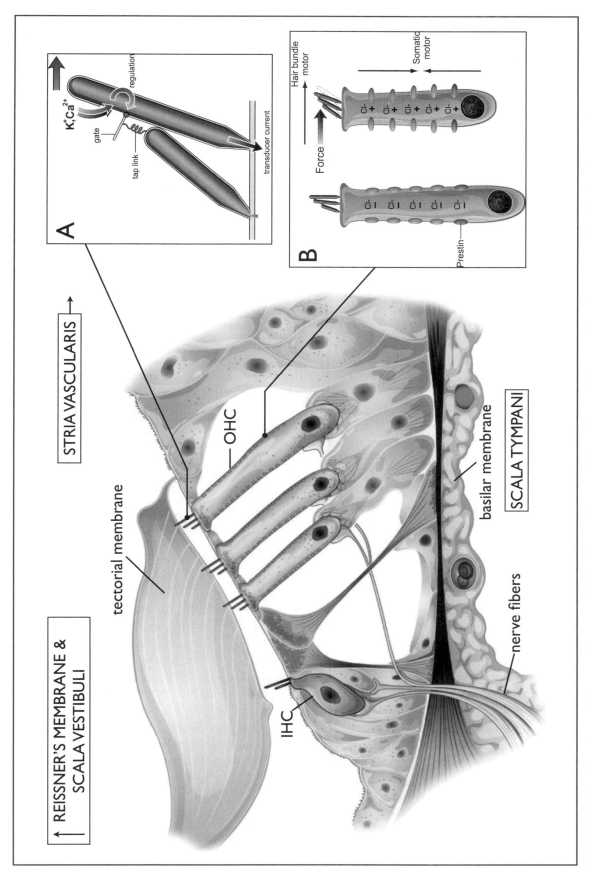

FIGURE 2–3. Structures in the cochlea constituting the cochlear partition. The cochlear partition is supported by the basilar membrane upon which sits the organ of Corti. The tectorial membrane forms the roof of the cochlear partition. Arrows and boxes are used to indicate the position of Reissner's membrane and the scala vestibuli above. Arrows and boxes are also used to indicate the position of the stria vascularis. The various types of supporting cells, which lend most of its weight to the cochlear partition are shown around the hair cells and in **B**. The boxes marked **A** and **B** display two candidates for the location of the cochlear amplifier and also the generator site for (at least distortion) OAEs.

arranged in rows, increasing in length with each row, and are connected to each other through delicate links near their tips and on the sides (Osborne, Cornis, & Pickles., 1984; Pickles, Cornis, & Osborne, 1984). The structural differences between IHC and OHC may already be evident from Figure 2–3. What is not obvious is that these two types of hair cells have completely different functions in the process of hearing. The IHC are primarily "sensory" in function as they mediate the final conversion of mechanical energy to neural signals. On the other hand, the outer hair cells perform a mechanical function by providing amplification to low-level signals (see Dallos, 2008, for a recent review). The gelatinous mass of the tectorial membrane forms the roof of the cochlear partition. It is hinged to the cochlear wall at its inner edge and by delicate filaments to the organ of Corti at the other edge. The tallest stereocilia of the OHC (*but not the IHC*) are embedded into the tectorial membrane.

The innervation to OHC and IHC is significantly different. The vast majority of the afferent auditory nerve fibers form one-to-one connections with single IHC after entering the cochlea through the central modiolus. A minority (~5%) of the afferent neurons cross the organ of Corti and connect with groups of OHC. It is universally accepted that afferent innervation of OHC and IHC has little, if any, impact on OAEs. However, OAEs are affected by the efferent fibers of the vestibulocochlear nerve. These fibers are part of a neural network whose cell bodies are in the brainstem. A small number of fibers (a few hundred) divides into numerous nerve endings on entering the cochlea, and terminate either on the afferent fibers underneath the IHC or directly on the OHC.

The final group of structures that is absolutely critical to the function of the cochlea and to the generation of OAEs consists of the stria vascularis and the spiral ligament. These structures are located on the outer wall of the cochlea and are responsible for maintaining ionic balance and the cochlear potentials. The stria vascularis is thought to be responsible for the secretion of endolymph and the generation of the endocochlear potential. The spiral ligament appears to be responsible for recirculating potassium ions from the organ of Corti to the stria vascularis.

Cochlear Processes and OAEs

In relation to OAEs, three aspects of cochlear function need to be understood. First, the passive mechanics of the cochlea that are driven almost entirely by the physical properties of the cochlear partition are important for the propagation of the stimulus energy to the region of the cochlea where the OAE is generated. The role of these mechanics in the propagation of the OAE back to the middle and outer ears is also important, and is currently under debate. Second, the electrical arrangement within the cochlea and organ of Corti is important in connecting the passive mechanics to the final aspect of active processes in the cochlea. These active processes are linked most directly to OAEs, but the exact nature of this link is yet to be fully understood. We will discuss the current status of knowledge in this area but the reader is warned that the discussion here is by no means inclusive of all details of cochlear physiology. The reader is directed to one of several recent reviews for a more complete discussion of these issues (e.g., Ashmore et al., 2010; Dallos, 2008; Robles & Ruggero, 2001).

The Traveling Wave

The basic understanding of energy propagation in the cochlea was largely derived from experiments conducted by the Hungarian biophysicist Georg von Békésy for which he received the Nobel Prize in Physiology or Medicine in 1961. In these pioneering experiments, Békésy used clever stroboscopic techniques to study the movement of the cochlear partition in human cadavers. The result was the classic image of a traveling wave with stimuli of different frequencies causing maximum disturbance along different lengths of the cochlear partition (tonotopicity) (Békésy, 1949, 1952). The traveling wave is a result of the pressure differential between the scala vestibule and tympani causing the cochlear partition to move either up or down. It is worth emphasizing that traveling waves in response to stimuli of different frequencies "travel" different distances along the cochlea partition before reaching the peak of their activity pattern due to the physical properties of the cochlear partition. The changing width, mass, and (mainly) stiffness of the cochlear partition dictates that stimuli of different frequencies find resonance at different locations along the length of the cochlea. We must clarify here that the traveling wave does not necessarily travel along the basilar membrane. A narrow cut (discontinuity) along the basilar membrane has no effect on the traveling wave. Rather, the traveling wave is an outcome of the pressure difference between the fluid chambers above and below the cochlear partition.

Today we know that the traveling wave patterns observed in cadavers lacked the very important element of active cochlear mechanics. That is, these dead cochleae would not have generated OAEs. Scientists suspected that passive mechanics leading to broad peaks of the traveling waves did not provide the complete story, as they could not be reconciled with the acute sensitivity of hearing and sharp frequency discrimination observed by psychophysicists. Rhode (1971) unearthed the first signs of active mechanics in the cochlea from measurements of basilar membrane vibration in squirrel monkeys. Evident for the first time in these recordings were critical facets of cochlear mechanics, such as compressive nonlinearity and also highest gain at lowest input levels constrained to a narrow region around the characteristic frequency place for a given stimulus. The search for the source of this nonlinear amplification is still ongoing, but strong candidate mechanisms have been identified. Before we can discuss these candidate mechanisms, we need to briefly discuss the electrical environment of the cochlea, which mediates the active processes.

Cochlear Potentials

We already have mentioned the uniqueness of endolymph (rich in potassium) among extracellular fluids. This richness in potassium leads to a positive charge or potential in the endolymph in comparison to any other part of the body. This is known as the *endocochlear potential*. The endocochlear potential can be detected by inserting an electrode (imagine a microscopic prong from a voltmeter used to test batteries) into the scala media. Similarly, *resting potentials* can be recorded by inserting a microelectrode into individual hair cells. The resting potential of the OHC is approximately –70 mV (Dallos, Santos-Sacchi, & Flock,

1982). Note that we are deliberately avoiding discussion about the resting potential of IHC. Of interest to us, the difference between the endocochlear potential and the resting potential of the OHC causes a gradient of approximately 150 mV. This gradient then constantly drives potassium ions from the endolymph into the hair cells. Displacement of the cochlear partition causes a shearing action between the basilar and tectorial membranes, resulting in a bending of the stereocilia of the OHC. Recall that the tallest of the stereocilia of the OHC is embedded into the cellular matrix of the tectorial membrane. The bending opens more ion channels in the stereocilia, increasing the flow of potassium ions into the OHC and depolarizing the OHC. Depolarization of OHC is the trigger for the activation of electromotility (shortening and subsequent elongation) of the OHC.

Other important cochlear potentials include the *cochlear microphonic* and the *summating potential*. Both the cochlear microphonic and the summating potential can be recorded from various recordong sites including intracochlear, round window, and ear canal locations by placing one electrode each in the scala vestibuli and the scala tympani. The cochlear microphonic is a fast response matched on a cycle-by-cycle basis with the stimulus, whereas the summating potential is recorded as an overall shift in the baseline voltage. Both potentials are generated primarily by the OHC (Dallos, 1975), but more recent evidence suggests that IHC activity contributes to the generation of the summating potential as well.

Active Cochlear Processes and Their Link to Cochlear Distortion

Overwhelming evidence suggests that there may be two candidates for the biophysical basis of the mammalian cochlear amplifier. One is motility in the stereociliary bundles or hair bundles (see Figure 2–3A) driven mainly by calcium currents (Martin & Hudspeth, 1999). The other is the voltage driven somatic motility (see Figure 2–3B) mediated by the protein prestin (Zheng, Shen, He, Long, Madison, & Dallos, 2000). In brief, current understanding appears to suggest that neither mechanism can fully account for the amplification characteristics observed in various mammalian species. Whether the hair bundle motor can deliver enough power to amplify the motion of the entire cochlear partition is questionable. On the other hand, it is not clear how a somatic motor can deliver amplification on a cycle-by-cycle basis at frequencies higher than 2 to 4 kHz due to the inherent low-pass characteristics of the basolateral membrane of the OHC. Consensus seems to be emerging that these two mechanisms somehow collaborate to bring about cochlear amplification, as we know it. Here in a brief paragraph, we have attempted to do the impossible, summarize the current knowledge on perhaps the most important question in hearing science. Needless to say, there is more to this issue and the reader is directed to reviews and opinion papers published recently that elucidate the main issues at hand in much greater detail (e.g., Ashmore et al., 2010; Dallos, 2008).

So which of these mechanisms is responsible for the generation of OAEs? Fortunately, the magic of genetic engineering has helped us start exploring this question. Genetically engineered mice lacking the motor protein prestin have been created and their phenotype studied extensively. The OHC in these mice cannot change length and the mice exhibit a 40- to 60-dB hearing loss. However, DPOAEs and other distortion products (in cochlear microphonics) can be recorded in these mice albeit with stimuli that are considerably higher in level (Cheatham, Huynh, Gao, Zuo, & Dallos, 2004; Liberman, Zuo, & Guinan, 2004). Imagine DPOAE input/output functions from these genetically modified mice to be shifted toward the left or higher input along the input axis. This resembles the expected outcome in conductive hearing losses where the signal is attenuated but function is restored once that attenuation is compensated for. In contrast, mice where another gene, stereocilin, has been inactivated do not produce DPOAE or distortion of the cochlear microphonic (Verpy et al., 2008). Lack of stereocilin disrupts the top connecters between stereocilia and makes them less organized and probably alters the stiffness of the bundle. These data appear to suggest that the actual distortion happens in the stereocilia and the somatic motor acts as the power amplifier resulting in distortion at lower input levels.

Reverse Propagation of OAEs

We have discussed the mechanisms that might be responsible for the generation of OAEs. Once their generation is complete in a certain region of the cochlea, the OAE energy still has to travel to the base of the cochlea, out through the middle ear, and into the external ear canal. The traditional view had been that if the stimuli enter the cochlea and arrive at their characteristic frequency region via a traveling wave, the OAE must also travel out of the cochlea via a *reverse* traveling wave. However, a reverse traveling wave may not be a viable mode of energy transport in the cochlea. For example, in Békésy's early experiments, he had attempted to stimulate the cochlea from within. However, he still observed a wave that appeared to be traveling from the base to the apex of the cochlea. Explicit experiments have been attempted in the recent past to understand the mode of propagation of OAEs.

The phase of basilar membrane vibration can be used to ascertain the direction of propagation of a traveling wave. Imagine a snapshot of basilar membrane motion at any given instant. Now imagine you were able to measure the phase of motion at points A and B along the length of the cochlea, where the point B is more apical. If the traveling wave were propagating toward the apex, the phase at point B would show a lead compared to the phase at point A. Another way of thinking about this would be to imagine two timekeepers sitting at points A and B along the cochlear length. As a stimulus wave is released at the base of the cochlea, these observers are asked to set their watches by radio communication. Then they each mark the time at which the traveling wave passes by their watch posts. In case of a traveling wave propagating toward the apex, it would pass by observer A before it passes by observer B. The opposite would be true if the traveling wave were propagating from the apex to the base. Thus, by comparing the time records of observers A and B or the recorded phases at multiple points along the basilar membrane, a determination can be made about the direction of propagation of a traveling wave. Using this logic in recording the phase of basilar membrane motion at multiple points has consistently shown a forward-traveling wave and not a reverse-traveling wave (He, Fridberger, Porsov, Grosh, & Ren, 2008). Interestingly, the results of a similar experiment conducted by measuring the pressure response in cochlear fluids do support the existence of reverse traveling waves (Dong & Olson, 2008).

The alternate to a reverse traveling wave would be the propagation of OAEs via an acoustic compression wave through the cochlear fluids (He, Fridberger, Porsov, & Ren, 2010). The critically important difference between these two modes of propagation of OAE energy is in the speed of transfer. A reverse traveling wave would be a much slower mode of energy transport that would require milliseconds for the emissions to arrive at the external ear from their generation site. In contrast, OAEs would arrive in the ear canal almost instantaneously if traveling via acoustic compression waves. The exact mode of propagation of OAEs may have a significant impact on theoretical models of OAEs and the clinical interpretation of

OAE data. These issues are discussed in detail in the next chapter.

CONCLUDING REMARKS

Even in this brief overview of the anatomy and physiology related to OAEs, it becomes clear that a complex chain of events has to be executed without the slightest glitch for the successful recording of OAEs. Both the external and middle ears play a double role in OAE recordings as the stimuli pass forward through these structures and the OAE have to return to the microphone using the same path. Although there is no question that OAEs are inextricably intertwined with the active cochlear processes in the OHC, the exact nature of these biophysical processes and the mode of propagation for OAEs are still not fully understood three decades after their discovery.

3

Classification of OAEs

INTRODUCTION

Writing this chapter is equivalent to telling a story that is yet to unfold fully. We have described the ongoing debate about the physiological mechanisms responsible for the generation and propagation of OAEs in Chapter 2. Critical issues related to the physiology of OAEs remain to be settled. Differences in the details of these processes in different species are becoming more evident. Interestingly, the acoustic outcome appears to be grossly similar across species even when the underlying mechanisms leading to that outcome could be very different. In Chapter 4, we argue that the vast majority of experimental OAE data have been recorded using calibration techniques that could be improved upon. Thus, the full story of OAEs is yet to be told. In this chapter, we outline the current thinking about different OAE types from two perspectives. First, we describe the more traditional classification system for OAEs based on the type of stimulus used to record them (Figure 3–1). Then, we discuss a more mechanism-based approach in classifying OAEs. In this category, we highlight the phenomenological differences between different OAE types that provide insight into the responsible mechanisms

and discuss the models behind this new type of classification. Although the models we refer to have very rigorous and formal representations in the literature, our approach is to build them as "thought" models. It is important to point out that discussion about these models and mechanisms will be confined to mammalian ears. The goal is to convey the spirit of the thinking behind the models and provide the clinician an entry into world of biophysics where these models are born and bred.

STIMULUS-BASED CLASSIFICATION

The traditional method of classifying otoacoustic emissions is based on the stimuli utilized to record them in the ear canal. These common types of OAEs are shown in Figure 3–1. The reader should note that the list of evoked emissions is based on the types that have been explored in the literature, some of which are in clinical use today. However, the variety of stimuli that can be used to evoke OAEs is essentially infinite. This method of classifying OAEs is well aligned with the procedures used in the clinic and allows us to discuss the common properties of each of these OAE types.

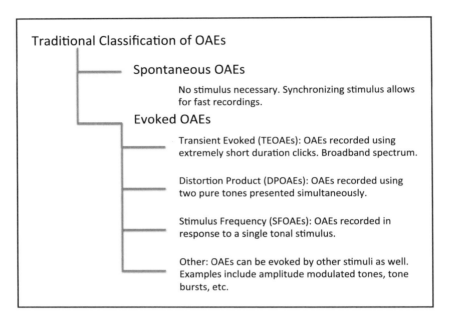

FIGURE 3–1. Schematic of the traditional classification system for OAEs. Based on the stimulus used to generate and record different OAE types, this system aligns with procedural considerations. Note that the enumeration of different evoked OAEs is essentially incomplete as they can be evoked by stimuli of infinite variety. The types shown in the figure are explored extensively in the literature and commonly used in the clinic.

Spontaneous OAEs

Spontaneous (S)OAEs are low level tonal signals that can be recorded in the ear canal without any external stimulation. The spectrum calculated from a relatively long (3 minutes) recording from the ear canal of a normal-hearing young adult female is displayed in Figure 3–2. The results of two recordings, 1 week apart, are displayed using solid and dashed lines. Each peak in the spectrum represents one SOAE. Note the stability of the spectra over a week; even though the peak levels of individual SOAEs fluctuate between the two recordings, an overwhelming majority of the SOAEs are present at approximately the same frequencies on both days. SOAEs that were present only on one of the two recording days are marked by asterisks on the figure.

The large number of SOAEs in one ear displayed in Figure 3–2 is certainly rare. However, estimates of the prevalence of SOAEs in normal-hearing human ears has grown over the years from less than 40% to approximately 80% (e.g., Kuroda, 2007; Strickland, Burns, & Tubis, 1985), with more SOAEs recorded in the ears of newborns than in adults. This increase in the estimated prevalence of SOAEs is most likely tied to the improvement in instrumentation leading to lower noise floors and hence the identification of SOAEs of lower levels that otherwise would be concealed by the noise floor of the recording equipment. In our laboratory today, we expect most if not all normal-hearing young adults to have detectable SOAEs, the SOAEs to be between −15 and 0 dB SPL in level (higher level SOAEs, although rare, are observed), and the SOAEs to be present between 0.8 and 4 kHz (SOAEs at higher frequencies are observed but more infrequently). When SOAEs are detected in an ear, there usually are three to four identifiable SOAEs.

SOAEs are thought to reflect the activity of the cochlear amplifier. In fact, Thomas Gold famously predicted the existence of SOAEs based on his expectations of the cochlear amplifier (see Chap-

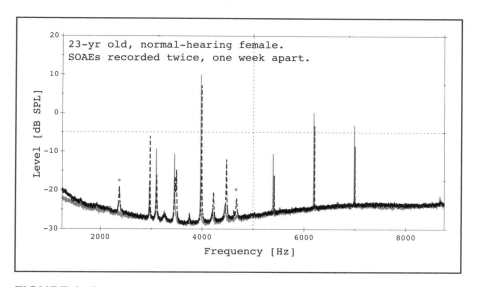

FIGURE 3–2. Spectrum of SOAEs recorded on 2 days, 1 week apart, from a normal-hearing young adult female subject. The SOAE spectra from the 2 days are represented by the solid gray and the dashed black lines, respectively. Note the multiple SOAE peaks present on both days. The levels of each SOAE fluctuate from day to day. However, the SOAEs are very stable in their frequency locations. Two SOAEs that are recorded on one day but not observed on the other are marked by asterisks on the figure.

ter 1 for details on the history of OAEs). Any and all insults to the cochlear amplification process, including chemotoxins (Kuroda, Chida, Kashiwamura, Matumura, & Fukuda, 2008; Long & Tubis, 1988) and age (Bonfils, 1989) also adversely affect SOAEs. However, the mere presence of SOAEs does not necessarily guarantee that a given ear will have better hearing sensitivity than a control group of ears without SOAEs. The functional relevance of the presence of SOAEs is far from clear. A few reports do suggest that ears with multiple SOAEs tend to have better thresholds than those without (e.g., McFadden & Mishra, 1993). Sharper psychophysical tuning curves have also been reported in ears with SOAEs (Micheyl & Collet, 1994). SOAEs have also been shown to be associated with hormonal levels (McFadden, 1993; Penner, 1995). Although these results would argue that ears with SOAEs are in superior physiological condition than those without, a considerable body of literature also portrays SOAEs as markers of cochlear damage. These paradoxical results are discussed next.

In a very interesting case study, Penner (1996) reported serial SOAE recordings from one subject over a period of approximately 3 months following attendance of a live music concert. The subject did not have a recordable SOAE before the event. But multiple SOAEs that appeared and disappeared at different frequencies on different days could be recorded over nearly 3 months, following which the SOAEs did not appear again. The author argued that the emergence and subsequent disappearance of these SOAEs signified a temporary degradation of the normal cochlear amplification processes. Indeed, such "noise-induced" SOAEs have been reported in the chinchilla (Clark, Kim, Zurek, & Bohne, 1984). Large SOAE also have been recorded in other animals and often been associated with regions of cochlear damage (e.g., Nuttall, Grosh, Zheng, de Boer, Zou, & Ren, 2004; Ruggero, Kramek, & Rich, 1984). Although the association of cochlear damage with SOAEs and the link between the cochlear amplifier and SOAEs may seem paradoxical, these may well represent two classes of SOAEs. The SOAEs recorded in normal-hearing ears are a result of the active cochlear mechanisms and a phenomenon of global resonance in the cochlea (discussed later). On the other hand, isolated, high-level SOAEs associated with cochlear damage could be a result of artificial punctate boundaries produced along the cochlear partition, which act as barriers or reflectors creating SOAEs. In that sense, this latter class of SOAEs would be similar to what Gold had envisioned—an amplifier gone astray leading to uncontrolled oscillations.

Recording SOAEs

Most laboratory recordings of SOAEs reported in the literature are truly spontaneous, that is, they are recorded without any external acoustic stimulation. A second method in which the ear is stimulated using clicks can also be used to record SOAEs. In this method, the click-evoked OAE is allowed to subside before evaluating the signal in the ear canal. Typically, this requires discarding the first 20 millisecond of the poststimulus inter-

val. What remains in the ear canal after the decay of the click-evoked emission is energy related to SOAEs synchronized by the click stimulus. These recordings, where the responses to a train of clicks can be averaged in the time domain, can be obtained much more quickly than when no stimulation is used. Typically, SOAEs obtained in this fashion are referred to as synchronized (S) SOAEs. SSOAEs and SOAEs have been shown to be approximately equivalent in the literature (Prieve & Falter, 1995; Wable & Collet, 1994). The levels and frequencies of SOAEs and SSOAEs recorded from eight normal-hearing young adults are compared in Figures 3–3A and 3–3B. The frequency and level of two SSOAE recordings are compared to the level and frequency of SOAEs recorded in the same subject at approximately the same frequency. Note that the data points in Figure 3–3B are aligned with the diagonal showing near equivalence between the frequencies of SSOAE and SOAE. The data points in Figure 3–3A show a little more variation. However, the difference in SOAE and SSOAE levels is no greater than

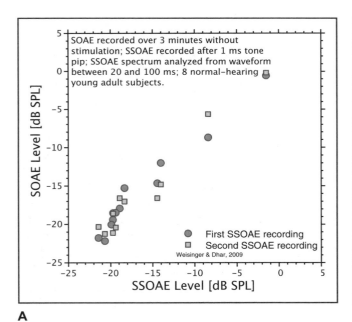

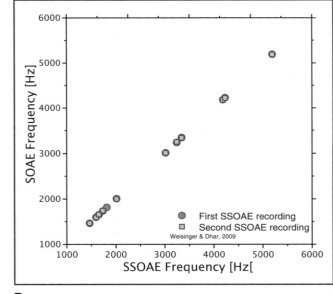

A **B**

FIGURE 3–3. The relationship between SOAE and SSOAE level and frequency in eight normal-hearing young adult subjects. **A.** The relationship between SOAE and SSOAE levels are presented where two measures of SSOAE level are compared to the corresponding SOAE level. **B.** The frequency estimates of two SSOAE recordings are compared to the corresponding SOAE frequency. These data were recorded by Coryn Weissinger during her AuD Capstone research project at Northwestern University.

the two level estimates of the SSOAEs. These data along with those previously presented in the literature give us confidence that SSOAEs can be used as an alternate measure of SOAEs.

SOAE Spacing

When multiple SOAEs are recorded in any given ear, the frequency spacing between adjacent SOAEs typically shows an interesting pattern. Note the spacing between adjacent SOAEs between 3 and 5 kHz in Figure 3–2. SOAEs in this frequency region appear to be evenly spaced. This spacing between adjacent SOAEs can be quantified by dividing the center frequency between two neighboring SOAEs by the frequency difference between them. That is, the frequency spacing between adjacent SOAEs is normalized for the center frequency between them, thereby accommodating the logarithmic frequency mapping of the cochlea. Computed in this fashion, the spacing between adjacent SOAEs appears to be remarkably stable over a large population of SOAEs taken from various reports in the literature (Shera, 2003). A modal value around 16 for this ratio of center frequency and frequency difference between adjacent SOAEs suggests that SOAEs are approximately equally spaced along the basilar membrane. There certainly are frequency regions where SOAEs appear to be farther apart. This is the case above 5 kHz in Figure 3–2. However, it could be argued that these are cases of dropped SOAEs. That is a missing SOAE creates a spacing that is essentially twice or another multiple of the nominal spacing. Although this discussion about SOAE spacing may seem esoteric at this point, we will return to it later in the chapter. We will connect this concept of spacing to the fundamental relationship between OAEs and cochlear mechanics and this will shape our discussion of a mechanism-based classification system for OAEs. Now we turn to evoked OAEs.

Evoked OAEs

Using a generic definition, any OAE recorded after the ear is stimulated by an external stimulus is an evoked OAE. We are most familiar with transient (click)-evoked OAEs and distortion product OAEs, as these are currently utilized heavily in the clinic. Entire chapters of this text are devoted to TEOAEs and DPOAEs. Thus, we will merely define TEOAEs for the sake of completeness of this classification scheme and will discuss certain aspects of DPOAEs that fit the scope of this chapter better. We do devote a portion of this chapter to discussing stimulus frequency (SF) OAEs, which are an important research tool at this time but may well become a part of the clinical arsenal in the near future. Readers should note that a variety of stimulus types, beyond those described above, have been used successfully to record OAEs. These include, but are not limited to, tone bursts of various frequencies (e.g., Epstein, Buus, & Florentine, 2004; McPherson, Li, Shi, Tang, & Wong, 2006; Zhang, McPherson, & Zhang, 2008) and amplitude modulated signals (e.g., Bian & Chen, 2011).

Transient Evoked OAEs

Transient evoked OAEs or TEOAEs are synonymous with click-evoked OAEs or CEOAEs. Measurement and analysis techniques usually employed to record TEOAEs in the clinic are discussed fully in Chapters 5 and 6. Briefly, clicks of very short duration (~80 microseconds) are used in combinations of click level and polarity such that averaging a combination set cancels out the click stimulus and leaves the emission response to be analyzed and displayed. The first few milliseconds of the poststimulus window typically are discarded before transforming the data to the frequency domain. The level of the TEOAEs in comparison to the noise floor is analyzed in narrow frequency bands to ascertain the physiological condition of the test ear.

Stimulus Frequency OAEs

Stimulus frequency (SF)OAEs are recorded using single pure tones. Although SFOAEs have not made their way into a regular clinical battery yet, they have played a very important role in our understanding of the mechanisms responsible for the generation and propagation of OAEs. Perhaps a significant roadblock to their incorporation into a clinical protocol is the difficulty associated with accurate measurements of SFOAEs. The problem should be self-evident—the OAE is generated at the exact same frequency as the stimulus tone. How can the OAE be reliably differentiated from the stimulus tone in the ear canal? Here we discuss the logic behind the methods employed to measure SFOAEs, but leave a deeper discussion about their role in exposing OAE mechanisms for later in the chapter.

Recording SFOAEs. Various methods have been used in the literature to extract the SFOAE from the ear canal signal. Here we discuss two methods that are not only common but also embody the general principles employed in extracting SFOAEs from the ear canal signal. Because the main hurdle is in the fact that the stimulus tone and the OAE are at the same frequency, the SFOAE is essentially extracted by comparing two snapshots of the ear canal pressure under different conditions. By design, the only difference between these two snapshots is the SFOAE of interest. Thus, the difference between the two snapshots yields the level and phase of the SFOAE. The two common methods by which this can be accomplished can be categorized under the general headings of suppression and compression.

Suppression. In the suppression method, the first recording is made using the probe tone at the frequency of interest, say 1 kHz. Usually, the probe tone used is at a low or moderate level, such as 20 or 40 dB SPL. Next, another recording is made but a suppressor tone, very close in frequency to the probe tone, is added to the signal. For example, the suppressor tone could be at 0.96 kHz and 60 dB SPL. The basic idea is that the presence of the suppressor tone removes the SFOAE from the second recording and all that is left is the probe tone alone at 1 kHz. Now vector subtraction is performed between the two recordings to extract the phase and magnitude of the SFOAE.

Compression. In the compression method, the first recording is obtained with a low or moderate probe tone (say, 20-dB SPL) at the frequency of interest (say, 1 kHz). Next another recording is obtained at 1 kHz but using a higher level tone (say, 60-dB SPL). The 60-dB SPL tone is 100 times larger in magnitude than the 20-dB SPL probe

tone. The assumption is made that, as the probe tone grew linearly between the two recordings, the SFOAE did not. The SFOAE is assumed to have been saturated and not changed between the two probe levels or to have grown in a compressive manner. The complex pressure in the ear canal recorded for the 60-dB condition is then scaled down 100 times. Now the scaled-down version of the signal is an exact replica of the probe alone for the 20-dB condition. Thus, vector subtracting the scaled-down version from the complex signal recorded for the 20-dB condition yields the phase and magnitude of the SFOAE.

The two methods of SFOAE extraction described above (as well as another method) have been shown to yield approximately equivalent estimates of SFOAE level and phase (Kalluri & Shera, 2007). As was mentioned before, SFOAEs are yet to find regular use in the clinic. However, initial attempts to use them for clinical purposes certainly show promise (Ellison & Keefe, 2005).

Distortion Product OAEs

Distortion product (DP)OAEs are recorded by stimulating the cochlea simultaneously with two pure tones. The cochlea generates additional tonal signals at frequencies arithmetically related to those of the stimulus tones. If the stimulus tones are designated to be at frequencies f_1 and f_2, respectively, a healthy cochlea can generate several DPOAEs at frequencies such as $2f_1$-f_2, $3f_1$-$2f_2$, $2f_2$-f_1, $3f_2$-$2f_1$, f_2-f_1, and so on. An example of a family of distortion products generated in a healthy human ear is displayed in Figure 3–4.

Multiple DPOAE orders are evident in the spectrum displayed in the upper panel of Figure 3–4. Note the family of DPOAEs lower in frequency to the stimulus tones (lower side band) and another family of DPOAEs higher in frequency to the stimulus tones (higher side band). The first DPOAE, just lower in frequency to the stimulus tones is at the frequency $2f_1$-f_2 and is most commonly used in clinical practice. In the remainder of this chapter, we will use the term DPOAE to refer to the particular DPOAE at the frequency $2f_1$-f_2. Other DPOAE orders will be specified by their actual frequency relationship with the stimulus tones.

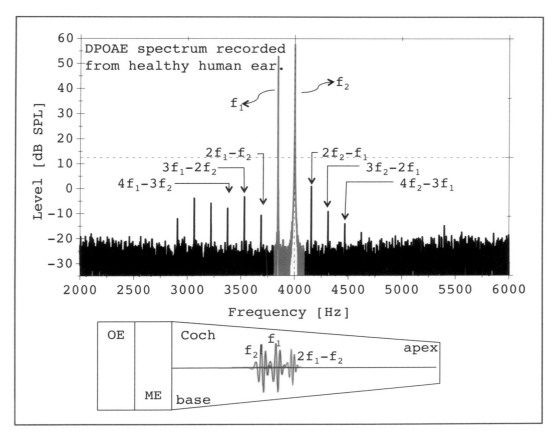

FIGURE 3–4. DPOAE spectrum recorded from a healthy human ear above. Schematic of traveling wave patterns in the characteristic frequency regions of the stimulus tones and the DPOAE at $2f_1$-f_2 below. Several DPOAE are evident in the spectrum and they all maintain a predictable arithmetic frequency relationship with the stimulus tones. Note the characteristic frequency region of the DPOAE at $2f_1$-f_2 is apical to those of the stimulus tones. OE = outer ear; ME = middle ear; Coch = cochlea.

The DPOAE at the frequency $2f_1$-f_2, and indeed all other lower side band DPOAEs share a common feature. The characteristic frequency (CF) regions for these DPOAEs are at regions apical to that of the stimulus tones. This is a direct outcome of the tonotopic frequency mapping of the cochlea where the base is most sensitive to high-frequency signals and the apex is most sensitive to low-frequency signals. It is well accepted at this time that DPOAE energy is generated in the region of overlap between the mechanical disturbances created by the two stimulus tones. The CF region for lower side band DPOAEs being apical to this overlap region, the mechanical gradient of the basilar membrane directs some of the DPOAE energy toward the CF regions of the respective DPOAEs. As we will see later, this complicates the DPOAE energy recorded in the ear canal.

Recording DPOAEs. We are most familiar with DPOAE recordings where DPOAE levels are displayed as a function of stimulus frequency in the form of a DPgram. Typically, these recordings are made using stimulus tones at various frequencies while the frequency ratio between them is held constant. This paradigm of DPOAE recordings is often referred to as the "fixed-ratio" method. However, this is not the only recording paradigm that can be used to record DPOAEs. Either the lower (f_1) or the higher frequency (f_2) stimulus tone can also be held at a fixed frequency while the frequency of the other is altered. When the lower frequency stimulus tone is held at a fixed frequency, the recording paradigm is often referred to as the "fixed f_1" or "swept f_2" paradigm. Conversely, when the higher frequency paradigm is held at a fixed frequency, the recording paradigm is referred to as the "fixed f_2" or "swept f_1" paradigm. Yet another recording paradigm is when the frequency of the DPOAE is held fixed and the frequencies of both stimulus tones are varied. In this case, the frequencies of the stimulus tones as well as the frequency ratio between them are varied to allow the DPOAE frequency to remain fixed.

MECHANISM-BASED CLASSIFICATION

Although it is convenient to classify OAEs based on the stimuli used to generate them, a different approach has gradually gained popularity over the last two decades. This second approach segregates OAEs into different types based on the mechanism responsible for them. At the core of this line of thinking is the following question: how is the reverse propagation of OAEs initiated in the cochlea?

By construction the basilar membrane in particular and the cochlear partition in general is built to aid signal propagation from the base to the apex. How then do OAEs make their way back to the ear canal? What triggers the reverse transmission of energy that eventually is recorded in the ear canal as OAEs? In the mechanism-based approach to thinking about OAEs, the trigger to this reverse propagation determines how OAEs are classified. Two distinct mechanisms are recognized: (1) reflections from random perturbations that are fixed in place along the basilar membrane and (2) reverse transmission induced by distortion caused by the stimulus or stimuli. Thus, the two mechanisms are thought to be either place fixed or fixed to the stimulus wave (Kemp & Brown, 1983; Knight & Kemp, 2000, 2001). In a parallel view, the two mechanisms are thought to be either reflection or distortion based (Shera & Guinan, 1999). Place-fixed and reflection-source emissions are essentially the same category as are wave-fixed and distortion-source emissions. Rigorous math-ematical models of these mechanisms have been developed and presented in the literature (Mauermann, Uppencamp, van Hengel, & Kollmeier, 1999; Shera & Guinan, 1999; Talmadge, Tubis, Long, & Piskorski, 1998). Here, we will discuss the basic concepts behind these models. Let us start with a thought model of reflection emissions.

Reflection Emissions

A highly schematized representation of the model of reflection emissions is presented in Figure 3–5. The basilar membrane is represented as a flat surface with many sources of perturbations on it. A schematized representation of the stapes marks the base of the cochlea. Silhouettes of outer hair cells represent randomly distributed irregularities or roughness that serve as reflectors along the basilar membrane. It is important to note that the outer hair cells do not have to serve as reflectors, although some evidence exists to support the role of irregular distribution of hair cells as the source of roughness (Lonsbury-Martin, Martin, Probst, & Coats, 1988). The traveling wave pattern in the background reminds the reader that the magnitude of vibration of the basilar membrane grows from base to apex and peaks near the CF region for that particular stimulus. Reflections (arrows in Figure 3–5) occur from any and all perturbations. However, the phase of these reflections depicted by the direction of the arrows is dependent on the phase of the stimulus at the reflector. The strength of the reflections is proportional to the magnitude of the incoming traveling wave. Therefore, the strongest reflections occur near the peak of the traveling wave. When a few of these strong reflections from near the peak of the traveling wave happen to have similar (coherent) phase, reverse transmission of energy is initiated and a portion of this energy escapes the cochlea to be recorded in the ear canal as OAEs. It should be noted that there may well be reflections from parts of the cochlea farther away from the traveling wave peak that are coherent in phase to those emerging near the peak of the traveling wave. Thus, reflectors farther from the peak of the cochlea theoretically can contribute to the ear canal OAE. However, because the strength of these reflections is small, their contribtion to the ear canal signal is not likely to be immense.

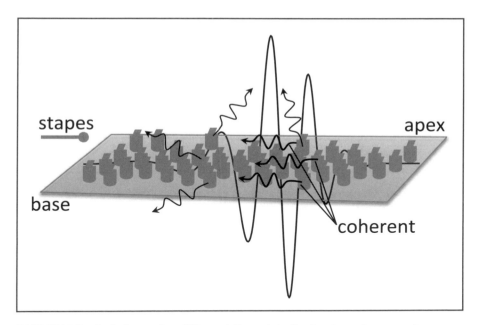

FIGURE 3–5. Schematic of "thought" model of reflections from random perturbations on basilar membrane (following Shera, 2003; Shera & Guinan, 1999). A simplified picture of the cochlear partition is presented with the stapes representing the source of energy input into the cochlea. The traveling wave pattern in the background reminds the reader that the magnitude of vibrations of the cochlear partition increases from base to apex, peaking at the CF region. The silhouettes in the shape of outer hair cells represent random perturbations that act as "reflectors" of this incoming energy. Note that the phase of the individual reflections (*represented by direction of the arrows*) are random. A few reflections from around the CF region happen to be in the same direction (have coherent phase).

SOAEs: Pure Reflection Emissions

Recall our discussion about SOAEs being evenly spaced along a log-frequency axis. This property can be modeled to be an outcome of the reflection-based mechanism of OAE generation. Imagine physiologic noises or even environmental noises acting as sources that set the basilar membrane in vibration thereby supplying the forward traveling energy along the basilar membrane. This forward traveling energy encounters the random roughness that was introduced above and the reflectors initiate reverse propagating waves that encounter another impedance mismatch as they arrive at the boundary between the cochlea and middle ear. As shown in Figure 3–6, these reverse propagating waves are then reflected back into the cochlea and travel toward their CF regions. In some locations, the total phase accumulation in this forward and backward propagation is a full cycle or multiples thereof and standing waves are set up (shown by solid arrows in Figure 3–6). In most other locations, this phase relationship is not met (shown in dashed arrows) and a standing wave cannot be sustained as the forward and reverse propagating waves cancel each other. In the locations where a standing wave can be sustained, the multiple internal reflections lead to a gradual increase in the magnitude of the reverse propagating wave and ultimately a portion of it leaks out into the middle and outer ears as an SOAE.

According to this view of SOAEs being an outcome of standing waves in the cochlea, the distance between regions along the basilar membrane where these standing waves can be sustained drive the frequency distance between adjacent SOAEs. In turn, this separation between

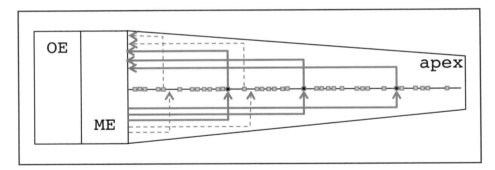

FIGURE 3–6. Schematic of global resonance model of SOAEs. Several perturbations are depicted using small boxes on the basilar membrane. Only a few reflectors (*filled black*) are in positions such that reverse and forward propagating waves add in phase creating standing waves in the cochlea that leak out as SOAEs. In locations where this phase relationship is not sustained, SOAEs cannot be formed (*shown in dashed arrows*).

SOAEs is then tied to the mechanical phase characteristics of the basilar membrane (Kemp, 1979; Shera, 2003; Zwicker, 1990).

Distortion Emissions

Distortion emissions are defined as those where the perturbation that causes the reverse propagation of energy is created by the stimulus itself. The two stimulus tones used to record DPOAEs do just that in the region of overlap between their mechanical vibration patterns on the basilar membrane. Nonlinearities in the mechanics of the cochlea create regional mechanical nonlinearities that cause reverse propagation of distortion energy that are ultimately recorded in the ear canal as DPOAEs. The distinguishing feature of this type of emission is that the perturbation causing the reverse propagation is not place fixed but is fixed to the wave of the stimulus itself. As we will see below, this has profound implications on the phase characteristics of distortion-source or wave-fixed emissions.

Mixed Emissions

Our discussion would not be complete if we did not map the different emission types discussed in the stimulus based classification system into this

mechanism based classification system. So far, we have accounted for SOAEs and DPOAEs (apical DPOAEs are complicated by multiple sources as discussed below). What about the other emission types? It turns out that both TEOAEs and SFOAEs receive contributions from both reflection- and distortion-based mechanisms. In the case of TEOAEs, the main mechanism appears to be that of reflection. However, TEOAE energy can be detected at frequencies outside those present in the click stimulus suggesting that intermodulation distortion between the various tonal elements of a click contributes to the TEOAE recorded in the ear canal (Yates & Withnell, 1999). Similarly, the role of nonlinearity has been modeled (Talmadge, Tubis, Long, & Tong, 2000) and experimentally documented in the case of SFOAEs (Siegel, 2005).

Apical DPOAEs

DPOAEs that are lower in frequency than the stimulus tones pose an interesting case. The distortion energy is certainly generated by wave-fixed mechanisms in the overlap region between the two stimulus tones. A portion of this distortion energy travels back toward the ear canal. However, as the CF regions of these DPOAEs are apical to the region of generation, a portion of the energy also travels to the CF region and is returned from there as a reflection emission. Thus, two components of DPOAE arrive at the ear canal, from different parts of the cochlea, generated by different mechanisms. This dual component, dual mechanism model is schematized in Figure 3–7.

This two-source, two-component model of apical DPOAEs has been confirmed experimentally by various groups (Heitmann, Waldmann, Schnitzler, Plinkert, & Zenner, 1998; Mauermann et al., 1999; Talmadge, Long, Tubis, & Dhar, 1999). The difference in the mechanisms responsible for the two DPOAE components manifests in differences in phase behavior of the two components as a function of frequency. In fact, phase behavior as a function of frequency is the key distinguishing characteristic between OAEs generated by the two mechanisms.

In the case of a wave-fixed or distortion-source OAE, the perturbation that triggers the reverse propagation of energy moves along with the stimulus wave as the frequency of the stimulus changes. Thus, the phase of any distortion source emission does not change appreciably as a function of frequency. On the other hand, the phase of a reflection-source emission is dependent on the phase of the incident wave at the reflector. As the frequency of the stimulus is changed, the location of the reflector remains fixed, and hence the phase of the new stimulus at the reflector changes dramatically. Thus, the phase of reflection-source or place-fixed emissions changes dramatically as a function of frequency.

Now, let us return to the ear canal where an apical DPOAE is being recorded in the form of a DPgram. This being an enhanced DPgram, DPOAE level and phase are being recorded every few Hz and not just at a few points per octave. The ear canal DPOAE is essentially a mixture of two OAE types, each with a very distinct phase characteristic. The phase of the DPOAE component from the overlap region does not change with frequency. On the other hand, the phase of the DPOAE component from the CF region rotates rapidly. Thus, as stimulus frequency changes, the phase relationship between the two components alternates causing alternating peaks and valleys in DPOAE level. Peaks are observed at frequencies where the two components are in phase. Conversely, valleys in DPOAE level are observed when the two components are out of phase. Such a pattern of peaks and valleys recorded from a healthy young adult human ear is presented in Figure 3–8.

As a confirmatory experiment we ask if this pattern of peaks and valleys can be abolished if one of the two components is eliminated from the signal in the ear canal. This is achieved by presenting a suppressor tone close in frequency to the DPOAE frequency simultaneously with the stimulus tones. The effect is evident in Figure 3–8. The dashed line in Figure 3–8 represents DPOAEs recorded in the presence of a suppressor. The DPOAE component from the CF region is eliminated, thereby eliminating the alternating pattern of peaks and valleys. This recording represents DPOAE energy generated in the overlap component only, to the extent that the suppression of the second component was complete.

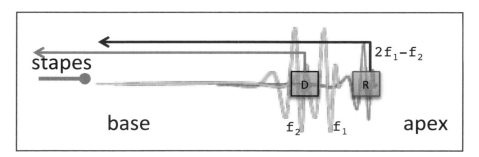

FIGURE 3–7. Schematic of two-component model of DPOAEs. See text for details.

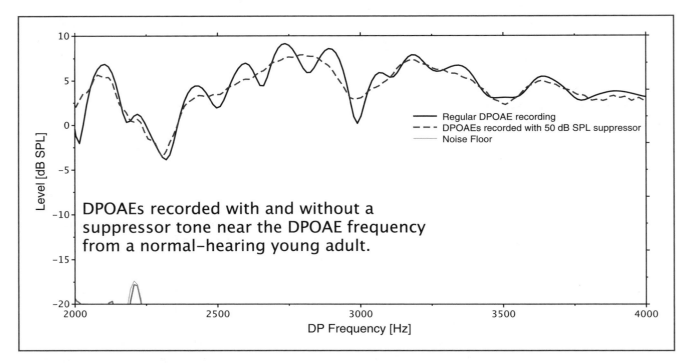

FIGURE 3–8. DPOAE level recorded at closely-spaced frequencies (*solid line*) showing a typical pattern of peaks and valleys in a normal-hearing young adult. The peaks and valleys are eliminated in the same ear when a 50 dB SPL suppressor tone is introduced along with the stimulus tones. Experiment conducted at Northwestern University by Ryan Deeter and Dr. Gayla Poling.

CLOSING COMMENTS

We have discussed two classification systems for OAEs. The stimulus-based classification system is convenient and straightforward. The mechanism-based classification system represents newer thinking in the field. According to this system, SOAEs represent reflection-source OAEs. Most if not all other OAEs are essentially mixed in nature. It could be argued that SFOAEs and TEOAEs obtained with very low stimulus levels could represent reflection-source emissions. Similarly, it could be argued that when the DPOAE component from the overlap region is isolated, it is a pure representation of a distortion-source emission.

In presenting these complex mechanisms and models, we have taken significant liberties and glossed over details. The reader is encouraged to refer to the primary literature to gain a full understanding of the models. Perhaps the most important fact to keep in mind when considering OAE generation is that OAEs are generated over distributed regions of the cochlea and not by isolated hair cells or other point sources. Thus, the behavior of the OAEs observed in the ear canal represents the collective properties of this entire array. The participating region in the cochlea is most likely stimulus level dependent with larger cochlear areas recruited in contributing to the ear canal OAEs with increasing stimulus levels (e.g., Martin, Stagner, & Longsbury-Martin., 2010). Thus, the frequency specificity of OAEs should be interpreted with care paying heed to the possibility of extended generation regions, the presence of multiple components, and the contribution of multiple mechanisms.

‖‖ 4 ‖‖

Instrumentation and Calibration

INTRODUCTION

We suspect that many of you might be like us—you wear many hats in a week, if not a day. You may be a clinician, a teacher, a manager, a scientist, a student, or chief troubleshooter for your clinic or your laboratory—sometimes all of them within a short span of time. The value of understanding the basics of instrumentation and calibration is of great importance to you in each one of these roles. In one instance, your chief concern may be to understand exactly what stimulus is being delivered to a subject's ears. In another instance, the concern may change to being certain that a test result is not a product of measurement artifacts. In yet another instance, the task may be to troubleshoot which of the several components of your OAE system may be malfunctioning. The ideas discussed in this chapter should help you in each of these situations.

INSTRUMENTATION

Clinical equipment for recording OAEs is ever evolving, much like ongoing technological advances in other medical devices. Thus, any attempt to describe OAE hardware or software available today in exact details is futile. The hardware and software in use today will soon be rendered obsolete, and replaced with more advanced, reliable, and, possibly, smaller and smarter versions. In this chapter, we describe the basic principles on which these devices are constructed. As the forms and functions of the devices change, these principles will remain. Specific functionality of clinical equipment is discussed to some extent in various other chapters.

Understanding calibration techniques and their consequences are important from two perspectives. First, proper calibration guarantees that the desired signal is being presented to the patient's ears. In the case of OAEs, calibration also guarantees that the recorded signal is a reasonable approximation of the physical signal that was present in the ear canal. Second, calibration gives us confidence in our test results. This is especially important when we are comparing test results for the same patient across time or across clinics. More generally, calibration also allows us to take the test results from an individual and be able to compare them against the results from a population with confidence.

Standards and Regulations

As clinicians, we are end users of electronic devices that are capable of providing the physiologic information needed to make informed decisions about our patients and their conditions. We count on standards, certifications, and regulations to ensure that the equipment we use adheres to the basic standards and requirements. Typically, clinical equipment is built to perform according to standards specified by the American National Standards Institute (ANSI) and the International Electrotechnical Commission (IEC). Both these organizations are non-profit, nongovernmental organizations dedicated to preparing and publishing standards. In essence, these standards are codification of practice in a field that is agreed upon by a contextually knowledgeable group of individuals. Currently, there are no ANSI standards for OAE equipment and calibration. The IEC standard 60645-6, published in 2009, provides guidelines for both screening and diagnostic OAE equipment. The IEC-60645-6 standards will be discussed at various points in this chapter.

All OAE devices marketed in the United States of America must be approved by the Food and Drug Administration (FDA). The FDA is an agency of the United States Department of Health and Human Services charged with the protection and promotion of public health through regulation and certification of, among other things, medical devices. In Europe, medical devices are regulated through directives issued by the European Union that define the operational requirements of the device. Additionally, devices are expected to obtain markings such as the "UL" or the "CE" marks that guarantee that the device meets basic standards of consumer safety.

HARDWARE

The basic instrumentation required to make OAE measurements is, in principle, similar to that used by any electronic musician to generate and record sounds. In that sense, a generic "sound card," either external or internal to a computer, and appropriate peripherals (speakers and microphones), should suffice for an OAE device. However, the demands of small size, wide dynamic range, and relatively distortion free performance make hardware used for OAE measurements unique and sophisticated, and, in turn, expensive.

The System

Any OAE system requires four stages as depicted in Figure 4–1. For the purposes of our discussion, let us call these stages the computer, the input/output (I/O) device, the signal conditioning stage, and the acoustic stage. The computer needs no introduction. The I/O device is a digital-to-analog converter (DAC) and an analog-to-digital converter (ADC). The signal conditioning hardware ensures that the speakers in the probe are optimally driven and the signal recorded by the microphone is adequate to be digitized by the ADC. Finally, the OAE probe houses the speakers that deliver the stimuli to the ear and the microphone that records the OAE in the ear canal. In a handheld OAE device, the first three stages are often built into a common housing. However, inside the housing, the functions of each are still separately performed by different electronic components. Let us discuss each of these components in more detail.

The Computer

The computer on a diagnostic device is often a desktop or laptop computer that is the command center of the entire operation. The same computer is often multipurposed for other clinical or office purposes. In the case of handheld units, the computer is built on a board that fits within the housing. The essential purpose of the computer is to control the generation and recording of signals by communicating with the I/O device via an industry-standard electronic communication protocol. These communication protocols are evolving with time with increases in equipment speed and reliability. We expect clinical equipment to continue to take advantage of these gains as these new standards become available. One dramatic change already evident is the move from wired to wireless communication protocols. Clinical devices using Bluetooth technology to transmit and receive data are already on the market. Yet another interesting trend could be the use of miniature handheld computers such as smart phones or tablets. In conversations with industry engineers and executives, it appears that, dedicated devices similar to the handheld units available today will be the norm, at least for the near term.

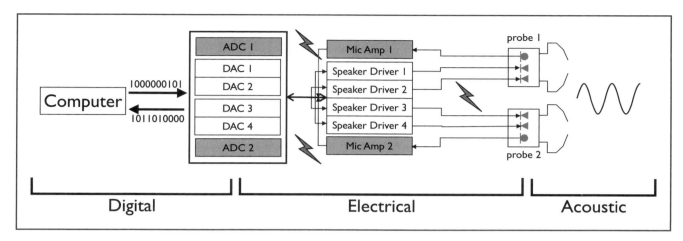

FIGURE 4–1. Schematic of the instrumentation necessary for the measurement of OAEs. Four main components or stages (described in the text) operate in three different domains (digital, electrical, and acoustic). The schematic shows two OAE probes that could be used to measure OAEs from both ears simultaneously. Alternately, the sound sources in the second probe could be used to deliver stimuli to activate the contralateral olivocochlear efferents (see Chapter 9 for details). Each probe is shown to house two miniature speakers and one miniature microphone. This configuration determines the need for four speaker drivers and two microphone preamplifiers. These components, in turn, are connected to independent digital-to-analog and analog-to-digital converters. These converters are controlled by the computer which houses the software.

The computer also analyzes and displays OAE data after they are recorded. Much of this functionality will be discussed later. However, we should mention here that the ability to do complex analysis quickly will improve as processors become faster and computers are built with more processors. Thus, we envision the improvements in computing power to translate to the availability of more complex analysis techniques leading to better results.

The Input-Output Device

The basic task of the I/O device is to convert signals between the digital and electrical domains. On the output side, the I/O device receives digital signals (often binary numbers) from the computer and translates them to electrical signals that can be fed to the speaker drivers. This is the process of digital-to-analog conversion (DAC). Typically, there are two DAC channels and both of them are utilized during DPOAE measurements, each generating one of the two stimulus tones. Four DAC channels are shown in Figure 4–1 to represent an OAE device that has two probes and allows simultaneous recordings from the two ears. A system

with two probes also would allow the delivery of a signal to the contralateral ear to evoke the efferent pathways (Chapter 9).

The I/O device performs the analog-to-digital conversion (ADC) for the signals recorded in the ear canal and delivers them to the computer for analysis and display. In this case, the I/O device receives electrical signals from the microphone preamplifier and sends the digital information to the computer. The sampling rate of the ADC is critical in determining the highest frequency electrical signal that can be reliably digitized. Per the Nyquest theorem, the highest frequency signal that can be accurately digitized by an ADC is *half* of the sampling rate of the device. Thus, for measuring OAEs in humans, a sampling rate of 32 or 44 kHz would suffice depending on whether we are interested in measuring OAEs up to 16 or 20 kHz.

The number of bits of an I/O device determines its dynamic range. A simplified way of thinking about the dynamic range issue is to imagine that each bit is a switch that can either be on or off. The switch in its off position represents a value of 0 and in its on position represents a value of 1. Thus, the amplitude of a signal, in this case, can be represented in only one of two states, either 0 or 1. Therefore, the dynamic range of this system can be represented as 2^n, where n is the number of bits (1 in this case). The dynamic range of this system will therefore be $20 * \log_{10}(2^1)$ or 6 dB. The I/O devices typically used for OAE measurements are either 16 or 24 bit systems, giving them a dynamic range of 96 or 144 dB, respectively.

One of the most critical properties of an I/O device for OAE measurements is synchronization between the ADC and DAC processes. Once a signal is sent from the computer to the I/O device to generate a signal, this also triggers the recording process. The time elapsed between the activation of the DAC and the ADC has to be strictly constant. It is perfectly acceptable if there is a delay between the two activations, but the delay has to remain constant. In cases where this synchronization between the DAC and ADC is not constant, all measures of OAE delay and phase are compromised. Additionally, when responses to multiple stimulus units have to be averaged, as in click-evoked or tone burst-evoked OAEs, improper synchronization causes errors in time-domain averaging leading to erroneous estimates of OAEs.

Speaker Drivers and Microphone Preamplifiers

Speaker drivers receive the electrical signal from the I/O device and pass it on to the speakers after modifications. These modifications may include amplification, impedance matching, frequency shaping, and so forth. Similarly, the microphone preamplifiers receive the electrical output from the microphone in the OAE probe and modify it before sending it to the ADC for digitization. Amplification is often necessary as the signal recorded in the ear canal is miniscule in magnitude. Compensation for the frequency response of the microphone also can be done at the microphone preamplifier.

OAE Probe

The OAE probe houses the speakers and the microphone(s). Typically, two speakers are used. The speakers could be physically housed in the probe assembly or external speakers could be used with their output routed to the ear canal through the OAE probe. More than one microphone is used in some OAE probes to improve the signal-to-noise ratio. Several aspects of the physical design of the OAE probe are important. First, the actual plumbing or tubing that carries the different output and input signals is critical. Suboptimal design of the tubing architecture can lead to *cross-talk* and other associated problems (see later discussion). The size and shape of the probe also is important. These features are critical in determining whether a probe is easy to fit in a variety of ear canals. The shape also determines whether a probe is comfortable for the patient. Finally, it is important that the OAE probe be designed with safeguards against easy entry of cerumen (earwax), vernix, or debris into the probe assembly. Similarly, it is important that the probe be easy to clean when its function is compromised by debris in one or more of the sound delivery or recording tubes and cavities.

Specifics of Standards

As mentioned before, ANSI standards for OAE equipment are not available at this time. However, IEC (60645-6) standards were recently published in 2009. Instruments are divided into two categories for screening and diagnostic purposes in this standard. The basic parameters and expectations of each class of equipment are summarized in Table 4–1.

Sources of Artifacts

Countless possible sources of malfunction and error in any typical OAE device may lead to the recording of artifacts. The device user may have few options for troubleshooting different components. The extreme example is that of a handheld unit where all components are built into a single case or enclosure. When possible, common sense troubleshooting techniques are applicable to OAE equipment. For example, if the device has multiple components, as depicted in Figure 4–1, ensuring that each component has power is the first step. It

Table 4–1. Select Specifications for Screening and Diagnostic OAE Equipment per IEC 60645-6

Specification	Screening	Diagnostic
Test Mode	Automatic	Automatic and Manual
Results Display	Pass/Refer	Full Details
Printout	Not Required	Required
Results Storage	Not Required	Required
TEOAE Frequency Range	1.5–3.0 kHz	0.5–4 kHz
DPOAE Frequency Range	1–4 kHz	0.5–8 kHz
TEOAE Signal Level	60–80 dB peSPL	30–90 dB peSPL
DPOAE Signal Level	50–65 dB SPL	0–70 dB SPL
Measurement Range	–20 to +30 dB SPL	–20 to +30 dB SPL

may also be worthwhile to power cycle each component independently. Finally, the output of each stage can be verified using appropriate means. The output of the acoustic stage can be checked through a listening check. In case the output of one of two channels is distorted, low, or absent, the cables to the probe can be switched between channels at the speaker driver stage. If the problem switches to the other channel, the malfunction is likely at the probe (one of the speakers may be malfunctioning). However, if the problem remains in the same channel as before, a malfunction in the speaker driver or I/O device is likely. Similarly, the output of the electrical stages (I/O device, speaker drivers, microphone preamplifiers) can be verified using a voltmeter or another appropriate device provided the connectors are accessible.

In our experience, a large proportion of malfunctions are due to debris (including cerumen and vernix) in the speaker or microphone tubes in the OAE probe. It is critical for the device operator to treat the probe with care and to thoroughly inspect visually the probe before and after every procedure. An otoscope may be handy to carefully examine the status of the sound tubes.

Next, we discuss two specific hardware related sources of artifacts that can compromise the accuracy of results.

System Distortion

Distortion due to hardware malfunction can masquerade as a biologic signal and lead to misinterpretation of results. System distortion usually occurs at the speaker but also can be due to a malfunctioning speaker driver or other components. The possibility of system distortion increases with increasing stimulus level. In fact, every OAE system distorts when driven at sufficiently high levels. Fortunately, system distortion can be quantified for each individual OAE system. The procedure is simple and is accomplished by conducting an OAE test in an appropriately sized cavity. This could be a 2-cc or smaller coupler. Syringes often are supplied with various OAE systems for this purpose. It may be even more accurate to conduct the test in a hard-walled cavity such as a hearing aid coupler or standard ear simulator. Once an OAE test is conducted, the software would report noise floor and some OAE levels. Under ideal conditions, no noticeable OAEs should be recorded above the noise floor. However, any apparent OAEs are certainly a sign of system distortion. In this case, the level of system distortion becomes the relevant noise floor and only OAE levels above this new noise floor should be considered biologic in origin. Because system distortion is dependent on output level, this test should be conducted at all stimulus levels (or level combinations) that one intends to use. Acceptable noise floors should then be established for each of the stimulus levels.

Cross-Talk

Yet another source of artifacts is cross-talk, defined as an undesired coupling of the output and input of the system. Stated differently, the microphone of an OAE system should register only the pressure signal in the ear canal. However, if the microphone and the sound source components are coupled in any other way, the output of the microphone would not represent the actual signal in the ear canal. This undesired coupling could be due to leakage of the signal inside the probe housing that is detected by the microphone. The microphone also could detect vibrations in any part of the sound delivery system. Problems with cross-talk in several commercially available OAE systems have been reported (Siegel, 1995). Measurement of cross-talk, and correcting the problem if detected, is not convenient for the clinician. Those interested in measuring cross-talk in their systems can follow the procedure described by Siegel (2007). Fortunately, in modern clinical devices, the levels of cross-talk are sufficiently low, at least in the frequency range below 8 kHz.

SOFTWARE

The software accompanying any clinical OAE device is custom designed by the manufacturer. However, all such software implementations share some common features in signal delivery, measurement, analysis, and display. The IEC standard (60645-6) is rather broad when describing requirements with only general guidelines for screening and diagnostic devices. For example, the standard requires that screening devices display only a "pass/refer" outcome. On the other hand, diagnostic devices are expected to display signal, OAE, and noise levels, along with the number of artifacts and the artifact rejection limit.

The digital signal processing involved in generating, recording, and analyzing signals involved in OAE measurements are generally hidden from the clinician. Similarly, different manufacturers implement different user interfaces for controlling the test process and displaying test results. However, there are some common elements to these user interfaces. For example, the interface allows the clinician to enter patient information, select the type of OAE to be measured (e.g., TEOAE or DPOAE), and determine the signal parameters (e.g., level, frequency, and frequency ratio). The clinician also is able to control the limit of artifact rejection during OAE data collection.

Variations in the implementation of these basic principles in different software systems make it impractical to discuss individual implementations. The clinician is encouraged to study the operator's manual and seek information from the manufacturer's representative about their specific equipment. Below, we describe a few interesting facets of software that are either unique to one manufacturer or common to several. These features are described so that the clinician can take full advantage of these innovative solutions.

Checking Probe Fit

Most clinical OAE software packets have built-in functions to evaluate the quality of the fit of the probe in the ear canal. Typically, probe fit is assessed with an analysis of the response to a broadband signal presented to the ear canal. Specific acoustic features are then identified to detect a leak in the fit of the probe, a blocked probe, or other anomalous conditions. The clinician is immediately given visual feedback about the problem, and further data collection is prevented by the software.

Sound Level Meter

At least one clinical OAE device includes a built-in sound level meter. Device software analyzes the sound level recorded by the probe microphone. Measurements can be initiated only if the detected ambient noise levels are acceptable for the conduction of OAE measurement. Detection of ambient noise conditions is especially useful when conducting OAE measurements in less than optimal environments, such as a hospital nursery or a newborn critical care unit.

Automatic Calibration or Leveling

Typically all OAE measurement software includes functions for calibrating stimulus levels and ensuring that the levels produced by two channels or speakers are comparable (applicable for DPOAE measurements). Calibration procedures are discussed in detail later in this chapter. Briefly, the level of a stimulus signal delivered in the ear canal and measured by the microphone built into the OAE probe is used to adjust the presentation level, arguably to accommodate individual differences in ear canal anatomy and insertion depth. Identical signals played through the two output channels and speakers are compared to ensure that the acoustic outputs of the two channels are approximately identical. In cases where minor errors or differences are observed, the output can be adjusted to compensate for these errors. However, when larger errors or differences are identified, the clinician is warned and often further progress in measurements is prevented.

Stopping Rules

The use of predetermined stopping rules is a potent tool for obtaining useful OAE measurements. Typically accessed through "Preferences," "Configuration," or "Setup" options on clinical OAE devices, these rules allow efficient data recording while mitigating the influence of ambient or physiologic noise. Several factors such as OAE amplitude level, noise floor level, signal-to-noise ratio, and maximum recording time can be set before data collection. Once programmed into the software, the OAE recording continues till the set criteria are achieved. To use the example of a DPOAE recording at any given frequency, the recording could continue until the DPOAE level reaches a predetermined level (say 10-dB SPL), or the noise floor reaches a predetermined level (say −20-dB SPL), or the signal-to-noise ratio reaches a predetermined level (say 6 dB), or a predetermined length of time has elapsed (say 12 seconds). When conditions are optimal, OAE levels are in or above the normal range, and the noise floor is sufficiently low, the recording time at each test frequency is reduced to the minimum necessary. Thus, the overall duration of the test is reduced and clinical efficiency is gained.

Artifact Rejection

OAEs are minute acoustic signals that are easily contaminated by ambient and physiologic noise within the external ear canal. Clinical equipment incorporates built-in functions to recognize and reject artifacts, thereby making the test process more efficient and the results more reliable. Each manufacturer may have proprietary techniques for detecting and rejecting artifacts, although information about some of these techniques may be available in the scientific literature or in patent databases. However, three general techniques for artifact rejection can be described.

First, ambient noise that is not within the frequency range of interest can be eliminated by filtering the incoming signal. This technique is usually implemented through a high-pass filter that is designed to reject all signals below a predetermined cutoff frequency (say, 300 or 500 Hz). Such a high-pass filter can be implemented in hardware between the microphone output and the input to the microphone preamplifier. A high-pass filter also can be implemented in software. In this scenario, the full bandwidth of the input signal is digitized. Unwanted portions of the recorded signal are then eliminated using a digital filtering algorithm.

Second, signals that are extraordinarily large in magnitude can be rejected prior to or after digitization. The premise is that the biologic signals of interest (e.g., OAEs) are small in magnitude and signals above a predetermined threshold are artifacts. The limit of this type of artifact rejection often is controllable by the clinician. The threshold of artifact rejection can be raised to allow more of the incoming signal into the recording. This modification is useful when the patient or the environment is noisy as it allows the test to be completed without compromising the quality of the data.

Finally, artifacts can be identified on the basis of repeatability. The incoming signal can be divided into two parts (buffers) and the averaged data in each buffer can be compared. In cases where there is a significant difference between the averaged signals of each buffer, the signal is interpreted to be an artifact. Often, the maximum allowable difference between the buffers is predetermined and cannot be altered by the clinician.

Automatic Decision Making

Equipment used for auditory screening with OAEs presents the results as a "pass/refer" decision. The rules that determine the outcome of the screening test are preset in the software, but often can be altered by the clinician. Typically, there are two steps in making a pass/refer decision. In the first step, a pass/refer decision is made for an individual test frequency. OAE level, noise floor level, and signal-to-noise ratio, either in isolation or in combination, are used to make a pass/fail judgment at a certain test frequency. Once decisions have been made for each test frequency, overall pass/refer decisions are made based on the proportion of test frequencies at which a pass or fail occurred. For example, a signal-to-noise ratio of 6 dB could determine a "pass" at each test frequency. Then, a "pass" decision in three out of four test frequencies may be required to make an overall recommendation of "pass."

CALIBRATION

At the outset of this chapter, we discussed the importance of calibration from two perspectives. First, accurately calibrated equipment provides confidence in our test results. Second, uniformly calibrated equipment across clinics allows us to compare test results against norms as well as across clinics, or even across measurements made on different days at the same clinic using the same equipment. We should note that standards for calibration of OAE equipment are not yet well established. Therefore, the clinician must be cognizant of the calibration method used in obtaining test results and also the calibration method used in obtaining any norms that will be used for analyses of test results for patients. Given the uncertainty in uniformly applied calibration procedures, we pragmatically advise clinicians to obtain a set of normative data from persons reflecting expected patient ages using their particular equipment in their typical test settings. When using OAE equipment with built in normative ranges, it becomes the clinician's responsibility to investigate and ensure that the normative data were obtained using the same calibration technique that is implemented in the instrument in question.

Goals

The goals of calibration in the case of OAE equipment are straightforward. The clinician needs to be reasonably confident that accurate stimuli are being delivered to the patient's tympanic membrane. The clinician also needs to be confident that the representations of the OAEs and the noise floor are reasonable approximations of the pressure signal recorded in the ear canal. Therefore, both the output (stimuli) and the input (OAEs + noise floor) of clinical OAE equipment need to be calibrated.

The clinician rarely actively calibrates either the input or the output portions of clinical OAE equipment. The input, including the microphone, is factory calibrated and the clinician's responsibility is to remain vigilant in monitoring calibration and detecting any obvious errors. The output, on the other hand, is calibrated "on the fly" for every patient measurement conducted by the clinician. Even in the case of output calibration, the process is automatic and entirely controlled by software. Therefore, the clinician cannot actively alter the process in any way. The clinician's role is to detect when an error in calibration occurs and to rule out any obvious malfunctions. In most cases, the malfunctions that can be alleviated in the clinic involve a bad connection in the hardware chain or calibration problems related to debris in either the speaker or microphone ports.

An accurate method for calibrating the microphone of an OAE system is described by Siegel (2007), and will not be discussed further here. Our discussion will focus on various methods available for calibrating the output. Again, the discussion is presented so that the clinician is aware of the principles underlying various calibration techniques and is able to ascertain the impact of these different techniques on the recorded OAEs.

Calibrating the Output

Let us begin by restating the purpose of calibrating the output of an OAE system—to deliver an accurate stimulus signal to the patient's tympanic membrane. Here the term "accurate" essentially covers three signal domains: (1) amplitude or level, (2) frequency, and (3) time. The temporal domain becomes most relevant in the case of OAEs recorded with brief stimuli (e.g., TEOAE). The domains of signal level and frequency are relevant to all stimuli used to record OAEs. Our discussion will focus on the complexities in delivering accurate stimulus levels across patients and across multiple measurements on the same patient. This focus is warranted because stimulus level is most variable across individuals and between tests. Intersubject variability is reflective of individual differences in ear canal geometry and volume, whereas intrasubject variability arises from differences in probe insertion and fit between tests.

We begin by describing a simple calibration method that is neither accurate nor utilized in clinical equipment. However, we will use this as a starting point to highlight the drawbacks of such a simple technique, and to introduce a discussion of more complex and arguably accurate methods.

Coupler Calibration

Imagine that there existed a standard coupler or ear simulator such as the Zwislocki coupler (ANSI S3.25-1989-R1995) or the IEC 711 ear simulator (IEC-60711, 1981) whose acoustic characteristics were identical to those of any individual's ear canal. Each of these couplers can be fitted with a microphone at the distal end that could serve as the surrogate of the tympanic membrane. With these tools at hand, let us embark on a calibration exercise. The goal of the exercise is to deliver a 65-dB SPL tone at 1 kHz to the tympanic membrane. We begin by driving one of the sound sources in the OAE probe with a known electrical signal of 1 kHz. We then measure the sound pressure level created in the ear simulator or coupler using the microphone at the distal end. The voltage of the electrical signal is then adjusted until we achieve the desired (target) 65 dB SPL measurement at the microphone. Calibration is complete because all we have to do is to supply the exact same voltage to the sound source every time we wish to deliver 65-dB SPL at 1 kHz.

Several complexities prevent this procedure from being the "gold standard" for calibrating OAE probes. First, we made an assumption about the "flatness" of the frequency response of the sound source. That is, we assumed that the electrical signal that produces 65-dB SPL at 1 kHz also will produce the exact same sound pressure levels at all other frequencies. The validity of this assumption is highly unlikely for all frequencies or over a very wide frequency range. Manufacturers of earphones, however, go to great lengths to ensure a reasonable frequency response for their sound sources. Usually, this characteristic is described in words as well as graphically in the specifications documentation for the earphones or the OAE equipment.

For the purposes of this calibration exercise, let us assume that we are working with a high-quality OAE probe that contains high-quality sound sources. The manufacturer may have used an equalization circuit in hardware or in software to ensure a reasonably flat response in the frequency range of interest as measured in one of the standard couplers. Although this solves the problem of generating the same sound pressure level for the same input voltage at all frequencies, the problem is solved only in the standard coupler.

The output of the sound sources used in OAE probes is heavily influenced by impedance of the load to which the probe is coupled. That is, the same drive voltage will produce different sound pressure levels in different ear canals (or couplers) depending on the impedance of the ear canal (the load). Thus, calibrating the OAE probe in a standard coupler would work perfectly only if the impedance characteristics of *every* patient's ear canal was identical to that of the coupler. The issue is further complicated by the fact that the physical shape of OAE probes is not designed to match any particular standard. Thus, probes of different manufacturers fit every ear canal differently with varying depths of insertion.

To summarize, differences in ear canal anatomy, probe fit, and depth of insertion are significant between patients and even between trials in the same patient. These factors lead to differences in the impedance characteristics of the ear canal experienced by the OAE probe, which in turn lead to differences in sound pressure levels generated by any given voltage drive. As a result, there may not be a convenient way to translate drive voltages calibrated in a coupler to generate desired sound pressure levels in a random ear canal. Perhaps the answer is to measure sound pressure in the ear canal and to adjust it to the desired level. We explore this avenue next.

Calibrating In The Ear Canal

Given the difficulties of calibrating signals necessary for recording OAEs in a coupler (as we do for measuring hearing thresholds), another possible and popular option is to perform a calibration procedure once the probe has been fit into the patient's ear canal. This calibration approach potentially solves the problems related to individual variations in ear canal anatomy, insertion depth, and probe fit. As it turns out, although these problems may be solved or at least reduced on the surface, a different problem emerges. The problem is related to the location of the measurement microphone relative to the plane where the signal needs to be calibrated—the tympanic membrane.

Given the goal of delivering a known signal to the tympanic membrane, the obvious choice would be to measure the sound pressure level at the plane of the tympanic membrane using a probe microphone. After all, clinicians are adept at making these real ear measurements during hearing

aid fitting. A practical problem becomes evident immediately. A probe microphone measurement prior to making each OAE measurement adds significant complexity, inconvenience, and cost. These problems are magnified several fold when the patient is a young child, particularly a newborn infant. Further, the estimation of the sound pressure level by a probe microphone depends critically on the location of the probe as sound pressure distribution in the ear canal is not spatially uniform (Siegel, 2007). These complications notwithstanding, scientists have used a probe tube placed strategically near the tympanic membrane under visual inspection to measure OAEs at extended high frequencies (Dreisbach & Siegel, 2001; Siegel & Dreisbach, 1995). The merits of this method, however, are outweighed by the impracticalities of its implementation in a busy audiology clinic in a variety of patient populations.

Using a probe microphone to calibrate stimuli for OAE measurements may be impractical, but every OAE probe is itself a measurement device with one or more microphones built in. Why not use this readily available microphone to measure the sound pressure in the ear canal for calibration purposes? Indeed, this is the procedure adopted by most OAE equipment as of this writing. Therefore, it is important to understand the complexities associated with this method.

The goal of the calibration procedure is to deliver a known signal to the tympanic membrane. The microphone of the OAE probe available in the sealed ear canal typically is located a few millimeters away from the tympanic membrane. For simplicity of discussion, let us assume that the sound sources and the microphone of the OAE probe are at approximately the same plane. We also can assume that the portion of the ear canal captured between the OAE probe and the tympanic membrane behaves acoustically as a cylinder with a compliant end (tympanic membrane) for signals up to approximately 10 kHz (Siegel, 2007; Stinson, Shaw, & Lawton., 1982). This assumption does not hold at higher frequencies where the complex geometry of the ear canal and the tympanic membrane have to be taken into account (Stinson, 1985). Both ear canal geometry and sound tube extension in the OAE probe turn out to be critical variables for the purposes of calibrating signals used in OAE measurements. The reader is directed to Siegel (2007) for a detailed discussion of these factors.

The calibration signal is generated at the OAE probe, and then propagates to the tympanic membrane. A portion of the signal is transmitted into the middle ear, and another portion is reflected back to the OAE probe. The incident and reflected waves interact in the ear canal and may create standing waves. At frequencies below approximately 2 kHz, the wavelength of the sound wave is considerably longer relative to the distance between the OAE probe and the tympanic membrane. As a result, the incident and reflected waves are "in phase" and sound pressure levels are similar at all locations in the ear canal. However, as the signal frequency increases and the wavelength decreases, the probability of the incident and reflected waves having different phases at any point in the ear canal increases. The problem is most prominent when the length of the cavity between the OAE probe and the tympanic membrane is ¼ of the wavelength of the sound wave. At these frequencies, the incident and reflected waves are out of phase at the OAE probe and cancel each other causing a pressure null (node). Thus, the sound pressure level measured at the OAE probe at this frequency is considerably lower than that measured at the tympanic membrane, where the incident and reflected waves are always in phase. This situation is schematically depicted in Figure 4–2.

The top panel of Figure 4–2 schematically shows an ear canal with the OAE probe inserted in one end and the tympanic membrane at the other. An imaginary microphone near the plane of the tympanic membrane is also depicted. This set up allows us to simultaneously measure the sound pressure level at the OAE probe and the tympanic membrane. In the middle panel, the dashed line represents the sound pressure level measured at the tympanic membrane as a function of frequency. The solid line represents the sound pressure level measured by the OAE probe. Note that, in the real world, only the latter measurement is available for calibration purposes.

Consistent with the above discussion on ear canal acoustics, the sound pressure levels measured by the two microphones at the two ends of the cavity are approximately similar at low frequencies. However, they divert from each other with increasing frequency with the biggest difference at the frequencies corresponding to the pressure nulls. The gray area between the two traces represents the difference between the two measurements. With deeper probe insertions, the residual distance between the OAE probe and the tympanic membrane is reduced pushing the ¼-wave nulls toward higher frequencies.

When the only measure available is the sound pressure level obtained from the OAE probe, the signal level has to be corrected (calibrated) based on this measure. The middle panel of Figure 4–2 demonstrates that the estimation made at the OAE probe is not an accurate representation of the sound pressure level at the tympanic membrane. However, when deviations from a nominal signal level are corrected based on this measurement, the delivered signal contains artificial corrections as seen in the bottom panel of Figure 4–2. These corrections can lead to "spikes" as large as 15 to 20 dB in the signal at the frequencies of the pressure nulls. Many clinical devices continue to use this form of convenient but compromised method of calibration. At least in one clinical device, corrections are made based only on the sound pressure levels measured up to 2 kHz, thereby avoiding the pitfall of incorrectly correcting for the pressure nulls.

The Future of Calibration for OAEs

The preceding discussion clearly confirms that standing waves in the ear canal are a major concern for calibration techniques using estimates of sound pressure in the ear canal. Standing waves lead to discrepancies between the sound pressure estimated at the tympanic membrane and the point of measurement, a few millimeters removed from the tympanic membrane. Recent thinking about calibration for OAE measurements, as well as other measures of auditory function, has revolved around using *power* or *intensity* calibration (e.g., Neely & Gorga, 1998). Although power or intensity is calculated from the sound pressure measured in the ear canal, these measures are not contaminated by standing waves.

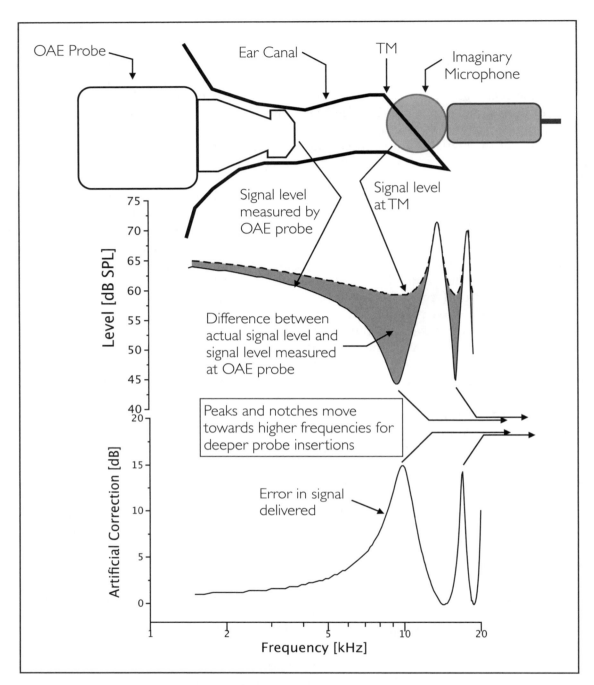

FIGURE 4–2. Schematic representation of a popular calibration method for OAE devices. *Top panel:* The ear canal with an OAE probe is shown along with an imaginary microphone at the approximate location of the tympanic membrane. This imaginary microphone serves as the reference. That is, the signal level at this microphone is considered to be the actual stimulus received by the middle ear and conducted to the cochlea. *Middle panel:* A nominal 65-dB SPL signal is represented as it is measured by the imaginary microphone at the tympanic membrane (*dashed line*) and by the OAE probe microphone (*solid line*). The gray area between the two lines represents the error in estimating the signal level at the tympanic membrane using the OAE probe microphone. *Bottom panel:* "Correction" made to the signal delivered when attempting to present a constant SPL signal to the tympanic membrane based on the measurement made by the OAE probe microphone. Figure inspired by published work of, and personal communications, with Jonathan H. Siegel (Siegel, 2007).

What follows is a grossly simplified description of the complex process of estimating stimulus levels at the tympanic membrane, free from contamination by standing waves. The reader is directed to the primary literature on these techniques for the technical and computational details (e.g., Keefe, 1997; Neely & Gorga, 1998; Scheperle, Neely, Kopun, & Gorga, 2008; Voss & Allen, 1994). The basic goal is to have precise knowledge of the impedance characteristics of both the source (OAE probe) and the load (ear canal and tympanic membrane) immediately before starting the OAE measurement. With this knowledge, several different metrics of sound intensity or pressure at the tym-

panic membrane (free of contamination by standing waves) can be computed. Thus, the problem, as depicted in the top panel of Figure 4–3, is the inability to differentiate between the incident and reflected signals in the ear canal. This confusion, in turn, is directly related to the lack of knowledge about the impedance characteristics of either the source or the load.

The first step in this calibration process is to ascertain the Thévenin-equivalent source properties of the OAE probe. Named after French engineer Léon Charles Thévenin, the Thévenin equivalent is a simplification of a complex electrical circuit that still represents the essential properties of the

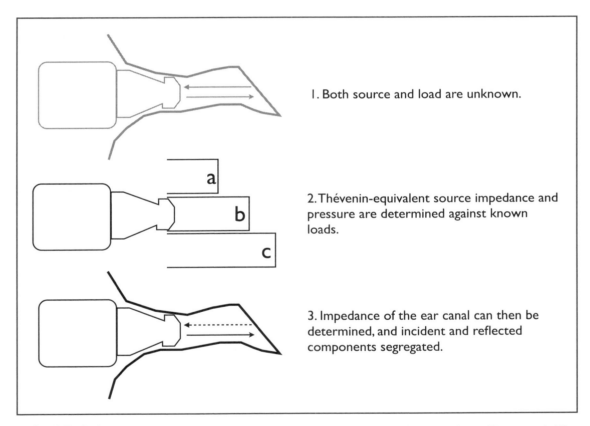

FIGURE 4–3. Schematic of newer methods of calibration for OAE probes. *Top panel:* The problem is in not being able to segregate the incident from the reflected sound wave at the OAE probe. The interaction between these two sound waves creates standing waves, which introduce a difference in the estimates of sound pressure at the OAE probe and the tympanic membrane. *Middle panel:* The Thévenin-equivalent characteristics of the OAE probe (source) are ascertained by evaluating the pressure response to a wideband signal in a series of known cavities. Because the acoustic properties of these known cavities can be computed, the properties of the OAE probe can be solved for. *Bottom panel:* Once the properties of the source are determined, the pressure response in an unknown cavity (the ear canal) is evaluated and its impedance characteristics determined.

circuit. In case of the OAE probe, the Thévenin-equivalent source pressure and impedance are ascertained by evaluating the pressure response of a wideband signal in a series of known cavities. As depicted in the middle panel of Figure 4–3, the probe is inserted in a series of metal tubes of known lengths, and the pressure response to a wideband signal evaluated in each cavity. Because these tubes are of known length and diameter, their impedance characteristics can be computed. The pressure response measured in each of these cavities is influenced by the impedance characteristics of both the OAE probe (unknown) and the cavity. By repeating this measurement in a series of cavities of varying impedance characteristics (due to the varying lengths), the Thévenin-equivalent source characteristics of the OAE probe can be solved for. That is, at the end of the process, the OAE probe is fully characterized. Practitioners beginning to use reflectance measurements in the clinic may already be familiar with the process of calibrating the OAE probe in multiple cavities to determine its Thévenin-equivalent characteristics. The exact lengths (impedances) of these calibration cavities as well as the minimum number of cavities necessary for accurate estimation of source characteristics are still being evaluated in various laboratories.

Once the OAE probe is adequately characterized, it is then inserted in the patient's ear canal. Again, the pressure response to a wideband stimulus in the ear canal is evaluated. At this stage, the acoustic characteristics of the source (OAE probe) are known but those of the load (ear canal and tympanic membrane) are unknown. Given that one half of the equation is known, the other half can be computed from these measurements. As depicted in the bottom panel of Figure 4–3,

once both the source and load characteristics are known, various metrics can be calculated that allow the estimation of sound pressure or intensity at the tympanic membrane. Exactly which of these measures is most appropriate for OAE measurements is not clear at this time. There appears to be significant momentum behind the use of forward pressure level (FPL), which is the estimate of sound pressure incident on the tympanic membrane. However, other measures such as pressure transmitted into the middle ear, or power absorbed by the middle ear can also be computed through the same procedure. Several groups are actively exploring the use of various measures, and the best measurement route should be evident in the near future. Initial experiments comparing the performance of FPL and SPL (traditional) calibration have not shown a significant improvement in DPOAE test performance at least up to 6 or 8 kHz (Burke et al., 2010; Rogers et al., 2010). The gains from these improved methods of calibration probably will be most evident at higher frequencies.

CLOSING THOUGHTS

Instrumentation used to measure OAEs certainly will undergo the same evolution as general signal processing electronics. We can expect hardware to get smaller, more powerful, and less demanding on power sources. Calibration methods also are on the cusp of a significant evolutionary step. It will remain the clinician's responsibility to be cognizant of these new developments and understand their effects on clinical applications of OAEs as well as their interpretation.

5

Clinical Measurement of OAEs: Procedures and Protocols

Introduction

At least four factors are critically important in the clinical measurement of OAEs: (1) The status of the external and middle ear; (2) The coupling (fit) of the probe assembly of the OAE system within the external ear; (3) Stimulus characteristics and stability; and (4) Noise (ambient and physiological) in the test environment and within the external ear canal. Each of these factors individually, or in some combination, can make or break successful, or even clinically usable, measurement of OAEs. Furthermore, failure to recognize the influence of any of these factors on OAE measurement, and failure to take these factors into account in analyzing OAE data for a given patient, can lead to misinterpretation of OAE findings. In turn, misinterpretation of OAE findings may result in errors in identification of hearing impairment or misdiagnosis of auditory dysfunction and, ultimately, to mismanagement of the patient. Pressures in a busy clinic often force the clinician to spend little time in preparation for OAE measurement. These essential preparations include inspection of the ear canal, verifying stimulus characteristics, and ensuring optimal test conditions.

Devoting the time required for proper preparation (generally less than one minute) will pay handsome dividends in time saved during OAE measurement. Once OAE measurement is under way, software for clinical devices incorporate "stopping rules" for determining automatically when test conditions are adequate and for statistical verification of the presence or absence of a response at any given test frequency. Although the specific algorithms vary among manufacturers and devices, the two most important criteria for stopping data collection are the amplitude of the OAE and noise level at corresponding frequencies. Test time will be shortest when OAEs are large and noise levels are low leading to a favorable signal-to-noise ratio. Of course, high amplitude OAEs and minimal noise levels are most likely when the four factors noted above are optimized.

From the start of a clinical OAE recording with a patient, the clinician must consider the possible influence of each of the four factors. Close inspection of the external ear and formal assessment of middle ear function should be a routine part of the OAE test protocol, particularly if OAE findings are not entirely normal. Selected test conditions, such as the probe fit, stimulus parameters (e.g., intensity level), and noise within the external ear canal, should be monitored consistently during OAE recording, with modifications in test technique made as indicated. Troubleshooting problems during clinical OAE measurement often is necessary to ensure the highest possible quality in OAE test outcome.

We will now review in more detail each of the four measurement factors. Of course, successful OAE measurement and meaningful analysis and interpretation of OAE findings are dependent on other nonpathologic and pathologic factors. These factors include the OAE test protocol, additional audiologic findings, and the skill and clinical experience of the tester. All of these topics are reviewed in this chapter and the next, as well as in hundreds of articles on OAEs in other textbooks devoted to OAEs (e.g., Hall, 2000; Robinette & Glattke, 2007). The application of OAEs and other electroacoustic and electrophysiologic measures in identifying and diagnosing auditory dysfunction in children is the focus of another Plural textbook entitled *Objective Assessment of Hearing* (Hall & Swanepoel, 2010). The Otoacoustic Emissions Portal (http://www.oto emissions.org), a Web site organized by Dr. Stavros Hatzopoulos, is also a comprehensive and up-to-date resource on the topic.

Status of the External Ear Canal and Middle Ear

External Ear Canal

OAE measurement invariably requires placement of a relatively rigid probe assembly into the external ear canal. A foremost consideration with probe insertion is to avoid causing the patient discomfort or harm. Risk to the patient is minimized by first examining the external ear canal for debris, foreign objects, excessive cerumen, and pathology. In addition to these health concerns associated with the external ear canal, there are important technical concerns. The external ear canal plays two major roles in OAE measurement. The stimulus is presented to the cochlea with a probe assembly located within the external ear canal and, in addition, a microphone housed in the same probe within the ear canal captures OAE activity returning from the cochlea. As summarized in Table 5–1, a variety of factors involving the external ear canal, some nonpathologic and some pathologic, can influence OAE measurement. In view of the twofold contribution of the external ear canal to OAE measurement, potential problems associated with the ear canal must be anticipated, identified, and, if possible, solved before or during data collection. The most important task prior to beginning OAE measurement is to rule out the presence of debris or foreign objects that might

Table 5–1. Nonpathologic and Pathologic Conditions Involving the External Ear Canal and the Middle Ear Possibly Influencing OAE Measurements

Factor	Influence
External Ear Canal	
Nonpathologic	
Probe tip condition	• Improperly placed rubber or foam probe tip on probe assembly may partially or completely occlude the ports for stimulus delivery or microphone and preclude detection of OAEs
Probe insertion	• Ill-fitting (too small or too large) probe tip will not adequately seal ear canal, thus affecting stimulus intensity and ambient noise levels
	• With tortuous (very curvy) external ear canal, probe tip may push against the wall and occlude ports for stimulus or to microphone

Table 5–1. *continued*

Factor	Influence
Standing waves	• For DPOAEs, interference between incident and reflected stimulus waves within the ear canal may affect stimulus calibration and valid measurement of DPOAEs for stimulus frequencies greater than 3000 Hz
Cerumen or debris	• May occlude one or more ports within the probe assembly
	• May block passage of the stimulus to the inner ear and/or the OAE from the inner ear to the external ear canal
Vernix caseous	• May occlude one or more ports within the probe assembly preventing valid measurement of OAEs
Leak in probe fit	• May cause the noise floor to rise in the low frequencies
Pathologic	
Fungal infection	• The probe tip should be discarded after OAE measurement to eliminate the possibility of spread of infection.
External otitis	• Probe insertion may cause discomfort or pain for the patient. The probe tip should be discarded after OAE measurement to eliminate the possibility of spread of infection.
Stenosis	• The probe may not adequately fit into the ear canal. The ear canal may not be patent with a complete passageway to the tympanic membrane.
Middle Ear	
Nonpathologic	
Mesenchyme	• Within the first few days after birth, uninfected liquid within middle ear space may preclude valid OAE measurement. Mesenchyme usually dissipates spontaneously
Ventilation tubes	• OAEs can be recorded in patients with ventilation tubes (also referred to as grommets) if the middle ear is otherwise normal. OAE amplitude may be reduced for lower test frequencies.
Pathologic	
TM* perforation	• OAEs can be recorded in patients with perforation of the tympanic membrane if the middle ear is otherwise normal.
Eustachian tube dysfunction	• Eustachian tube resulting in negative middle ear pressure will result in reduction of OAE amplitudes and/or absence of recorded OAEs
Negative middle ear pressure	• Negative middle ear pressure will result in reduction of OAE amplitudes and/or absence of recorded OAEs. Compensating for negative pressure during OAE measurement enhances amplitude and detection of OAEs
History of ear disease	• OAEs may be reduced in amplitude at selected frequencies, or absent, in patients with a history of chronic middle ear disease even in the absence of audiometric evidence of active ear disease
Otitis media	• OAEs are typically not recorded in patients with otitis media. Evidence in the literature confirms the sensitivity of OAEs in the identification of otitis media
Other middle ear disease	• OAEs are usually not recorded in patients with middle ear disease (e.g., otosclerosis and discontinuity of the ossicular chain).

* TM = tympanic membrane (eardrum).

Note: The literature describing OAE findings in ear pathology is reviewed in Chapter 7 for children and Chapter 8 for adults.

confound probe operation (e.g., excessive cerumen) or put the patient at risk for injury during probe placement (e.g., foreign objects). Otoscopic inspection of the external ear canal prior to probe insertion is advisable for most patients. Two exceptions to this policy are patients who have just undergone otologic inspection in a medical facility and newborn infants undergoing OAE hearing screening who are unlikely to have pathology, foreign objects, or cerumen within the ear canal.

The tonal signals used for measuring DPOAEs create a special concern due to the possibility of standing wave formation. Standing waves are created due to the interaction between the stimulus ("primary") tones delivered to the external ear canal and their reflections off the eardrum. A cancellation null can be formed near the microphone (probe) end of the closed ear canal at certain frequencies determined by the depth of insertion of the OAE probe. Standing waves that cause pressure nulls near the microphone in the closed ear canal are problematic when the pressure estimated by the microphone is used to correct the stimulus level in the ear canal. This indeed is the practice in many OAE devices. The pressure null near the microphone "tricks" the measurement system into thinking that the stimulus level is much lower than desired and then an appropriate correction factor is applied. This correction factor erroneously raises the signal level to compensate for the null near the microphone thereby causing the signal level at the eardrum to be higher than desired. Problems with standing waves generally are limited to stimulus frequencies above 3000 Hz in adults. The frequency of the notch or the null increases with the depth of probe insertion. In other words, the smaller the cavity size between the OAE probe and the eardrum, the higher the frequency of the null. For example, in young children, the null usually is above 8000 Hz.

Standing wave interference in the ear canal can lead to a discrepancy between the actual signal level within the ear canal and the target signal level of up to 15 or 20 dB (Siegel, 2007). If the end of the probe could be placed very close to the tympanic membrane, standing wave interference would be pushed to frequencies higher than those used in OAE measurements today. Unfortunately, that is not possible in clinical OAE measurements. A more detailed explanation of standing wave interference is found in a discussion of stimulus calibration in Chapter 4, along with the recent emergence of calibration techniques that help avoid the problem all together. How can standing wave interference be handled clinically? Highly atypical and unreliable amplitudes at one or more high frequencies in a DP-gram should be viewed with suspicion. DP calibration and measurement should be repeated after the probe is removed from, and then repositioned in, the external ear canal. Unreliable amplitude data at any frequency should not be analyzed or included in the analysis of clinical DPOAE measurements.

Probe Fit

Guidelines and recommendations for fitting the probe of the OAE device to the ear canal, and verifying adequacy of the probe fit, vary from one manufacturer to the next. The reader is strongly advised to review the manual provided with the OAE system before conducting clinical measurements. Some general statements regarding probe fit, however, can be applied to all TEOAE and DPOAE devices. Some probe fit considerations are included among the nonpathologic conditions involving the external ear canal, shown earlier in Table 5–1. At the least, stimulus intensity level (s) should approximate the target stimulus intensity level (s) within a specific tolerance (e.g., ±1 dB SPL), stability or consistency of the stimulus intensity level(s) should be verified, and an adequately low noise floor must be confirmed. Additionally, prior to TEOAE measurement, it is important to verify a crisp transient stimulus temporal waveform with little "ringing" and a relatively flat stimulus spectrum.

Adherence to a few straightforward guidelines for probe tip selection and insertion will increase the likelihood of a good probe fit within the external ear canal. The probe tip should be large enough to fill the ear canal, yet small enough to be placed rather deeply within the ear canal. Some probe tips are designed to compress easily as they are inserted into the ear canal, and then expand once they are within the ear canal. Probe tips consisting of firm rubber should be selected by size, much as they are for tympanometry. In this era of universal precautions for the prevention of transmission of infection, rubber probe tips should be either discarded after use with a single patient or, if in compliance with institutional policy, properly disinfected before reuse. A wide selection of sizes should be available to accommodate patients of all ages and ear canal sizes. If foam probe tips are used, the user should wait for the foam to expand fully inside the ear canal before starting the calibration (check fit) or test procedures. Waiting for between 30 and 60 seconds usually ensures a properly expanded tip.

As explained in later in this chapter, problems with ambient noise in OAE recording will be reduced by deeper insertion of the probe tip within the ear canal, and tighter probe tip coupling with the ear canal wall. Secure placement of a probe tip deep within an ear canal is facilitated by first straightening the external ear canal immediately before insertion. For older children and adults, this is accomplished by grasping the superior pinna of the external ear and pulling gently but firmly up and out (away from the head). For infants and younger children, the same objective is achieved by grasping the earlobe and pulling downward and outward. Admittedly, the design of the probes and probe tips for some devices seems to facilitate simple probe insertion and consistent probe fit within the ear canal for all patients, including newborn infants. In contrast, with other devices, more effort and technical skill is required to obtain and maintain an adequate probe fit throughout OAE measurement. Prior to purchasing an OAE device, a prospective user (or users) should take a "dry run" by performing sample OAE measurements with different types of patients (e.g., newborn infants, older children, and adults) to develop a clear sense for the probe characteristics and ergonomics of the device.

Stimulus Verification

The minimum stimulus criteria just summarized for the probe fit should be monitored throughout OAE measurement. With cooperative older children and adults, deviations from the initial stimulus properties documented during the probe fit are unlikely. However, for younger children or

physically active patients of any age, the probe assembly may become dislodged partially or slip entirely out of the ear canal during measurement. A less than optimal coupling of the probe with the ear canal usually affects stimulus intensity level and allows increased ambient noise into the ear canal. Periodically, the operator should view the probe to be sure it remains well-seated in the ear canal and, alternately, view the stimulus and noise values displayed on the screen of the OAE system. All clinical devices permit constant monitoring of stimulus intensity and ear canal noise levels either through visual inspection by the operator or by automated display of special symbols. If either stimulus or noise conditions change during OAE measurement and, in particular, if the tolerance for stimulus accuracy and stability or noise levels is exceeded, measurement should be paused to permit troubleshooting and correction of any problems. In some cases, the best solution is to abort the OAE recording, to make necessary technical adjustments, and to begin again.

OAE test protocols are described below. However, it is appropriate at this juncture to mention the impact of the measurement configuration, including the stopping rules noted already. Usually, clinical OAE devices are marketed with a handful of default measurement configurations developed for different clinical applications and/ or patient populations. The various configurations utilize different criteria for determining when data collection will stop for a specific stimulus set. Common stopping rules include a combination of one or more of the following parameters: (1) noise levels (maximum or minimum) during measurement, (2) absolute OAE amplitude in dB SPL, (3) differences between the OAE amplitude and noise level (e.g., signal-to-noise ratio), and (4) test duration as defined by either test time in seconds or the number of stimuli presented or averaged. A configuration developed by the manufacturer with OAE data collected in a sound-treated chamber from cooperative adult patients may work very well, and quickly under those optimal test conditions. OAE measurement in a restless child in a non-sound-treated setting with the same default protocol can require an unacceptably long test time. A protocol and configuration designed by the manufacturer for use with young children in relatively noisy test settings, and then evaluated with clinical trials, would be a better choice for OAE measurement under less than optimal conditions. Clinicians are advised to experiment with different default protocols for an OAE device before the protocols are actually applied clinically to verify that adequate results are obtained within an acceptable time period.

Noise

OAEs are low level (often subthreshold) signals measured within the ear canal and associated with energy produced by the outer hair cells. Confident detection and accurate analysis of OAE activity is invariably enhanced when extraneous noise levels in the external ear canal are as low as possible. Conversely, increased noise levels in the ear canal adversely affect OAE measurements. Indeed, excessive noise levels can make it impossible to detect OAE activity (at least for some test frequencies) and may preclude valid OAE mea - surement. Systematic and diligent efforts to minimize extraneous noise in the ear canal during OAE recording is by far the most important single step a clinician can take to optimize test outcome. Some noise in the ear canal, at least within the low frequency region (<1500 Hz) is inevitable. Essential bodily functions, such as breathing and cardiovascular activity (e.g., blood flowing through vessels within and near the ear) are physiologic sources of noise that can be quantified in the ear canal. Indeed, the absence of any noise in a patient's ear canal during OAE recording would be highly unusual, prompting the concern about the patient's vital signs and the possible need for CPR!

Some practical steps for minimizing the two general sources of sound, ambient and physiologic, are summarized in Table 5–2. Good clinical practice involves taking as many of these steps as necessary, and as soon as possible, to achieve noise levels at each test frequency that are lower than the upper limit for normal noise. Normal noise levels typically are determined by analyses of noise data during the collection of normative data for an OAE protocol. A common definition for the upper acceptable limit of noise is the 90th or 95th percentile for the normal subject group.

Of course, normal noise levels will vary considerably as a function of OAE device and test frequency. Noise levels also vary considerably as a function of patient population (e.g., adults versus infants) and test setting (e.g., sound-treated room, quiet clinic room, or newborn intensive care unit). Most clinicians or clinics do not collect normative data for large numbers of subjects prior to the application of an OAE device and/or protocol. However, it is important to ensure that the normative data for amplitude and for noise level used in clinical OAE measurement are appropriate for a specific device and for a specific patient age group. The appropriateness of a normative data set can be determined in two steps. First, OAEs are recorded from a small sample of subjects with confirmed normal auditory function. Then, the findings (e.g., mean OAE amplitude and mean noise level) are compared to the equivalent values within a normative database. If the OAE data for the small group of normal subjects are consistent with the normative OAE data, then the database can be used confidently for clinical analysis of subsequent patient findings.

Status of the Middle Ear

In almost all audiologic procedures, except tests involving bone-conducted stimulation, the middle ear is a vital link for delivery of stimuli to the inner ear. Uniquely, however, OAE measurement is doubly dependent on the status of the middle ear. That is, in OAE measurement, stimulus energy must be transmitted through the middle ear to the cochlea and also OAE-related energy from the cochlea must be propagated outward through the middle ear to the ear canal. You will recall that the tympanic membrane and middle ear system serve an important impedance-matching function, essentially preventing a loss of more than 25 dB as energy travels from the ear canal to the cochlea. That is, for inward propagation of sound, the middle ear actually prevents the loss of stimulus energy due to the impedance mismatch between the outer and inner ears. Unfortunately, with OAE measurement you cannot (as the saying goes) have it both ways.

Table 5–2. Simple Steps for Minimizing the Effect on OAE Measurement of Noise Within the Ear Canal

Ambient Noise

- Eliminate extraneous noise sources in test room or setting, including:
  - Equipment that is not being used
  - Talking among testers, patient, or family members
- Close the door to test room
- Place a "Quiet! Hearing Test in Progress" sign outside the test area
- Select a probe size that is appropriate for the patient
- Insert the probe deeply within the external canal
- Verify a tight coupling of the probe tip and ear canal walls
- Secure the probe cord if it moves during OAE measurement
- Position the test ear as far away from the OAE equipment as permitted by the length of the cord. With the patient is sitting in a swivel chair, the test ear can easily be turned away from the equipment.
- Replicate OAE recordings to verify the presence of OAE activity at each test frequency
- Modify the test protocol as needed to include only test frequencies ≥ 2000 Hz as most ambient noise is occurs in the frequency region below about 1500 Hz

Physiologic Noise

- For older children and adults, instruct the patient to
  - Remain quiet (no talking) and still
  - Avoid chewing or other jaw movements
- For infants and younger children, encourage minimal physical activity by conducting OAE measurement while the child
  - Drinks from a bottle, uses a pacifier, or is nursing
  - Sleeps
  - Is amused by parents, tester, or assistants
- Monitor the noise level throughout OAE recording. Repeat the above steps whenever the noise level exceeds the 95th percentile for normal noise.
- Modify the protocol to limit test frequencies to those ≥ 2000 Hz

Even under the best of circumstances (i.e., normal middle ear function), much of the OAE energy is lost on the return trip from the cochlea to the ear canal. The recorded OAE amplitude rarely exceeds 10 dB SPL in adult normal hearers with normal cochlear function. Neonates who are products of term deliveries may produce larger amplitude OAEs (15 to 20 dB SPL). The best estimates from experiments on laboratory animals suggest that the level of the OAE energy could be as much as 30 dB greater in the cochlea (see Chapter 2 for a detailed discussion). Thus, a significant amount of the OAE energy is lost in the reverse propagation to the ear canal. It stands to reason that any anomalies in middle ear function would magnify this problem, making it even more difficult to record OAEs in the ear canal. Presumably, the normal middle ear is the largest single variable responsible for the reduction of OAE amplitude as measured in the ear canal.

Even subtle middle ear dysfunction will reduce OAE amplitude measured in the ear canal, whereas clinically significant middle ear dysfunction prevents the transmission of OAE-related energy from the cochlea to the external ear canal. The clinical implication of this statement is clear. Whenever OAEs are abnormal (i.e., amplitudes are below normal limits), particularly within the low-frequency region, the status of the middle ear must be evaluated. Abnormal cochlear (outer hair cell) function cannot be inferred from abnormal OAE findings until functional integrity of the middle ear function is documented. In other words, middle ear dysfunction must first be ruled out. Patient history, of course, offers a clue to the possibility of middle ear disease and dysfunction. Audiologists, however, must perform formal measurement of middle ear function with either aural immittance or wideband reflectance to rule out or confirm a problem that might explain abnormal OAE findings. The sensitivity of OAEs to the status of the middle ear system can be exploited clinically for early detection of dysfunction. The rather large clinical literature on the topic of OAEs and middle ear function is reviewed in Chapter 7. Viewing the relation between OAEs and middle ear status from the opposite perspective, OAEs can serve as a quick and practical technique for ruling out clinically significant middle ear dysfunction, particularly with infants and young children who cannot be thoroughly or easily evaluated with behavioral techniques (air- and bone-conduction pure-tone audiometry). If OAE findings are entirely normal, including amplitude values within an appropriate normal region across a frequency region of 500 to 8000 Hz, then there is very little chance of middle ear dysfunction. Normal OAE findings argue strongly for normal middle ear status, and against the need for additional aural immittance or other audiologic measures.

RECORDING TEOAEs

Test Parameters

In the early years following the discovery OAEs in the late 1970s, virtually all clinical research and clinical application of TEOAEs was conducted with a single device (the ILO 88 instrument marketed by the British manufacturer Otodynamics Ltd.). Now multiple manufacturers market instruments for recording TEOAEs. Selected commercially available devices are listed in Table 5–3. The following brief review focuses on features common to most TEOAE devices, rather than a detailed description of any single device. The discussion covers important stimulus and acquisition parameters and options, using as examples various TEOAE devices available clinically. The format for display of stimulus and response information varies from one device to the next, as seen in the figures inserted within the text. Users of a specific instrument are encouraged to consult the manual provided by the manufacturer for details on hardware, including the probe design and contents, on device specifications, and for guidelines on operation of the device. The next discussion focuses exclusively on measurement of TEOAEs, with the following section devoted to DPOAEs. The closely related topic of TEOAE and DPOAE analyses is covered in Chapter 6.

Stimulus Parameters

Stimulus parameters important in TEOAE measurement include: (1) type, (2) spectral characteristics (spectrum), (3) intensity, (4) stability, (5) temporal characteristics, (6) polarity, and (7) rate. Each of these stimulus parameters will be discussed briefly here, with reference to appropriate screens from several clinical TEOAE devices. Examples of typical stimulus parameters as depicted on several TEOAE systems are shown in Figures 5–1 and 5–2. However, emphasis is placed on three stimulus parameters that are most often selected, manipulated, and/or initially documented and closely monitored by the tester, with TEOAE measurement. They are the: (1) type of stimulus (e.g., click or tone burst), (2) spectrum of the stimulus, and (3) stimulus intensity level. As expected from the terminology used to describe them, transient OAEs are evoked by very brief stimuli. Most often clinically the *type of stimulus* is a click. Because it is so brief (on the order of 0.1 to 0.2 ms), the click contains a broad range of frequencies. In fact, the spectrum of the click stimulus, that is, the lower and upper frequency cutoffs, is limited by the

Table 5–3. Commercially Available Instruments for Recording Transient Evoked Otoacoustic Emissions (TEOAE) and Distortion Product Otoacoustic Emissions (DPOAEs)

Manufacturer	Device (s)	TEOAE	DPOAE
Biologic Systems (NATUS)	AuDX devices	X	X
	Scout	X	X
	ABaer	X	X
	Fischer-Zoth Echoscreen	X	X
Etymotic Research	ERO Scan	X	X
	ERO Scan Pro*	X	X
GN-Otometrics (Madsen Electronics)	Echo-Screen	X	
	Capella	X	X
	Hortmann AB devices	X	
	AccuScreen	X	
	Echomaster**	X	X
Grason-Stadler	GSI 60	X	
	GSI 70 screener	X	
	AudioScreener	X	X
Intelligent Hearing Systems (HIS)	SmartOAE	X	X
Interacoustics	TEOAE-25	X	
	DPOAE-20	X	
	OtoRead	X	X
Labat Biomedical Instruments	EchoLab Eclipse	X	X
Maico Diagnostics	ERO scan	X	X
Otodynamics	Echoport ILO288	X	X
	Echoport ILO292	X	X
	Echocheck	X	X
Sonamed	Clarity system	X	
Starkey-Labs	DP-2000	X	X
Vivosonic	Integrity	X	X

*EroScan Pro includes TEOAE, DPOAE, and tympanometry (with high-frequency probe tone option).

**Option also for tympanometry.

Note: Many manufacturers also include instrumentation for auditory brainstem response measurement, plus options for automated TEOAE, DPOAE, and ABR, and also multiple products (e.g., hand-held and laptop devices) within a single device category. Adapted from Otoacoustic Emissions Portal (http://www. otoemissions.org).

FIGURE 5–1. A TEOAE system (EchoPort device courtesy of Otodynamics Ltd.) showing data collection and analysis screen. Several stimulus characteristics are displayed, including stimulus spectrum and intensity level (*lower left portion*).

properties of the transducer. Although it also is possible to evoke a transient OAE with a tone-burst stimulus and the literature includes at least a dozen articles describing various response parameters (e.g., latency, amplitude, reproducibility) as well as test performance in the identification of hearing loss (e.g., Lichtenstein & Stapells, 1996), for some reason, this approach is not popular clinically. There also are a handful of papers reporting TEOAE measurement with chirp stimulation, although this technique also typically is not applied clinically.

Closely related to stimulus type is *stimulus spectrum*. Again, by definition, a click stimulus has a broad spectrum and, of course, tone-burst stimuli have energy primarily at a designated and limited frequency. Without doubt, the most important spectral feature of the click stimulus used to evoke transient OAEs is a flat frequency response. It is highly desirable in clinical TEOAE measurement to present stimuli that contain relatively constant amount of energy across the entire

frequency range of the stimulus. Often, the spectrum of the transient stimulus is displayed on the screen of a TEOAE device so the user can initially inspect, and then periodically verify, that the spectrum is adequate (refer again to Figures 5–1 and 5–2). If initial inspection of the stimulus spectrum during the premeasurement "probe fitting" routine reveals a jagged pattern, with sharp peaks and valleys at various frequencies, then the probe must be repositioned in the ear canal or the probe tip replaced in an attempt to produce a flat frequency response. Even in an entirely normal ear (normal cochlea and outer hair cells), the response recorded (the TEOAE) is only as good as the stimulus presented. Put another way, if there are frequency regions of reduced stimulus energy or, conversely, increased stimulus energy, the spectrum of the resulting TEOAE will reflect rather precisely these inconsistencies in stimulus frequencies. It certainly is possible to record abnormal TEOAEs even when the stimulus spectrum is optimal, due to dysfunction of the cochlea or

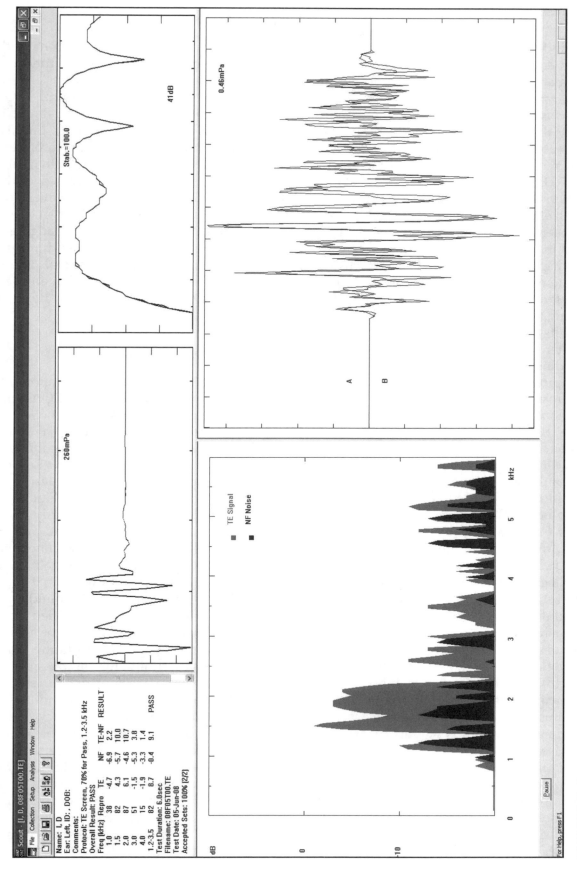

FIGURE 5–2. Data collection and analysis screen for a TEOAE system (Scout device courtesy of BioLogic Systems, Inc.). Several stimulus characteristics are displayed in the top portion, including stimulus temporal waveform and spectrum.

even the middle ear. However, it is not possible to record an entirely normal TEOAE if energy is missing or markedly reduced in some portions of the stimulus spectrum.

Stimulus intensity is another essential stimulus property that should be verified, as TEOAE recording is about to begin, and then monitored throughout recording. Manufacturers inevitably recommend TEOAE measurement with a default stimulus intensity level (e.g., 80 or 86 dB SPL) that is effective in eliciting a robust response from normal ears yet sensitive also for detection of cochlear dysfunction in abnormal ears. Stimulus intensity usually is displayed on the analysis screen for TEOAE systems (refer again to Figures 5–1 and 5–2). The default stimulus intensity level may differ depending the patient population (e.g., infants versus adults). As in DPOAE measurement, stimulus intensity can be varied systematically when recording TEOAEs. Although findings for TEOAE input/output functions have been reported in formal investigations, stimulus intensity rarely is manipulated in clinical application of TEOAEs. The term *stimulus stability* is used to describe consistency of stimulus intensity throughout the period of data collection, from calibration of the stimulus intensity at the start to the last stimulus presented during the signal averaging process. Stimulus stability usually is described as the percentage of variation in intensity level. Obviously, stimulus stability of 100% indicates no change in intensity level throughout data collection. A stimulus stability value of 90%, however, would imply that, although the target stimulus intensity level was 80 dB SPL, the actual intensity level of some of the stimuli presented during the averaging process were as low as 76 dB SPL or as high as 84 dB SPL (8 dB is 10% of 80 dB). Higher stimulus stability values clearly are a good thing, but clinicians should consult the user's manual for guidance about acceptable ranges of stimulus stability for specific TEOAE devices.

The *temporal waveform* of the stimulus is the last of the parameters that is likely to be documented, manipulated, and periodically verified by the tester. The temporal waveform is the amplitude or intensity of the stimulus in the patient's ear canal within milliseconds after it is delivered, as measured by a miniature microphone within the probe assembly and displayed as a function of time. The temporal waveforms of transient stimuli as displayed by two typical clinical TEOAE devices also are illustrated in Figures 5–1 and 5–2. Because very brief duration (transient) stimuli are used to elicit the TEOAE, the goal is to present a crisp and brief temporal waveform, with minimal "ringing" or extension of the stimulus over time. With some devices, the temporal waveform is achieved with an optimal probe fit in the ear canal, and can be manipulated by changing the probe tip size, and the angle and depth of probe insertion within the external ear canal. Because users of TEOAE devices rarely alter default selections for stimulus polarity and rate, these two parameters are not reviewed further.

Acquisition Parameters

Acquisition or response parameters important in TEOAE measurement include: (1) TEOAE magnitude or amplitude, (2) Background noise level, (3) TEOAE spectral characteristics (spectrum), (4) Analysis time (windowing), (5) Reliability (reproducibility) and (6) TEOAE latency. Response parameters also are displayed on an analysis screen of TEOAE devices, as illustrated in Figures 5–1 and 5–2. TEOAE latency is not regularly quantified with current clinical systems. The *amplitude* or *magnitude* of the TEOAE as a function of frequency usually is depicted graphically across the frequency region, almost always in the same display as the spectrum of background noise. TEOAE amplitude is also quantified in dB SPL somewhere on the display screen, again as a function of frequency. As amplitude cannot be displayed numerically for all possible frequencies within the spectrum of the response, the dB SPL values for TEOAE amplitude are given only for representative frequency regions, or bands of frequencies (e.g., a band of frequencies centered around 800 Hz, 1000 Hz, or 2400 Hz). As indicated in Figures 5–1 and 5–2, numerical display of amplitude as a function of frequency is referred to as the "response data summary." Of course, the format used for summarizing response data varies slightly among clinical TEOAE devices.

One of the most important response variables in any OAE measurement is *background noise level*. Non-stimulus-related sound level (i.e., noise) in the ear canal during the time of the recording is critical because the response being measured, the OAE, is itself sound. The challenge, then, is to distinguish the stimulus-related sound from the ongoing and inevitable background noise. Recall that noise in OAE measurement has both physiologic (patient) and ambient (environment) origins. The screens shown in Figures 5–1 and 5–2 include a display of the noise spectrum, that is, the noise levels as a function of frequency. Conventionally, noise levels are depicted in black or a dark color, whereas OAE amplitude is indicated in the same graph by a lighter or brighter color. Notice that noise levels tend to be greater for lower frequencies and may deviate considerably across the frequency region included in the spectrum.

Another response or acquisition parameter that is regularly viewed by testers is the TEOAE *temporal waveform*. It is an entirely different parameter than the stimulus temporal waveform discussed already. Since the temporal waveform is, as the term indicates, a display of the TEOAE throughout the *analysis time*, both response parameters will be discussed together. Shown rather prominently in Figures 5–1 and 5–2, the temporal waveform is a reflection of the amplitude of the response (i.e., the TEOAE) over time. The convention with clinical devices is to plot TEOAE amplitude over the course of about 20 ms. Actually, that statement is not entirely accurate because as you will notice in the figures (Figures 5–1 and 5–2), OAE amplitude is not displayed within the first few milliseconds. Each temporal waveform display instead begins with a flat line. Stimulus energy remains in the ear for 2 to 3 ms after the stimulus is presented, making it technically impossible to confidently quantify TEOAE activity independent of residual stimulus activity. The temporal waveform of the stimulus in reality is a graphic representation of signal level coming from the cochlea into the ear canal. Naturally, the first portion of the cochlea that is stimulated (closer to the basal region) will produce a response first and, conversely, the final portion of the cochlea activated (the apical region) will be the last to produce a response. Consequently, high-frequency cochlear activity is displayed to the left (toward the beginning) of the graphs and lower frequency activity is shown later (toward the right side). This general principle of temporal waveform analysis, however, must be qualified. The spectrum of TEOAE activity is rather limited at each extreme. Recall that the spectrum of the click stimulus usually does not include energy below about 100 Hz or above about 5000 Hz. The OAE evoked by this stimulus is similarly constrained by the bandwidth of the transducer. In addition, because of the problem with residual stimulus in the ear canal for several milliseconds after the stimulus is presented, it is not possible to detect high-frequency cochlear activity that reaches the ear canal within that early time period. For this reason, the upper limit for TEOAE frequency detection is about 5000 Hz and the temporal waveform begins at about 2 ms after stimulus presentation.

The final response parameter to be mentioned is "reproducibility" or "correlation." Very close inspection of the temporal waveform graphs in Figures 5–1 and 5–2 will reveal the presence of two different waveforms, usually designated "A" and "B." Within the same recording period, TEOAE activity for alternative click stimuli is stored separately. If a robust TEOAE is present, and background noise is negligible, then the correlation between these two "copies" of the TEOAE response should approximate 100% and, in the graph, the two waves will overlap almost totally. On the other hand, if there is little or no TEOAE activity and/or excessive measurement noise, then the correlation (or reproducibility) will be much lower. Analysis of TEOAEs is reviewed in greater detail in the next chapter.

Test Protocols

Typical default parameters for a TEOAE test protocol are summarized in Table 5–4. Recommendations for the actual values for each stimulus and acquisition parameter vary for different manufacturers. As noted already, the user of a specific TEOAE device should be familiar with the default parameters, and the possibilities for parameters, that are available for the device. Although the user has the option and flexibility to alter any of the measurement parameters, this option should be

Table 5–4. Typical Parameters for Recording Transient Evoked Otoacoustic Emissions (TEOAEs)

Stimulus	
Type	Click (less often tone bursts)
Duration	80-μsec pulse
Intensity	80 to 86 dB SPL (e.g., 0.3 Pa)
Rate	50/sec
Polarity	alternating paradigm
Number of stimuli	260
Acquisition	
Analysis time	~20 ms
Data points	512
Frequency scale	0 to 6000 Hz
Frequency resolution	49 Hz (for fast Fourier transform)
Bandwidth for analysis	1000 Hz
Temporal resolution	40 μs per data point
Noise rejection threshold	47 dB SPL
Sets of averages	Two (A and B) averaged simultaneously to alternate stimuli

Note. Specific default parameters and values vary among manufacturers and devices.

exercised cautiously when TEOAEs are applied for clinical purposes (e.g., infant hearing screening). Modification of parameters may well invalidate manufacturer recommendations based on clinical trials, and may radically alter reported test performance (e.g., sensitivity and specificity) of the device in certain patient populations and/or test environments (e.g., newborn nursery setting). Nonetheless, there may be clinical indications for modification of TEOAE test parameters. To minimize test time under selected conditions, for example, recording a robust (>20 dB SPL) TEOAE from a sleeping infant in a quiet environment, as few as 20 to 50 stimulus presentations may be necessary to achieve an adequate signal-to-noise ratio. On the other hand, under less than optimal recording conditions that include increased ambient and/or physiologic noise levels, it would be reasonable to extend the averaging process well beyond the default 260 stimulus presentations, perhaps to 520 or even 1040 stimuli. The result of this simple modification of the test protocol is invariably an enhanced signal-to-noise ratio, and more confident identification of the TEOAE. Research evidence supports the strategy of manipulating, with certain clinical applications (e.g., infant hearing screening), the number of averaged responses to optimize detection of TEOAEs while minimizing test time (e.g., Korres et al., 2006).

Another common modification of the TEOAE test protocol for newborn hearing screening purposes is to shorten the analysis time from 20 to 10 ms. With the earliest TEOAE device, the protocol employing the shorter analysis time is known as the "Quickscreen." Two clinical advantages are associated with decreasing the analysis time. As noted above, the temporal sequence of the recorded waveform reflects the site of generation of the TEOAE along the basilar membrane in the cochlea. TEOAE energy from the basal end of cochlea is found within the first portion of the analysis time, whereas energy from more apical regions is found in the latter portion of the time period. Because noise levels tend to be greater within the lower frequency region and because sensory hearing loss is more likely to occur for higher frequencies, nothing is lost and much is gained by limiting the analysis time to the first 10 ms (i.e., eliminating analysis for the time period of about 10 to 20 ms). One of the practical dividends of this simple modification in TEOAE protocol is faster test time, as less processing of data is required and noise levels are generally lower in the region of OAE analysis. Kei and colleagues (2003) confirmed significantly larger signal-to-noise ratios and superior test performance for the QuickScreen versus default TEOAE test protocols, as least with a group of adult subjects. The clinician should be cognizant of the loss of sensitivity to middle ear issues when the low frequencies are eliminated from the TEOAE protocol.

RECORDING DPOAEs

Test Parameters

Even though OAEs were discovered in 1978, almost 20 years passed before the earliest widespread clinical applications of distortion product OAEs. The introduction of user-friendly devices for recording DPOAEs in the mid-1990s was followed by a rapid increase in clinical reports describing the value of findings in varied patient populations. For example, in 1996 alone, dozens of published group studies and case reports described a curious pattern of auditory findings that included normal OAEs in patients with markedly abnormal ABR results, usually with no ABR responses. Among these publications was the article by Arnold Starr and colleagues (1996) in which the term "auditory neuropathy" initially appeared. We now immediately recognize the combination of normal OAEs and an absent ABR as the initial signature of auditory neuropathy spectrum disorder, or ANSD. Literature pertaining to OAEs in ASND is reviewed in Chapter 7. Hundreds of other published studies provided evidence in support of multiple clinical applications of DPOAEs, including newborn hearing screening, pediatric diagnostic assessment, monitoring potential ototoxic hearing impairment, and others.

Initially, DPOAE parameters and protocols were not evidence-based but, rather, were developed and applied somewhat arbitrarily. As an example, intensity levels for the f_1 and f_2 stimuli (L_1 and L_2) at first were as high as 70 to 75 dB SPL, with $L_1 = L_2$. Although high stimulus intensity levels usually produce robust OAEs, there also is a higher likelihood of detecting an artifact rather than an actual OAE and of recording obvious OAEs in patients with sensory hearing loss (false-negative type errors). Before long, however, findings from experimental research with animal models and formal clinical investigations and trials contributed to efficient and effective clinical protocols. Regarding parameters for stimulus intensity, experimental and clinical research clearly confirmed that a stimulus intensity paradigm of $L_1 = 65$ dB and $L_2 = 55$ dB, and $L_1 - L_2 = 10$ dB, had the twofold beneficial effect of increasing the amplitude of DPOAEs while also enhancing sensitivity to cochlear auditory dysfunction. The following discussion highlights current test parameters and test protocols for clinical measurement of DPOAEs. New directions in OAE instrumentation and test protocols are noted in Chapter 10.

Stimulus Frequency

There are three frequencies to consider in DPOAE measurement, and then combinations of the three frequencies. DPOAEs are activated by two pure tones, referred to as the "primary" tones and abbreviated as f_2 (the higher frequency) and f_1 (the lower frequency), that is $f_2 > f_1$. Stimulus frequencies in DPOAE measurement are clearly visible in the analysis screens of clinical devices, as shown in Figures 5–3, 5–4, and 5–5. The relation, or difference, between the two frequencies is critical for activation of distortion products within the cochlea. The relation, defined by the ratio expression f_2/f_1, is usually about 1.20. Amplitude of the DP varies slightly for ratios within the range of 1.15 through 1.30, and as a function of the absolute frequency of either f_2 or f_1. In theory, the use of optimal combinations of f_2/f_1 ratios and $L_1 - L_2$ would yield higher DPOAE amplitudes and, therefore, more effective test protocols for identification and diagnosis of hearing loss (e.g., Fitzgerald & Prieve, 2005; Johnson et al. 2006; Moulin, 2000).

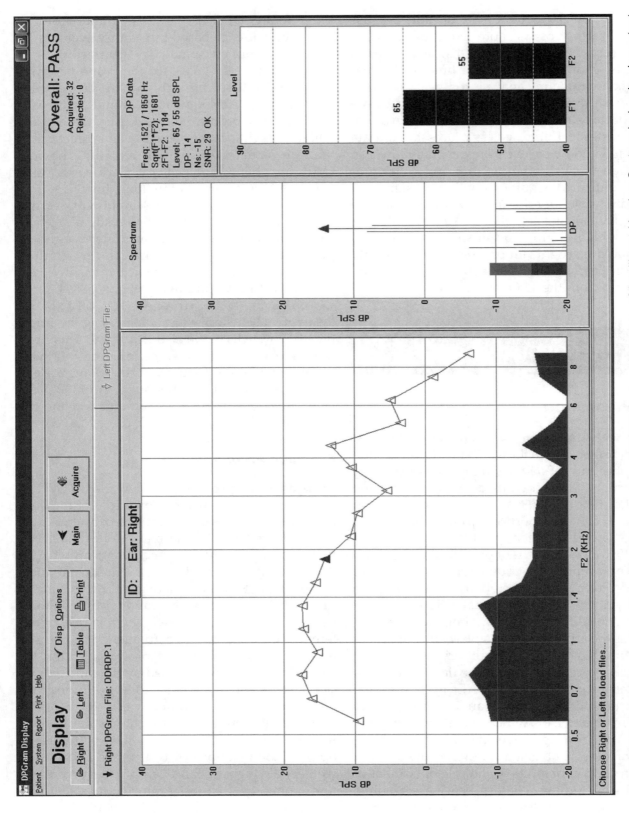

FIGURE 5–3. Data collection and analysis screen for a DPOAE system (courtesy of Intelligent Hearing Systems) showing important stimulus and response parameters. Also displayed in the right portion of the figure is energy at the f_2 and f_1 stimuli, the distortion product $(2f_1 − f_2)$ frequency, and the relative intensity levels of L_1 and L_2.

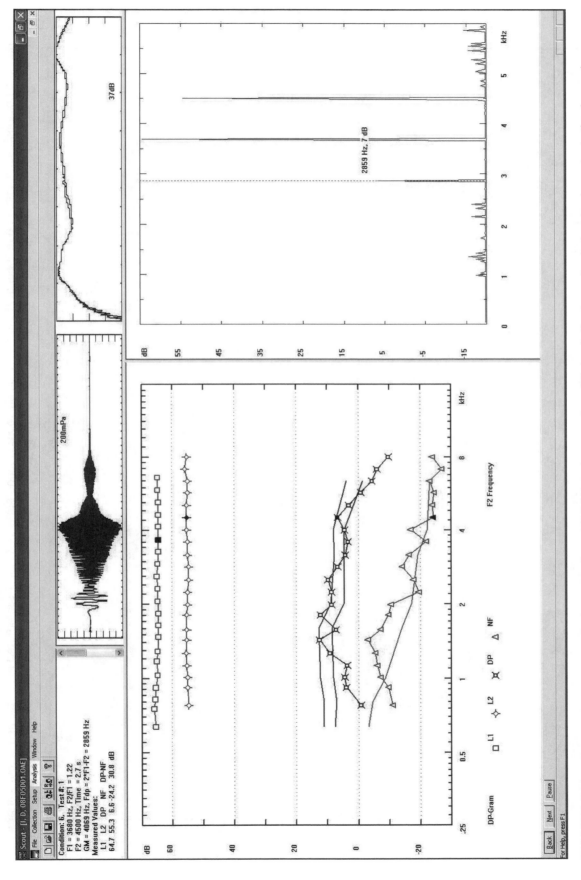

FIGURE 5–4. Data collection and analysis screen for a DPOAE system courtesy (BioLogic Systems, Inc.) showing important stimulus and response parameters. Energy at the f_2 and f_1 stimuli, the distortion product ($2f_1 - f_2$) frequency, and the spectrum of noise recorded simultaneously in the external ear canal is shown in the right portion of the figure.

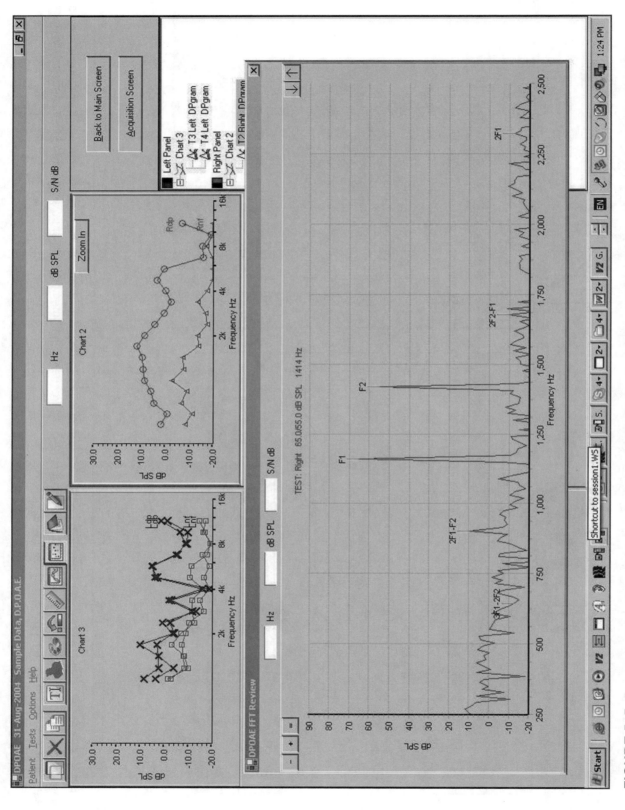

FIGURE 5–5. Data collection and analysis screen for a DPOAE system (courtesy of Grason-Stadler Incorporated, GSI) showing important stimulus and response parameters. Displayed in the bottom half of the figure is energy at the f_2 and f_1 stimuli, the distortion product ($2f_1 - f_2$) frequency, and the spectrum of noise recorded simultaneously in the external ear canal.

The optimal ratio, and the relation between L_1 and L_2, is likely to be different for very high stimulus frequencies (>8000 Hz) versus conventional frequencies (Dreisbach & Siegel, 2005). The optimal f_2/f_1 ratio also varies as a function of age in children, and for children versus adults (e.g., Vento et al, 2004). However, because a default value for the ratio of 1.20 to 1.22 is generally effective for most patients and frequency values, it remains a typical default value for clinical DPOAE instrumentation.

In clinical DPOAE measurement, multiple sets of stimulus frequencies (f_2 and f_1) are presented within each octave. Diagnostic DPOAE protocols applied with relatively cooperative patients include three or more test stimuli within each octave across the range of 500 to 8000 Hz, and even higher frequencies. For specific clinical applications of DPOAEs, such as early detection of drug-induced (ototoxicity) or noise-induced hearing impairment, the protocol may include as many as five to eight frequencies per octave to provide a high degree of frequency specificity. Protocols for hearing screening with DPOAEs, on the other hand, may be limited to a modest number of stimulus frequencies (e.g., three or four) within one or two high frequency octaves (e.g., 2000 to 4000 Hz) well above the dominant spectrum of ambient and physiologic noise (>1500 Hz). The user of DPOAE devices can modify frequency parameters easily in creating customized protocols for different clinical applications.

The third, and obviously important, frequency is the actual distortion product (DP). The DP frequency is illustrated in Figures 5–3, 5–4, and 5–5. The largest DP evoked by tonal stimulation in the human ear is defined by the simple expression "$2f_1 - f_2$." Although numerous other DPs theoretically are possible, and several others usually are generated in human ears during DPOAE measurement, the most robust DP is found at the frequency defined by $2f_1 - f_2$. Algorithms incorporated in clinical devices are designed to detect energy at the DP frequency, and distinguish it from noise within the ear canal in the same frequency region (e.g., ±30 to 40 Hz). As calculated from the equation, the DP frequency is lower than either of the stimulus frequencies. For example, if the f_2 frequency is 2000 Hz, and the f_1 frequency is 1600 Hz, then inserting those values in the equation 2 (1600 Hz) – 2000 Hz equals 1200 Hz. Given the low levels of the DPOAEs of interest, noise within the ear canal is detrimental to their recording. The stimulus tones are relatively unaffected by noise due to the fact that they are considerably higher in level. Moreover, the clinician's concern is much more about the purity of the DPOAE signal. In some instruments, the frequency range around the DPOAE signal for the calculation of the noise floor can be manipulated. Care should be exercised when manipulating this window and it should never be expanded to include the frequencies at which stimulus tones will be presented for any given DPOAE.

A clinically important concept is the relation of the three DPOAE frequencies to the audiogram, and to hearing loss at different frequencies. With the stimulus intensity paradigm typically employed clinically ($L_1 - L_2 = 10$ dB), the bulk of the DPOAE energy is suspected to be produced near the f_2 stimulus region. This area of DPOAE generation becomes more diffuse if stimulus level is increased or if much higher level stimuli are used. The geometric mean of the stimulus frequencies often is used as the nominal DPOAE generation frequency in those cases. Although a portion of the DPOAE recorded in the ear canal may have been generated near the $2f_1 - f_2$ region of the cochlea, this contribution usually is small compared to the contribution from around the region of the basilar membrane with a characteristic frequency of f_2. The practical implication of these statements is that hearing thresholds, and hearing loss, at audiogram frequencies (e.g., 4000 Hz) are most likely to be correlated with the DPOAE recorded using an f_2 stimulus of 4000 Hz.

Stimulus Intensity

Intensity levels of stimuli in DPOAE measurement can be described in absolute and relative terms. Effective absolute intensity levels for eliciting DPOAEs are within the range of 40 to 70 dB SPL. Lower intensity levels are insufficient to produce DP levels above the noise floor in most normal hearing persons without the use of specialized recording techniques. The source of DPOAEs recorded with stimulus levels above 70 dB SPL can be questionable. At such high levels, it becomes difficult to rule out instrumentation and other artifacts. Within the range of 40 to 70 dB SPL, increases in absolute stimulus intensity normally are associated with growth of DP amplitude. Plotting DP amplitude as a function of stimulus intensity for one frequency is referred to as a "DP input/output function." Examples of DP input/output functions for a variety of test frequencies (12,000 Hz down to 2000 Hz) recorded as a function of the f_1 stimulus frequency from one ear of a normal hearing young adult are illustrated in Figure 5–6. Stimulus intensity can be either increased from a low level (e.g., 0 dB SPL) or decreased

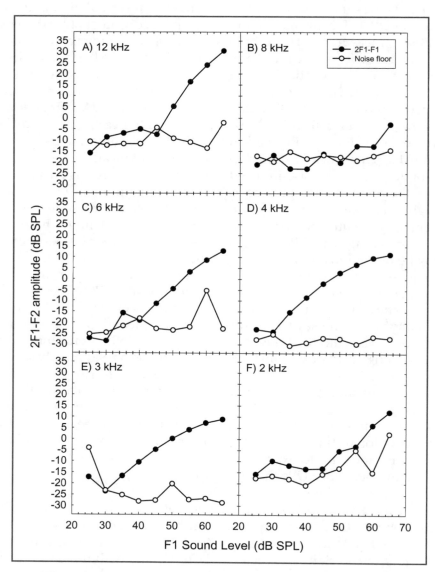

FIGURE 5–6. Illustration of input-output functions for distortion product otoacoustic emissions elicited from a normal hearing college-age subject at multiple high test frequencies ranging from 12,000 Hz (*A*) to 2000 Hz (*F*). The distortion product at $2f_1 - f_2$ is plotted as a function of the f_1 frequency. Input-output functions were recorded with the Mimosa Acoustics HearID otoacoustic emissions device. Figure courtesy of Dr. Colleen LePrell, University of Florida.

from a high intensity level (e.g., 75 dB SPL), usually with either 5 or 10 dB increments. Stimulus presentation and averaging continues until the criterion for a DP, or a pre-established time limit, is reached. Criterion for the presence of DP activity is amplitude of 6 dB above the noise floor (NF). As soon as either a time or DP-NF criterion is met at a given intensity level, stimulus intensity is changed (increased or decreased) and data collection begins at the new intensity level. With DP input/output measurement, stimulus intensity is changed systematically while frequency is held constant.

Intensity levels of stimuli in DPOAE measurement are abbreviated with the letter "L." The intensity levels of the f_1 and the f_2 stimuli are, respectively, L_1 and L_2. Optimal DP amplitudes are recorded when L_2 is less than L_1 by 10 to 15 dB. Initial clinical studies typically utilized equal stimulus intensity levels, such as L_1 and $L_2 = 70$ dB SPL. Although it is not intuitive, experimental and clinical evidence clearly confirms that the amplitude of DPOAEs actually increases (by about 3 dB) when the intensity level of one of the stimuli is decreased (e.g., Stagner et al., 1995; Whitehead, Lonsbury-Martin, & Martin, 1992). In addition to slight increase in DPOAE amplitude with $L_1 > L_2$, animal and clinical studies showed enhanced sensitivity to cochlear deficits. Reflecting an appreciation of this animal and clinical research, the common stimulus paradigm for clinical DPOAE measurement is now $L_1 = 65$ dB SPL and $L_2 = 55$ dB SPL, that is, $L_1 - L_2 = 10$ dB or $L_2 - L_1 = -10$ dB. There is recent evidence, however, suggesting that "individually optimized stimulation" with L_1 and L_2 values selected to produce the highest amplitudes and lowest thresholds may enhance the specificity of DPOAEs in diagnosing middle ear and cochlear auditory dysfunction (e.g., Kummer et al., 2006).

The display of DPOAE amplitude at different frequencies for fixed intensity levels is referred to as a "DP-gram." In contrast to the input/output functions described above, the frequencies of the stimulus tones are varied while holding their levels and the frequency ratio constant in case of the DP-gram. Displays of DP-grams for three different clinical devices were shown previously in Figures 5–3, 5–4, and 5–5. Of course, DP-grams can

be recorded at different sets of stimulus intensities. For the sake of simplicity, the DP-grams displayed in the figures are not replicated. As stressed in Chapter 6. OAE Analysis and Interpretation, documentation of amplitude repeatability is an important step in clinical DPOAE measurement. The DP-gram can begin at the lowest frequency and proceed to higher frequencies or from the highest frequency and proceed to lower frequencies. In noisy test settings or with patients who are physically active, there is a distinct advantage to beginning with the highest frequency. Ambient and physiologic noise levels invariably are lower for the highest frequencies. Due to the increased likelihood of large signal-to-noise ratios (e.g., large differences between DP amplitude and the noise floor) when absolute noise levels are low, criteria for detection and confirmation of DPOAEs are met more readily for higher stimulus frequencies. DPOAE measurement therefore gets off to a quick start. In some cases, test conditions improve with ongoing measurement as the patient relaxes and becomes more accustomed to the procedure. Infants who initially are disturbed by probe placement tend to become quieter as testing continues. Beginning a hearing screening protocol at the highest frequency often quickly leads to a "pass" outcome, as criteria are met for the majority (but not the lowest) of stimulus frequencies. Screening beginning at a low stimulus frequency could very well be delayed for an unacceptable length of time.

Other Parameters

A range of other parameters may be selected for DPOAE measurement, although the options and specific terminology vary from one device to the next. Examples of screens from several devices displaying miscellaneous parameters are shown in Figures 5–7, 5–8, and 5–9. In addition to the requisite stimulus parameters, such as intensity level of each frequency (L_1 and L_2), the starting and ending frequency, the frequencies per octave, the f_2/f_1 ratio, the user also may select the number of sweeps for each frequency presentation, the number of sampling points, the sampling rate, artifact rejection threshold, and a variety of stopping criteria or rules for that control DPOAE measurement and, in addition, have a substantial

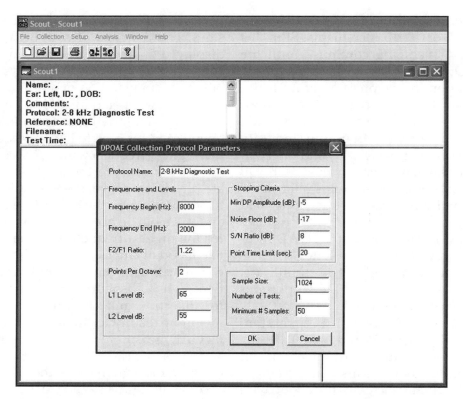

FIGURE 5–7. Selection screen for a DPOAE device (courtesy of Bio-Logic Systems, Inc.) showing options for parameters used in DPOAE measurement.

impact on test time. The stopping criteria include the amplitude of the DP, the amplitude of the noise floor, and perhaps the DP to noise floor difference. Criteria also may incorporate the number of stimuli presented at each frequency and the amount of time allotted for averaging at each frequency, before the device moves on to the next set of frequencies. Various criteria employed for automatically determining "pass" versus "fail" outcomes for when OAEs are applied in hearing screening are discussed in Chapter 6.

Test Protocols

DPOAE test protocols can be modified for specific clinical applications. A protocol included within a test battery for diagnostic assessment of auditory function would not be appropriate, or even feasible, for newborn hearing screening. The diagnostic protocol includes test frequencies in two octaves below 2000 Hz, and many test frequencies per octave. Each of these features of the diagnostic test protocol would create serious problems for newborn hearing screening. High levels of ambient and physiologic noise would adversely affect data collection in the low-frequency region, and the price for frequency specificity (many frequencies per octave) would markedly extend test time. Early clinical trials with various OAE devices (see Hall, 2000, for review) quickly revealed the clinical disadvantages encountered when newborn hearing screening was attempted with diagnostic test protocols. Conversely, there are good reasons to go beyond simple DPOAE protocols in conducting a diagnostic assessment of auditory function. Hearing screening protocols yield little specific frequency information on cochlear function, and data analysis does not adequately differentiate among degrees of cochlear dysfunction. Unfortunately, clinicians have a tendency to rely on simple hearing screening test protocols and

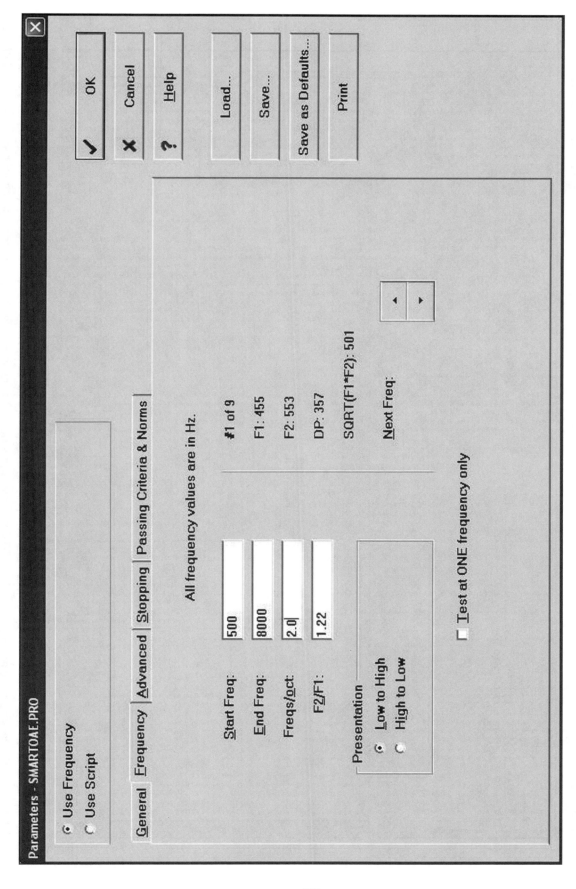

FIGURE 5–8. Parameter selection screen for a DPOAE device (courtesy of Intelligent Hearing Systems) showing options for parameters used in DPOAE measurement.

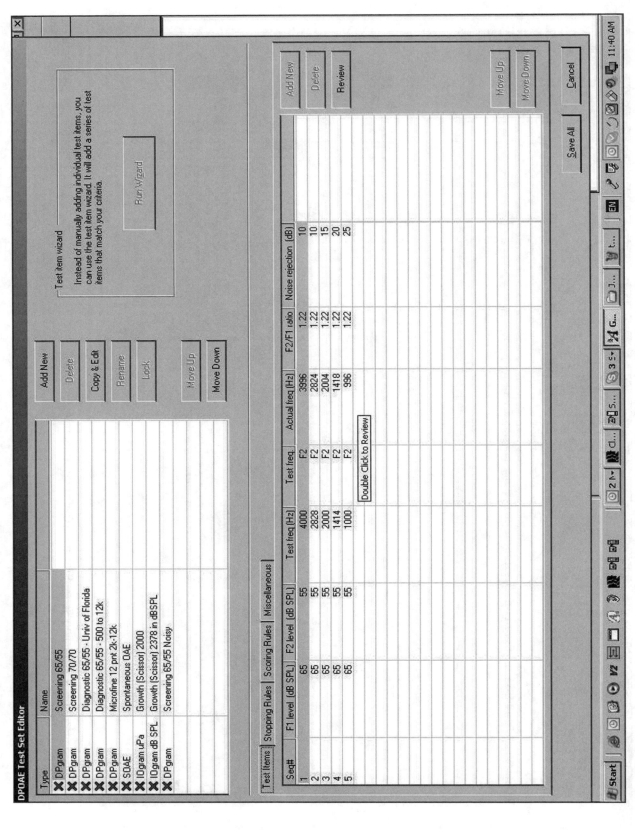

FIGURE 5–9. Parameter selection screen for a DPOAE device (courtesy of Grason-Stadler, Inc.) showing options for parameters used in DPOAE measurement.

analyses strategies even when DPOAEs are applied in diagnostic test batteries. The equivalent practice in general audiologic assessment would be to routinely conduct a hearing screening with three or four pure-tone stimuli at a fixed intensity level (e.g., 20 dB HL) over a limited frequency region (e.g., 1000 to 4000 Hz). Of course, no audiologist would consider routinely taking this simple hearing screening approach for audiologic assessment, rather than performing a full pure-tone audiologic assessment.

Three practical protocols for clinical DPOAE applications are summarized in Table 5–5. Several parameters are commonly found in most DPOAE test protocols, including those applied in hearing screening and diagnosis of auditory dysfunction. In previous figures, we saw many parameters that can be quickly and easily modified by entering a different value into an existing field. Examples of these parameters are the f_2/f_1 ratio (always approximately 1.20) and the initial stimulus intensity paradigm (generally an L_1 and L_2 of 65 dB SPL and 55 dB SPL. A cursory review of Table 5–5 reveals that the obvious differences among protocols involve the nature and extent of the frequency region, and the number frequencies per octave. Protocols customized for other applications can be created quickly by the user of OAE devices, and then saved for easy access on a moment's notice. For example, protocols can be tailor-made for diagnostic assessment of persons with suspected endolymphatic hydrops or Ménière's disease (an emphasis on lower frequency DPOAE measurement) or noise-induced cochlear dysfunction (an emphasis on the frequency region from about 2000 to 6000 Hz). Customized screening protocols might be considered also for preschool and school-age screening (covering the frequency region of 2000 to 8000 Hz) or even ongoing screening for possible drug-induced hearing loss (ototoxicity) after a diagnostic assessment confirmed normal cochlear function (DPOAE measurement for test frequencies of 4000 to 8000 Hz, or even higher frequencies). With most devices, the upper frequency limit is affected by the sampling rate used to detect OAE data. With a sufficiently high sampling rate (e.g., 32,000/sec), it is possible with some devices to record OAEs up to 10,000 to even 16,000 Hz. As a general rule, when DPOAEs are applied in monitoring for possible drug-induced

Table 5–5. Parameters and Protocols for Measurement of Distortion Product Otoacoustic Emissions (DPOAEs) in Different Patient Populations and for Three Selected Clinical Applications

Clinical Application Parameters	Infant Hearing Screening	General Diagnostic	Ototoxicity Monitoring		
Stimulus					
Intensity level					
L1	65 dB	65 dB	65	55	45
L2	55 dB	55 dB	55	45	35
Frequency ratio (f_2/f_1)	1.22	1.20	1.22		
Frequency range	2000 to 4000 Hz	500 to 8000 Hz	2000 to 8000 Hz*		
Frequencies/octave	4 or 5	5	8		
Acquisition					
Repetitions	More	Standard	Less		
Noise algorithm	High Noise	Standard	Low Noise		
Other					

*Higher frequencies if possible.

cochlear dysfunction, it is desirable to extend measurement to the highest possible frequency.

As noted already, other differences for protocols applied in different populations or for different purposes can be found in the acquisition parameters, that is, the algorithms or configurations that determine test time, criteria for response detection, and how or how much noise is reduced during data collection. So called "stopping rules" are criteria selected by the operator of a device, or the supervisor of a hearing screening program, to govern data collection and when data collection is automatically terminated by the device. Often the settings in the screen essentially consist of a series of choices that are either "yes/no" questions or "if this, then that" questions. Of course, once a set of criteria is selected for a specific clinical purpose and perhaps population, the program can be given a logical "name." The protocol then can be saved and, later, quickly retrieved in the clinical setting. Stopping rules for newborn hearing screening with OAEs can be triggered by a "pass" outcome (data collection stops as soon as criteria for a pass outcome are met), or data collection can continue until there is no statistical chance of a pass outcome. The manufacturers of DPOAE devices or independent laboratories or clinics in collaboration with manufacturers of clinical instruments usually include with the devices a small number of default DPOAE protocols with varied acquisition parameters designed for specific (e.g., screening or diagnostic) clinical applications. The user of a newly purchased OAE device is advised to take few hours for some practice recordings on cooperative adults (e.g., yourself, colleagues, family members, or other convenient normal hearers) to become familiar with, and to evaluate the usefulness and appropriateness of, default protocols for different clinical purposes.

CONCLUDING COMMENTS

In this chapter we tried to highlight the "universals" in recording TEOAEs and DPOAEs. As mentioned in the text and demonstrated in the figures, specific implementations of these universals vary from manufacturer to manufacturer and from device to device. However, the fundamentals remain unchanged and the clinician well versed in these fundamentals will be able to quickly learn the specific variations in a given device. There certainly is no source more reliable than an instrument's manual for a complete tutorial on the nuances of that particular device. We suspect that as design of OAE instruments progresses and we develop even better grasp on the important aspects of OAE recording and analysis, there will be considerable convergence of features and nomenclature. Again, a clinician well versed in the basics will have no trouble adapting to these changes.

Understanding the underlying effects of the stimulus and measurement parameters is vital to the successful integration of any OAE measurement device in a clinic. Although a majority of clinical situations can be handled by the default sets of test conditions that are preprogrammed in the instrument by the manufacturer, knowing which parameters to alter for the atypical situation will go a long way in fully utilizing OAEs in the clinical setting. To that end, gathering normative data and even a smaller set of data with altered stimulus and measurement parameters allows a clinician to become familiar with a new piece of equipment and to develop confidence in manipulating the instrument as necessary to maximize its utility.

‖‖ 6 ‖‖

OAE Analysis

INTRODUCTION

General Principles

The analysis of OAEs is greatly simplified and enhanced if the OAEs are first recorded with good technique under optimal measurement conditions. That is, when an appropriate test protocol is selected, and OAE recordings are made and then replicated in a quiet setting from a quiet patient, analysis is rather straightforward and the accurate analysis of the outcome is almost inevitable. For this reason, the reader is advised to review the preceding chapter on "OAE Measurement" before beginning this chapter. Of course, the general statement that, "good technique leads to good outcome" applies to most auditory procedures. Unfortunately, even experienced clinicians using proven test protocols will encounter unexpected problems and challenges in OAE measurement. The likelihood of trouble increases substantially when OAEs are recorded from a potentially uncooperative patient outside of a controlled test environment. Perhaps the best, and most common, recipe for problems is when OAE measurement is made in a neonatal nursery setting with a restless infant. Under these challenging test conditions, a host of variables may conspire to disrupt data collection and complicate data analyses.

The following discussion of OAE analysis presumes that good test practices and techniques were employed with appropriate test protocols and instrumentation. However, recognizing the unpredictable nature of clinical audiology, we devote a section of the chapter to troubleshooting measurement problems, with the goal of assisting the reader in finding solutions to the difficulties most often encountered clinically. We also review the many diverse factors that influence OAEs and their analysis. Several factors, most notably noise, have a serious impact on the quality of OAE measurement and analysis. Because noise is one of the few factors that can be manipulated, and the effects of noise on OAE recordings can be minimized, it seemed reasonable to devote detailed discussion to the topic. The effects of other factors on OAEs, such the nonpathologic subject factors age and gender, obviously cannot be modified or manipulated. Still, these factors can almost always be well defined in advance and have, for many years, been extensively studied and reported in the literature. Finally, there are some factors that are less well appreciated and understood, and their effects less clear. Our objective is to at least make the reader aware that these factors exist and they may exert some influence on the outcome of OAE measurement.

Screening Versus Diagnostic Applications

The distinction in application of OAEs for identification versus diagnosis of auditory dysfunction is readily apparent. Identification of auditory dysfunction is hearing screening of a population (e.g., newborn infants, preschool children, or school-age children). The objective is to separate persons with auditory disorder that might interfere with communication from those who do not have a communicatively important auditory disorder. The goal in diagnosis of auditory dysfunction, on the other hand, is to provide as much information as possible regarding the type, configuration, and the site of auditory dysfunction. One would expect the distinction between these two OAE applications to be appreciated by most all audiologists. Yet, clinical audiologists tend to apply the same rather simple approach to inspection and analysis of OAE used in newborn hearing screening to OAEs applied within a diagnostic audiologic test battery. Infant hearing screening was among the first clinical applications of OAEs, and even now remains important for identification of hearing loss in otherwise healthy babies. With today's instrumentation, automated technologies and statistical algorithms usually are relied on for stimulus presentation and response analysis when screening with OAEs. However, the simple categorization of OAE findings as "present" versus "absent," commonly used in hearing screening, is inadequate for diagnostic applications of OAEs. Diagnostic measurement of OAEs yields a wealth of information on auditory function that cannot be untapped with a simple dichotomous categorization of findings.

Diagnostic application of OAEs deviates from the use of OAEs in hearing screening in two main ways. First, when OAEs are applied diagnostically, the goal is to obtain a high degree of frequency specificity on outer hair cell functioning, that is, independent information for many more test frequencies usually is sought. For school-age children and adults, it's clinically feasible to record a DPgram for 5 or even 8 frequencies per octave over the range of 500 to 8000 Hz. TEOAEs can be recorded for all available frequency bands, usually covering the range of 0 to 5000 Hz. Second, in contrast to the screening categorization of

OAEs as either present or absent (pass versus fail), diagnostic OAE findings for each of the many test frequencies are minimally defined as: (1) normal (amplitudes within a designated normal region), (2) present but abnormal, or (3) not present (absence of OAE activity ≥6 dB above the noise floor). Analyses of TEOAEs and DPOAEs next are reviewed in more detail and also presented in a step-by-step approach to facilitate clinical application of the information.

Initial steps in the general analyses of OAEs (transient or distortion product) for diagnostic applications are summarized in Table 6–1. The first goal is to determine whether measurement conditions were adequate. Was the stimulus intensity at or very close to the target intensity level throughout the recording? Was noise during measurement within normal expectations (i.e., not excessive or not exceeding the upper limit for noise levels in a normal population)? If noise was sufficiently low, then was OAE activity present (≥6 dB above the noise floor) at any frequency or within any frequency region? Finally, what was the distribution of OAE outcome (normal, abnormal, or absent) across the frequency region included in the test protocol? The steps in analysis will now be reviewed more fully for transient evoked and distortion product OAEs.

As reviewed in some detail in Chapter 2 (Anatomy and Physiology) and in Chapter 3 (Classification of OAEs), it is highly likely that different mechanisms underlie generation of transient evoked versus distortion product OAEs. Recall the terms "linear reflection" versus "nonlinear interaction" and "wave-fixed" versus "place-fixed" used to distinguish the sources and mechanisms of OAE generation. Recall from Chapter 2, that TEOAEs are dominated by "reflection" or "wave fixed" emissions whereas DPOAEs in most cases are dominated by "nonlinear" or "place fixed" emissions. Keep in mind also that outer hair cell motility may be more important for generation of TEOAEs whereas nonlinearity in outer hair cell physiology (e.g., transduction) may play a critical role for generation of DPOAEs. Fundamental differences in the mechanisms of these two types of OAEs presumably have implications for their clinical application in various etiologies of cochlear dysfunction (e.g., monitoring ototoxicity, early detection of

Table 6–1. Initial Steps in the Analysis of Otoacoustic Emissions (OAEs) (Transient or Distortion Product) for Diagnostic Applications

- Perform analysis of OAE amplitude and noise floor (NF) level at all test frequencies. Avoid generic statements (e.g., OAEs are present or absent). The goal of analysis is to categorize OAE findings as:
  - Normal
  - Abnormal but present
  - Absent
- Verify adequately low noise floor (<90 or 95% of normal limits for noise)
- Verify the reliability of OAE
  - TEOAE reproducibility value >90%
  - DPOAE amplitude (±2 dB) from at least two runs
- OAE analysis
  - Is the OAE - NF difference ≥6 dB?
  - No, there is no evidence of OAEs *or*
  - Yes, OAEs are present
  - If present, are OAE amplitudes within normal limits?
    - Yes? OAEs are normal
    - No? OAEs are abnormal (but present)

noise induced cochlear changes, diagnosis of inner ear pathophysiology, etc). The following discussion separately addresses the application of transient and distortion product (OAEs) in identification and diagnosis of auditory dysfunction. Combining both types of OAEs, however, could enhance the clinical value and expand the clinical usefulness of OAEs, particularly for diagnosis of varied etiologies of auditory dysfunction. Combined TEOAE and DPOAE measurement in clinical audiology is reviewed in Chapter 10 (New Directions in Research and Clinical Application).

ANALYSIS OF TRANSIENT EVOKED OTOACOUSTIC EMISSIONS

Screening Applications

With the discovery of active processes within the cochlea that produced energy measured as sound in the external ear canal (e.g., Kemp, 1978), TEO-AEs were recognized as a potential technique for hearing screening. Soon after, reports of newborn hearing screening with TEOAEs appeared in the literature (e.g., Johnsen, Bagi, & Elberling, 1983). Although the clinical feasibility and value of TEOAEs as a screening technique was supported almost immediately by research evidence, instrumentation was rather cumbersome and data analysis was tedious. TEOAE equipment consisted of large and heavy "desktop" computers that were slow and limited in capacity in comparison to technology today. Probes used for presentation of stimuli and housing the miniature microphone for detection of OAEs at that time were rigid, shaped like an otoscope speculum, and not designed specifically for use with tiny infant ear canals. Analysis of TEOAE amplitude always was performed manually (not automatically). In fact, with one system, the user actually was required to calculate TEOAE amplitude or signal-to-noise level with a millimeter ruler pressed against the computer screen. The excursion of the TEOAE on the monitor screen in millimeters was then converted to dB SPL by consulting a chart. Screening outcome (i.e., pass versus fail) also was determined by analysis of reproducibility of the overall TEOAE (whole reproducibility), reported in percentage. There were, however, no evidence-based guidelines for the optimal reproducibility value for differentiating a pass versus fail outcome. Screening centers and testers used as cutoff values as low as 50% and as high as 80 or 90%. And, because the reproducibility was based on the entire TEOAE, relatively high reproducibility certainly was possible for infants with significant auditory dysfunction in a specific (e.g., high-frequency) portion of the TEOAE spectrum. False-negative screening outcomes, that is, passing children with hearing loss that might affect acquisition of speech and language, were a concern. Conversely, some of the early investigators of TEOAEs in newborn hearing screening reported unacceptably high failure rates (up to 50%).

In the 1990s, lessons learned from accumulated clinical experience and a diverse collection of technological advances led to widespread improvement in consistency and test performance of infant hearing screening. Failure rates dropped dramatically with the appreciation of the negative effects of vernix and noise (physiologic and ambient) on TEOAE measurement, and techniques for improving probe fit within the external ear canal. Without question, the development of handheld devices with algorithms for automated TEOAE measurement and analysis was the most important technological advance contributing to better TEOAE test-performance and to the advent and rapid expansion of universal newborn hearing screening. To some extent, automated technology has removed from audiologists the responsibility of selecting and properly applying appropriate and evidence-based test protocols for infant hearing screening with TEOAEs. Still, the audiologist should at the very least understand the principles underlying automated TEOAE hearing screening.

The specific algorithms for response detection and statistical verification vary among clinical TEOAE devices. Some devices analyze a number (e.g., 50 to 100) of individual data points within a defined time period (e.g., 5 or 10 ms) following the stimulus presentation, with the hypothesis that no signal time-locked to the stimulus is present. This hypothesis must be rejected with a rather high degree of certainty (e.g., 99%) for the device to yield a "pass" outcome. Furthermore, criteria the "pass" outcome must be met for a designated number of representative data points. Throughout the data collection process, artifacts generally are rejected automatically. This strategy is sometimes described as "signal statistics."

Automated TEOAE devices also may include algorithms for detection of a specific signal-to-noise ratio. In early years of hearing screening with TEOAEs, a popular criterion for a "pass" outcome was a SNR of 3 dB as failure rates were acceptably low. By the mid-1990s, however, clinical investigations showed that defining the presence of a TEOAE by a SNR of 3 dB increased the changes of false-negative screening outcomes. That is, the device appeared to detect a TEOAE response when, in fact, no response was present. Current automated devices require a SNR of ≥ 6 dB for detection of TEOAE activity (a "pass" outcome) and also, in most cases, other criteria to verify sufficiently low noise levels. In addition, all automated devices include "stopping rules" that control the minimum and maximum number of stimuli presented during data collection for screening purposes. Finally, during data collection, automated TEOAE devices monitor important test conditions, such as noise levels and stimulus level. Stimulus intensity level must be constantly verified to ensure that stimuli are presented at the target (desired) intensity level. If test conditions change, for example, stimuli deviate from the target intensity level, the device produces a colored light or verbal message on the screen to alert the operator.

Audiologists using automated OAE devices, or supervising the use by other screening personnel of automated devices, should be familiar with the formal clinical studies that were conducted to validate the algorithm for automated response detection, preferably, findings for validation studies are published in peer-reviewed scientific journals. We advise new users of automated TEOAE devices, really any OAE device, to take the time required to review the manual and to explore the measurement and analysis options available on the equipment.

Diagnostic Applications

In the diagnostic application of TEOAEs, analysis goes beyond simply confirming the presence of a response exceeding the noise floor by at least 6 dB somewhere within the frequency region. Rather, analysis of TEOAE for diagnostic purposes requires careful frequency-specific calculation of response amplitude with comparison of patient findings to normative data. Guidelines for extending the application of OAEs (transient evoked or distortion product) beyond screening to the diagnosis of auditory dysfunction are summarized in Table 6–2. TEOAE amplitude and SNR can be viewed in three or four different ways from the display of findings found on most clinical devices. As illustrated for three different clinical devices in Figure 6–1, Figure 6–2, and Figure 6–3, TEOAE data are almost always presented as: (1) amplitude as a function of time within the response temporal waveform, (2) absolute amplitude presented as a function of frequency within the response spectrum, (3) amplitude versus noise floor difference (i.e., signal-to-noise difference or ratio) presented as a function of frequency, and (4) summary of amplitude numerically, usually including absolute amplitude and SNR in dB

Table 6–2. Guidelines for Diagnostic Application of Otoacoustic Emissions (OAEs) Within a Test Battery

Current Limitation	*Recommendation*
Reliance on screening protocols	• Develop diagnostic protocols
Recording within limited frequency region	• Record OAEs for test frequencies over the widest possible range (e.g., 500 to 8000 Hz for DPOAEs) with multiple frequencies (5 or 8) per octave.
Simple "pass" versus "fail" outcome	• For each test frequency, categorize OAEs into one of three categories: (1) normal, (2) present but abnormal, or (3) absent.
Analysis limited to "present" or "absent"	• For each test frequency, perform close analysis of OAEs and noise floor rather than categorizing overall OAE as simply "present" or "absent."
Inconsistent analysis techniques	• For each test frequency throughout OAE measurement, verify that noise level does not exceed the 95th %ile for normal noise levels (see suggestions for reducing measurement noise elsewhere in chapter). – Calculate the difference (in dB) between OAE amplitude and noise floor (SNR) across the range of test frequencies. Presence of OAE is defined by SNR of ≥6 dB
Limited application of OAEs	• OAEs have clinical and diagnostic value in a variety of pediatric and adult patient populations (summarized elsewhere in the chapter). OAEs should not be applied only as a hearing screening technique with newborn infants
Why record OAEs if an audiogram is available?	• The audiogram is not alwauys in agreement with OAEs. OAEs provide diagnostic information not available from the audiogram. OAEs can be normal in patients with an abnormal audiogram and vice versa (as summarized in Table 3–8).

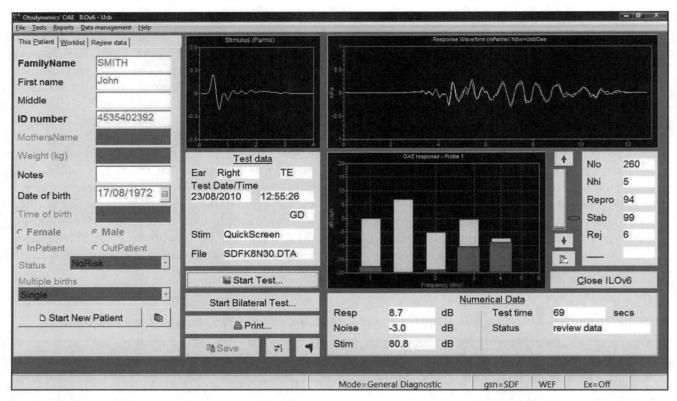

FIGURE 6–1. Data collection and analysis screen for a TEOAE system (courtesy of Otodynamics, Ltd.) showing OAE temporal waveform, TEOAE spectrum, and other test data.

SPL for bands of frequencies. The TEOAE screens shown in these figures will be familiar to you. The same figures were included with the discussion of stimulus and other measurement parameters in Chapter 5. Some modern devices also automatically display a message indicating whether TEOAE data meet criteria for a pass versus fail outcome.

Over 30 years of clinical experience and research documents the effectiveness of TEOAE in detecting sensory hearing loss. As already noted, an extensive literature describes the application of TEOAE in newborn hearing screening. In contrast, few publications report the results of formal investigation of the use of sophisticated TEOAE measurements in the diagnosis of auditory dysfunction in varied populations. TEOAE reproducibility and amplitude values generally are correlated with pure tone hearing thresholds in adults with normal middle ear status (Balat-

souras et al., 2004). However, objective estimation of hearing thresholds with TEOAEs as a clinical objective is not possible for several reasons. Absence of TEOAEs in persons with sensory hearing loss exceeding 35 to 45 dB HL is one of the biggest drawbacks in routinely predicting degree of hearing loss clinically. That is, when TEOAEs are not detected (assuming normal middle ear function), it is not possible to differentiate degree of hearing loss over a wide range, from mild to profound. Of course, the same statement generally applies also to distortion product OAEs.

As reviewed below, there are well-appreciated discrepancies between OAE (transient evoked and distortion product) findings and pure-tone audiometry. Put simply, OAEs can be abnormal in persons with entirely normal audiograms and, vice versa, OAEs can be normal in persons with significant hearing loss by pure-tone audiometry. Once recognized, the not uncommon disconnect

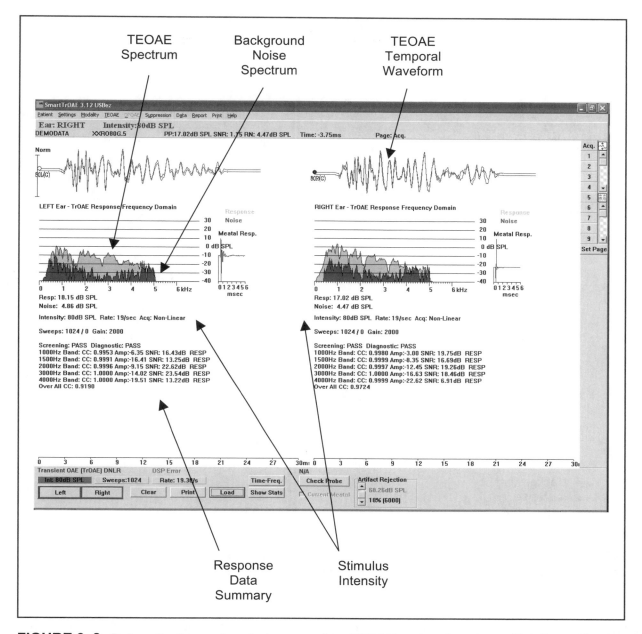

FIGURE 6–2. Data collection and analysis screen for a TEOAE system (courtesy of Intelligent Hearing Systems) showing important stimulus characteristics, including stimulus spectrum and intensity level.

in findings for pure-tone audiometry versus OAEs is not really a problem clinically. Quite the contrary, the discrepancy between OAEs and the audiogram can be exploited clinically. The complementary relationship of OAEs and the audiogram emphasizes the rather unique contribution of OAEs to the diagnosis of auditory dysfunction. The enhanced sensitivity of OAEs to cochlear dysfunction, even when the audiogram is normal, is the fundamental rationale for application of OAEs in early detection of drug- and noise-induced hearing loss. Systematic studies confirm

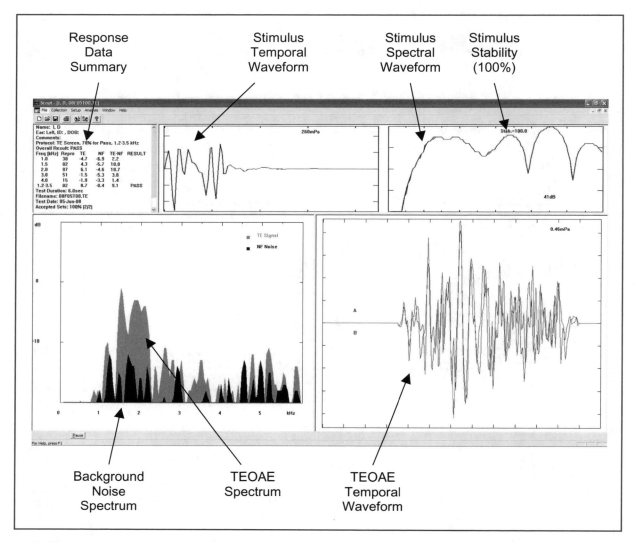

FIGURE 6–3. Data collection and analysis screen for a TEOAE system (courtesy of BioLogic Systems) showing important stimulus characteristics, including stimulus spectrum and intensity level.

that abnormal cochlear function in high-frequency (basal) regions affects the generation of TEOAEs for lower frequency regions, even when hearing sensitivity is normal by the audiogram (Murnane & Kelly, 2003; see Hall, 2000, for review).

Why are papers describing the diagnostic application of TEOAEs so few and far between? One likely explanation is the scarcity of readily accessible normative databases and the incorporation of normative data within clinical TEOAE devices. To be sure, published formal normative studies do exist (e.g., Glattke et al., 1995; Prieve et al., 1993; Robinette, 1992; Stover & Norton,

1993). However, most of the data reported were collected with now outdated analysis strategies (e.g., amplitude or reproducibility of the "whole" response, rather that specific frequencies or frequency bands). Almost without exception, TEOAEs were measured with a single clinical device (ILO 88) using default test parameters. Finally, despite the inclusion of automated signal detection algorithms in clinical devices for the purposes of hearing screening (presence versus absence of a response), normative data are not invariably included as an option for precise TEOAE analysis in diagnostic assessment of auditory dysfunction.

ANALYSIS OF DISTORTION PRODUCT OTOACOUSTIC EMISSIONS

Screening Applications

From the discovery of OAEs in 1978 until the mid-1990s, transient OAEs were used almost exclusively for newborn hearing screening. In fact, distortion product OAEs were viewed by some as a research tool rather than a clinical technique. However, beginning in the mid-1990s with the introduction of a variety of clinical devices, DPOAEs quickly assumed an important role, and were commonly applied by audiologists, in both hearing screening and diagnosis of auditory dysfunction. Initial studies were designed to provide evidence in support of a clinically feasible test protocol for newborn hearing screening. Diagnostic DPOAE protocols clearly were not appropriate for infant hearing screening. As reviewed by Hall (2000), research rather quickly confirmed that hearing screening was most efficient for a frequency region of about 2000 to 4000 or 5000 Hz, avoiding the often excessive noise levels in lower frequencies. Also, configurations with criteria for determining confidently when a DP was present, and for minimizing measurement noise without excessively extending test time, were developed from raw data collected from infants in well baby and intensive care nursery settings. Systematic study during this time period led also to the decision to employ f_2 and f_1 stimuli at moderate absolute and different relative intensity levels (e.g., $L_1 = 65$ dB SPL and $L_2 = 55$ dB SPL). Most current DPOAE screening protocols, and automated analysis strategies, are a direct outgrowth of these early clinical investigations.

Typical DPOAE protocols for hearing screening and for selected diagnostic applications are summarized in Table 6–3. The rationale for various parameters in DPOAE measurement was discussed in Chapter 5. As noted below, the user of

Table 6–3. Examples of Distortion Product Otoacoustic Emissions (DPOAE) Protocols Used for Hearing Screening and Diagnosis of Auditory Function

General Parameters	Hearing Screening	Ototoxicity Monitoring	Ménière's Disease	Noise-Induced Dysfunction	Diagnostic
Stimulus intensity (dB SPL)					
L1	55	55 or 45\	55	55	55
L2	65	65 or 55	65	65	65
Frequency Ratio					
f_2/f_1	1.21	1.21	1.21	1.25	1.21
Frequencies/Octave	4	8	4	8	5-8
Frequency Range (Hz)	2000 to 5000	2000 to ≥8000	500 to 2000	1000 to 8000	500 to 8000
Stimulus Presentations					
Averaging	Fewer	Fewer	More	Fewer	Conventional
Noise Reduction					
Algorithm	High Noise	Low Noise	High Noise	Low Noise	Conventional

The columns above are grouped under the spanning header **Clinical Application**, within which **Hearing Screening** stands alone and **Ototoxicity Monitoring, Ménière's Disease, Noise-Induced Dysfunction, Diagnostic** fall under **Diagnosis**.

clinical DPOAE devices can modify parameters for different protocols that are used for specific clinical purposes. In addition to the parameters summarized in Table 6–3, there are criteria for automated determination of a pass versus fail (refer) outcome. These criteria are defined or set either by the manufacturer as default settings or by the operator of the device, using screens with fields of relevant criteria. In this era of automated hearing screening, analysis almost always occurs following criteria and rules established within preprogrammed algorithms, rather than manually by the tester. For that practical reason, we will not review DPOAE analysis in screening applications further here.

Diagnostic Applications

The DPgram

A common approach for DPOAE analysis is illustrated with the replicated DPgram displayed in Figure 6–4. We urge the reader to apply the simple step-by-step process for analyses of OAEs for

diagnostic purposes listed previously in Table 6–1 to the findings shown in Figure 6–4. For all frequencies, first assess noise conditions, reliability of findings, and the possible presence of DPs above the noise floor. Then, determine whether DP amplitudes are normal, abnormal or absent. Of course, when noise floors are consistently low across the entire frequency region and DP amplitudes are reliably within the range defined by the 5th to 95th percentile lines, even a quick glance at the printout leads to the confident conclusion that findings are normal. In other cases, however, analysis will require very careful scrutiny of each of these response parameters, frequency by frequency, and perhaps consultation also with a numerical display of selected values. We urge the reader to practice analyses of DPOAEs by reviewing with this systematic approach DPgrams recorded clinically from various types of patients.

Figures 6–5, 6–6, and 6–7 show representative examples of DPgrams from three different clinical devices, along with a spectral display that includes the stimuli (f_2 and f_1, the DP ($2f_1$-f_2), and background noise during DP measurement.

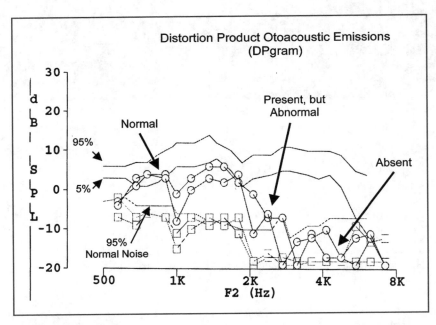

FIGURE 6–4. Basic steps in the analysis of distortion product otoacoustic emissions (DPOAEs).

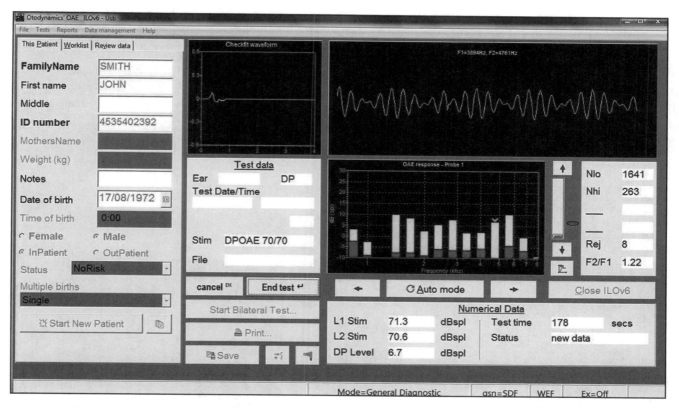

FIGURE 6–5. Data analysis screen for a DPOAE system (courtesy of Otodynamics, Ltd.) showing important stimulus and response parameters.

Notice that distortion products for other frequencies, either higher or lower in frequency than the $2f_1$-f_2 DP, also may be identified within the spectrum. Amplitude invariably is higher for the $2f_1$-f_2 DP. Symbols vary among manufacturers, but in each figure DP amplitude is shown as a function of the f_2 frequency and, a little lower on the amplitude scale, the upper limit for normal noise. Usually, DPgrams for right and left ear stimulation are available for viewing on a single screen, either next to or above and below each other. Each of the manufacturers represented in the figures offer the option for simultaneous display of DPgrams for the right and left ears. According to audiometric convention, right ear findings are displayed in red and left ear findings in blue. Also, with most clinical devices there is an option for superimposing multiple replications of DPgrams on a single screen to verify reliability.

Normative Data

Analysis of DPOAEs is greatly enhanced when an appropriate normative region is shown on the same screen as DP and noise floor amplitude. By appropriate, we mean that the normative region was developed with the same make/model for the clinical device and with the test protocol (e.g., stimulus conditions) actually used to record DPOAEs from patients. Normative data for adults almost always is supplied with OAE devices for default test protocols. In addition, clinical devices offer the option of including new normative databases collected by the user or taken from published or unpublished normative studies. Normal amplitude values, as defined by the lower and upper limits for normal subjects for DPs, and also normative data for noise at designated frequencies are entered numerically into a table.

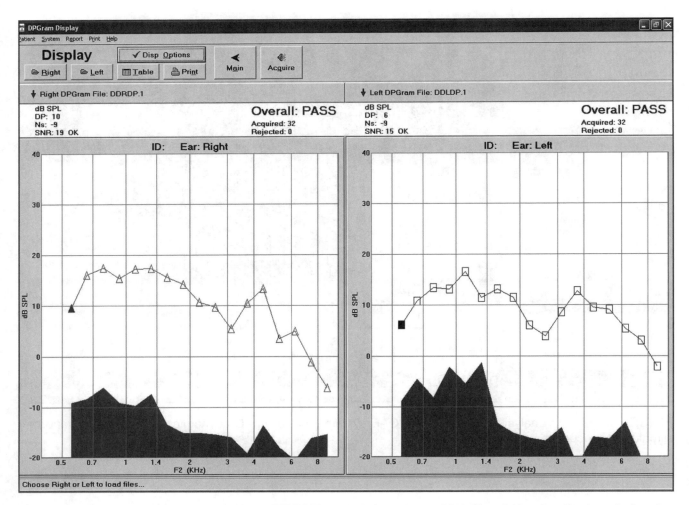

FIGURE 6–6. Data analysis screen for a DPOAE system (courtesy of Intelligent Hearing Systems) showing selected stimulus parameters, plus distortion product ($2f_1 - f_2$) amplitude and the spectrum of noise recorded simultaneously in the external ear canal.

Examples of a normative data table for a typical device are shown in Figure 6–8. With any clinical device, data for DP amplitudes and noise then can be displayed also on the actual DPgram screen for quick reference during on-line analysis as DPOAEs are recorded from a patient, as illustrated earlier in Figures 6–5, 6–6, and 6–7. The normative region may not be displayed on the DPgram when clinical devices are operated with a default (manufacturer provided) protocol. We encourage the user to immediately select, or create, a normative database before recording DPOAEs from the device. However, if a normative data region is lacking, a

convenient rule of thumb is to consider 0 dB SPL as the lower limit for normal DP amplitude. Inspection of published and manufacturer-provided normative data reveals that the lower limit for normal is generally about 0 dB SPL, although values vary somewhat as a function of frequency.

Input-Output Function

The DPgram is, as we have already reviewed, a plot of DP amplitude as a function of stimulus frequency (usually f_2) with stimulus intensity held constant. Although DPgrams are relied on almost

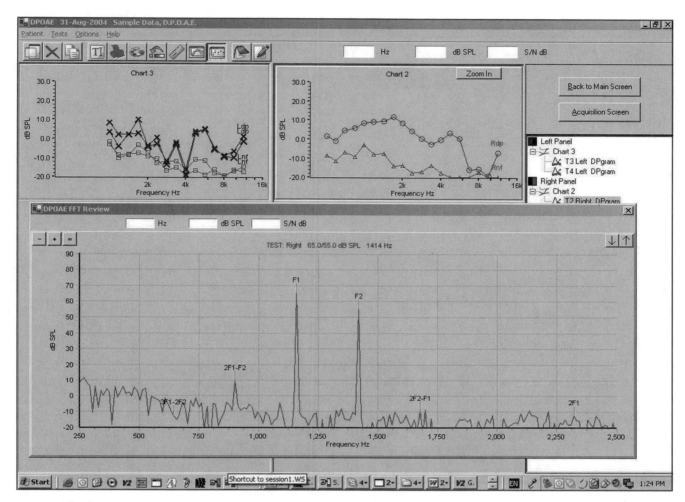

FIGURE 6–7. Data analysis screen for a DPOAE system (courtesy of GSI) showing DPgrams for the left and right ears. Also displayed (*below*) is energy at the f_2 and f_1 stimuli, the distortion product ($2f_1 - f_2$) frequency, and the spectrum of noise recorded simultaneously in the external ear canal.

exclusively for reporting DPOAE findings, more information is available from input-output (I/O) curves or functions. The I/O display consists of DP amplitude as a function of stimulus intensity level for a given stimulus frequency. A group of I/O functions for a normal hearer is shown graphically in Figure 6–9. With the I/O function, DP amplitudes for each intensity level are recorded separately using multiple stimulus presentations and an averaging process. The procedure can begin at a high stimulus intensity level and then progress for lower intensity levels or vice versa. As with the DPgram, noise levels are documented

under each stimulus condition (in this case each intensity level).

The I/O function, particularly for a wide range of intensity levels, provides a wealth of information about DP amplitude, including an estimation of the "threshold" of DP detection and the point of maximum amplitude. Because the findings for a single I/O function are limited to a single set of stimulus frequencies (f2 and f1 values), it is customary to record I/O functions for more than one frequency. Nonetheless, DPgrams invariably offer more frequency-specific information on cochlear function. A semblance of an I/O function can be

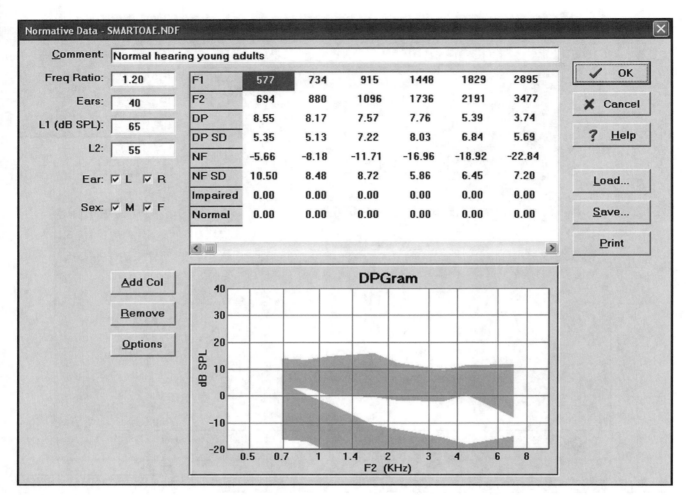

FIGURE 6–8. An example of a normative data table and normal range as displayed on a clinical distortion product otoacoustic emissions (DPOAE) device (courtesy of Intelligent Hearing Systems). Manufacturers usually provide a selection of databases, or the user of the device may enter data for normative databases customized for specific populations, protocols, and/or test settings.

obtained rather quickly for many frequencies by recording DPgrams for several different stimulus intensity levels. For a variety of technical and regulatory (e.g., FDA) reasons, input-output functions have, to date, been viewed and utilized mostly as a research tool. Recent studies (e.g., Janssen, Niedermeyer, & Arnold, 2006) refocus attention on the possible clinical advantages of I/O functions, among them predicting pure-tone thresholds from extrapolated I/O data and objectively documenting loudness recruitment. Reports of clinical applications with specific patient populations, such as adults with tinnitus or noise-induced hearing loss, are reviewed in Chapter 8.

Over the years, alternative analyses approaches, utilizing more than simple amplitude measures, have been investigated and proposed for clinical application. A variety of new strategies for DPOAE analysis are reviewed in Chapter 10. OAE findings for different pathologies are reviewed in Chapters 7 and 8. The relation between OAEs and hearing sensitivity, even in persons with pure-tone thresholds within clinically defined normal limits (e.g., <25 dB), are discussed briefly in the next section.

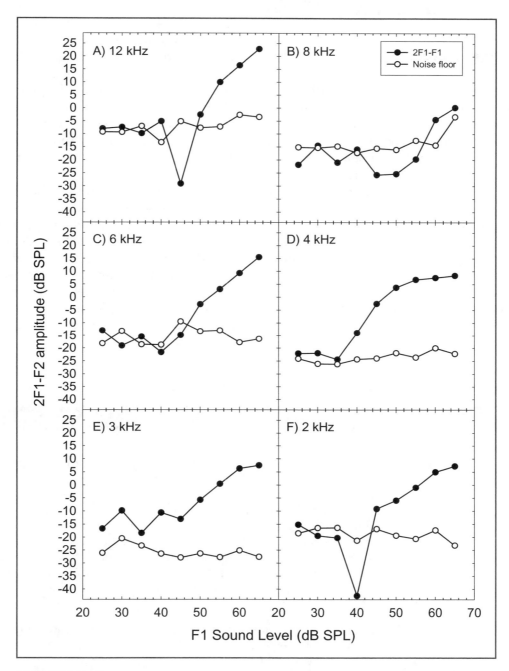

FIGURE 6–9. Distortion product otoacoustic emission (DPOAE) input-output functions recorded from a normal-hearing subject for different f₁ test stimuli.

FACTORS INFLUENCING OAE FINDINGS

Nonpathologic (Subject) Factors

Meaningful clinical analysis of OAE findings for a specific patient must take into account a diverse collection of nonpathologic or "subject" factors. The factors range from those that characterize each and every patient and can usually be documented, such as age (develop and aging), gender, and interear asymmetry, to those that always must be considered but may differ from one patient to the next (e.g., reliability), to those that occasionally need to be considered (e.g., nonpathologic middle-ear disorders), to those rarely recognized that sometimes exert a very important influence on OAEs. Selected factors in OAE measurement described in the literature are summarized in Table 6–4. Clearly, some variables have gener-

ated substantial research interest and consistent research findings. A good example is the difference in OAE findings for the right versus left ears. In contrast, research on other inevitable factors, like body position, attention, or race, has yielded somewhat conflicting findings. Given the ubiquitous potential influence of these factors on OAEs, further investigation would appear to be warranted. Complicating an application of the information on nonpathological factors is the apparent interaction among factors and the often differential impact of factors on transient versus distortion product OAEs.

Clearly, the clinician must have an appreciation of the many and varied factors on OAEs and a general understanding of the impact of major factors. Prior to analysis of OAEs under certain clinical conditions or with specific patient populations, it also may be necessary to develop separate normative data sets on the basis of known influences of a factor. For example, separate normative data for males and females and, within each gender,

Table 6–4. Factors That May Influence Otoacoustic Emissions (OAEs)

Factor	Influence
Nonpathological	
Time after birth	Referral rates (percentage of infants failing hearing screening) are highest within the first 24 hours after the infant's birth, and then decrease steadily over the next two to three days. Refer rates are lower when hearing screening with OAEs is deferred until at least 48 hours after birth, and lowest when screening is performed on the third or fourth day. The likely explanation is dissipation of vernix caseosa within the external ear canal and possibly fluid within the middle-ear space. In developing countries, hearing screening performed weeks or even 1 to 3 months after birth at a regular-follow up visit to a public health facility minimizes referral rates, and maximizes the proportion of babies who undergo screening and who are identified with hearing loss.
Time of bath (infants)	Significantly lower referral rates are reported when the time period from bathing an infant and OAE measurement is at least 7 hours (e.g., Marques et al., 2008). Moisture in the infant ear can contribute to false-failure in OAE hearing screening.
Age	OAE amplitude is highest at term birth (40 weeks gestational age). Amplitude is relatively decreased in premature infants, and also decreases systematically with age from term birth through childhood. Age effects presumably are secondary to maturation of the middle ear and outer hair cells. Effects of advancing age are minimal when age-related hearing loss is ruled out.

Table 6–4. *continued*

Factor	Influence
Gender	Transient OAEs are significantly larger in females than males at all ages, whereas there is no significant gender effect with distortion product OAEs.
Noise	OAEs consist of modest levels (<15 dB) of sound in the ear canal. Therefore, ambient noise in the test setting has a major impact on the detection of OAE activity, particularly for frequencies below 1500 Hz. All possible steps should be taken to minimize ambient noise within the external ear canal.
Ear probe placement	Depth and quality of ear probe coupling with the external ear canal affects the stability and intensity of the stimulus as well as the amount of ambient noise in the ear canal during OAE measurement. As a rule, a deeper and tight fit is desirable.
Ear canal status	Normal ear canal acoustics influence calibration and amplitude of DPOAEs for test frequencies in the region above 5000 Hz. TEOAEs are not affected by ear canal acoustics. Disorders and debris (e.g., cerumen, foreign objects) in the external ear canal can disrupt stimulus presentation and reception of OAE activity by the microphone within the probe.
Tester	
Background	The professional or educational background of the tester (e.g., audiologist, nurse, technician, volunteer) has no influence on screening outcome.
Experience	Infant hearing screening refer rates decrease directly with the experience of screening personnel. The efficiency of an infant hearing screening program is highest when many infants are screened by a few persons rather than many persons screening a few infants.
Instrumentation	Referral rates are influenced by different components of screening devices, such as the probe design (smaller and lighter is better), algorithms, and configurations for detection of the OAE and noise reduction). OAE referral rates typically are lower for DPOAE than for TEOAE technology.
Test protocol	Referral rates are lowest for test frequencies above 2000 Hz (reduced influence of noise), and lower for DPOAEs versus TEOAEs. Test settings, including stopping rules, signal-to-noise differences included in pass/fail criteria, influence referral rates and test times.
Pathological	
External ear canal	Stenosis and external otitis (infection) can affect ear probe placement and probe ports for stimulus presentation or OAE detection (microphone).
Middle ear disorder	Any middle ear dysfunction (e.g., negative middle ear pressure, otitis media) can minimize OAE amplitude or obliterate propagation of OAE activity from the cochlea to the ear canal. OAEs, particularly for high frequencies, can be recorded in patients with ventilation tubes if middle-ear status otherwise is normal.
Cochlear dysfunction	
Outer hair cell	Abnormal or absent OAEs characteristically are found in outer hair cell dysfunction.
Inner hair cell	OAEs typically are not affected by isolated inner hair cell dysfunction.
	Neural: Because they are generated before the first synapse in the auditory system (preneural generators), OAEs are not influenced by neural auditory dysfunction.

for the right and left ears contributes importantly to accurate analysis and clinical interpretation of transient evoked OAEs. Curiously, clinical analysis of distortion product OAEs is far less dependent on these two factors. Appreciation for some factors might lead to technologic advances in the instrumentation used for OAE measurement. The negative impact of negative middle-ear pressure on the amplitude and, in some cases, detection of OAEs suggests a potential clinical demand for devices that permit compensation for pressure imbalances prior to and during OAE measurement. Combination tympanometry and OAE devices are mentioned in Chapter 10. Finally, the influence of selected factors could form the rationale for new clinical applications of OAEs, such as clinically accessible and noninvasive documentation of pathologically high intracranial pressure (ICP) or even objective proof of acute alcohol intoxication.

Hearing Loss

Introduction

Although clearly a factor in the analysis and interpretation of OAEs, hearing loss cannot be confidently or accurately predicted with OAE findings. There are, as we have reviewed, some reasonably predictable influences of stimulus parameters (e.g., intensity levels) on the relation between the pure-tone audiogram and OAE amplitude. Even with very careful accounting for stimulus parameters, however, precise prediction of hearing thresholds with OAE data has proven elusive. Some productive and highly regarded groups of hearing scientists, who have published widely on OAEs, have conducted well designed clinical investigations of the relationship of OAE findings to hearing thresholds and, specifically, the precision with which OAEs predict the audiogram (e.g., Boege & Janssen, 2002; Dorn et al., 1999; Gorga et al, 2003; Martin et al., 1990). Dozens of investigations have been carried out with complex instrumentation in the laboratory setting, with precise manipulation of multiple stimulus parameters, repeated measurement of OAE input-output functions, examination of various OAE response parameters (e.g., amplitude, signal-to-noise ratio,

threshold of detection), careful control of experimental and subject variables, with detailed analyses of data to determine test performance (e.g., hit and false alarm rates, cumulative distributions of OAE findings), and using statistical methods for data analyses, such as clinical decision theory, receiver operating characteristics (ROC) curves and neural networks.

Summary of Research

Behavioral hearing threshold at the f_2 stimulus frequency explains much variability in DPOAE amplitude, even in persons with generally normal hearing sensitivity as defined clinically (Garner, Neely, & Gorga, 2008). With careful recording and analysis of DPOAE input-output functions, it might be possible to more accurately predict pure tone thresholds (e.g., Boege & Janssen, 2002; Gorga et al., 2003), but problems remain in the low-frequency region and when hearing loss exceeds 30 dB HL (Gorga et al., 2003). However, even with a rather complex analysis technique referred to as "feedforward artificial neural network," accurate prediction of hearing status was not possible for normal and hearing-impaired persons (e.g., de Waal et al., 2002). The complexity in generation of DPOAEs, specifically multiple generator sources, makes it difficult to accurately predict auditory thresholds from DPOAE data (e.g., Shaffer et al., 2003).

Two general approaches are taken in the reported investigations evaluating the estimation of hearing thresholds with OAEs. With one approach, some measure of the level of OAE activity, usually either the amplitude or the signal-to-noise ratio (SNR), produced by stimulation at a fixed stimulus intensity level (e.g., 80 dB SPL for TEOAEs or an L_1/L_2 stimulus paradigm of 66/55 dB SPL for DPOAEs) is plotted as a function of hearing loss. In studies utilizing DPOAEs, the OAE levels are plotted separately for a series of different stimulus frequencies (e.g., from 500 Hz up to 8000 Hz). The expected relation, of course, is decreased amplitude of the OAE (or the SNR ratio) as hearing loss increases. Unfortunately, for each stimulus frequency, the correspondence between OAE level and hearing loss is associated with considerable variability, even under ideal measurement conditions. For example, inspection of published data, including graphs plotting OAE level in dB SPL as a function of audiometric threshold, indicates that at any given OAE amplitude (e.g., a DPOAE amplitude of 0 dB SPL) for a stimulus frequency (e.g., 1000 or 4000 Hz), the predicted audiogram threshold data (plotted in 5 dB intensity increments) cover a range of 20 dB HL or more. Although data summarized statistically for relatively large groups of subjects appear to show promise for estimating the degree of hearing loss, for an individual subject (patient) the magnitude of error in estimating hearing loss from OAE level is clinically unacceptable.

Another approach for estimating hearing threshold is derived from analysis of the threshold for OAEs. First, input-output (I/O) functions (described above) are recorded for individual fre-

quencies. OAE amplitude (the output) is recorded as stimulus intensity is either decreased or increased over a range of about 80 dB. For example, beginning with a stimulus level of –10 dB SPL, The presentation level is increased in 5-dB steps and the TEOAE or DPOAE amplitude is recorded at each step up to a maximum stimulus intensity level of 70 or 75 dB SPL. OAE threshold is defined at the lowest stimulus intensity level that produces an OAE – NF (noise floor) difference of at least 6 dB. Figure 6–10 illustrates the relation of DPOAE threshold (in dB SPL) and behavioral hearing threshold (in dB HL), based on inspection of data reported by various investigators (e.g., Gorga, Neely, Dorn, & Hoover, 2003). The diagonal line extending from –10 dB to 90 dB for pure-tone and DPOAE threshold values indicates an ideal

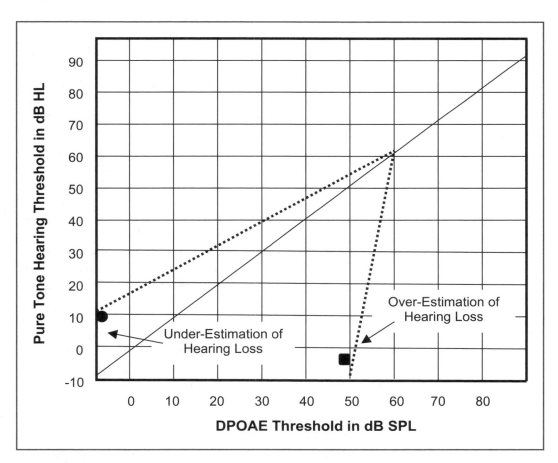

FIGURE 6–10. Relation of DPOAE threshold (in dB SPL) and behavioral hearing threshold (in dB HL), based on inspection of reported data (e.g., Gorga, Neely, Dorn, & Hoover, 2003).

correlation between the two variables. The limitation inherent in estimations of hearing threshold from OAE data is readily apparent, and highlighted by two hypothetical data points. The circle represents a modest underestimation of hearing threshold. That is, the DPOAE threshold of −10 dB SPL would imply better than average hearing level, but actual hearing threshold is about 10 dB HL. Although there is a discrepancy between the predicted and actual hearing level, it would probably not lead to misdiagnosis of sensory hearing loss or poor management decisions. Subjects in the above-noted investigations were carefully selected to ensure a sensory hearing loss.

As noted at the beginning of the chapter, patients with inner hair cell or neural auditory dysfunction may have normal OAEs in the presence of significant hearing loss by pure-tone audiometry. In other words, the trends illustrated previously in Figure 6–10 should not remove concerns about the possibility of serious and clinically significant underestimations of hearing loss by analysis of OAE findings. The more worrisome limitation associated with estimation of hearing loss by DPOAE (or TEOAE) thresholds, overestimation of hearing loss, is highlighted by the square symbol in Figure 6–10. Thresholds for OAEs may be markedly elevated or, in fact, OAEs may not be detected, in some persons with entirely normal hearing sensitivity. For the application of OAEs in hearing screening, the clinical consequence of this discrepancy is relatively minimal. An infant with an elevated OAE threshold presumably would undergo diagnostic audiologic assessment with a full test battery, including estimation of thresholds with frequency-specific ABR techniques (see Chapter 5). However, the not uncommon possibility of major overestimation of hearing loss by OAE thresholds argues strongly against abandoning pure-tone audiometry in favor of OAE measurement for defining the degree of hearing loss.

Despite these obvious limitations inherent in precise prediction of hearing loss from OAE findings, it is clinically useful to appreciate the expected relation between hearing level and amplitude of OAE, or the presence versus absence of OAEs. According to research conducted with well-defined patient groups, such as persons exposed to ototoxic drugs, DPOAE amplitude levels decrease on the order of about 0.3 dB SPL for each 1 dB HL increase in hearing threshold level (e.g., Ress et al., 1999). OAEs can be exploited in the diagnosis of auditory dysfunction with adherence to a rather systematic approach for analysis and interpretation, as summarized earlier in Table 6–1. We must begin with the premise, amply supported by both research evidence and clinical experience, that it is not possible to replace the information from an audiogram using OAEs. Then, we must assume that identification and diagnosis of auditory dysfunction would benefit even if OAE findings occasionally provided some guidance or direction regarding the degree and configuration of hearing loss.

The likelihood of recording normal OAE amplitude or detecting any OAE activity (exceeding the noise floor by at least 6 dB) at a particular frequency is shown schematically in Figure 6–11. In some respects, the relation depicted in Figure 6–11 highlights the clinical value of OAEs in hearing assessment. A few points are quite apparent. Normal OAE findings, that is, OAE amplitudes that are within an appropriate normative region (e.g., between the 5th and 95th percentile for OAE data collected with the same equipment and test protocol from a carefully selected group of persons with normal auditory function) are expected only from persons with hearing sensitivity of 10 dB HL or better. As hearing threshold worsens, even within a range usually defined clinically as "normal hearing sensitivity" (from 0 to 20 or 25 dB HL), the amplitude of OAEs steadily decreases. To be sure, OAEs still may be present, but unequivocally abnormal based on analysis of amplitude relative to a normative region. The standard deviation for pure-tone hearing threshold measurement is 5 dB. Hearing thresholds of 20 or 25 dB HL may be within the clinically normal region, implying for an adult patient at least that hearing sensitivity still is adequate for communication, and amplification or other management options are not indicated. However, from a statistical perspective, a hearing threshold of 20 dB HL is four standard deviations below normal mean value and, by definition, "abnormal." OAE amplitudes are abnormally decreased when hearing thresholds reach 20 to 25 dB HL, reflecting disruption in cochlear integrity, specifically, outer hair cell dysfunction.

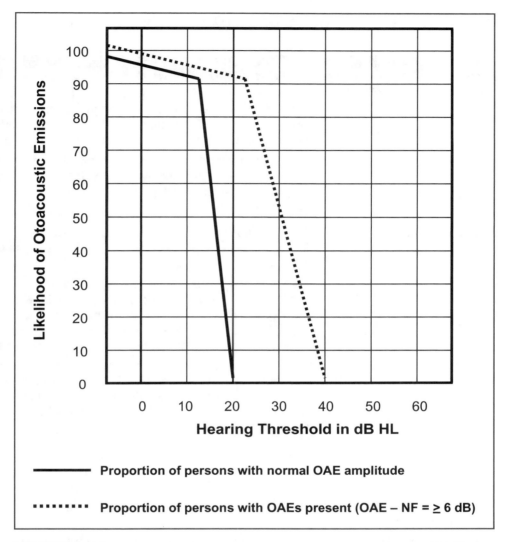

FIGURE 6–11. Likelihood of detecting a normal otoacoustic emission (OAE), that is, amplitude within normal limits, or the presence of any OAE activity (>6 dB) as a function of hearing threshold level.

Differences Between OAE and Audiogram Findings

As we have emphasized in this section, distinct discrepancies between the audiogram and OAEs commonly are encountered clinically. The discrepancy between OAE findings and the audiogram is not a reflection of a problem or weakness of either procedure. It's a good thing. Divergence among audiologic findings in general, and OAEs and the audiogram in particular, highlights the diagnostic value of individual audiologic procedures and the diagnostic value of the test battery approach in clinical audiology. If OAE findings simply mirrored the pure-tone audiogram, OAE would have minimal clinical usefulness. Without doubt, some of the most interesting patients are those who demonstrate a difference, even a marked disagreement, between the audiogram and OAEs. Some of the common discrepancies between OAE and audiogram findings, and brief explanations, are summarized in Table 6–5.

Let's consider this relation between OAEs and the audiogram from a clinical perspective. If,

Table 6–5. Two Clinically Common Discrepancies Between OAE and Audiogram Findings, with Brief Explanations

Discrepancy	Explanation(s)
Abnormal OAEs with hearing sensitivity within normal limits	• *Middle ear dysfunction:* Propagation of OAEs from the cochlea to the external is dependent on integrity of the middle ear. Slight middle ear dysfunction that doesn't cause a deficit in hearing sensitivity may entirely preclude detection of OAEs • *Inner ear (outer hair cell) dysfunction:* OAEs are more sensitive to outer hair cell dysfunction that the audiogram. OAE amplitude is typically reduced or OAEs are not detected with hearing sensitivity levels in the 15 to 25 dB region
Abnormal audiogram with normal OAEs	• *Inner ear (inner hair cell) dysfunction:* OAEs are not affected by inner hair cell dysfunction, but hearing sensitivity is dependent on inner hair cell integrityth • *Neural auditory dysfunction:* OAEs are not affected by neural dysfunction. The OAEs are preneural in origin, whereas the audiogram is dependent on neural integrity, including synapse between inner hair cells and afferent 8th nerve fibers, 8th cranial nerve, auditory brainstem, and even auditory cortical regions • *Behavioral factors:* Multiple listener variables, including attention, motivation, cognition, reliability, psychoacoustic factors (e.g., decision criteria) can influence pure-tone audiometry resulting in elevated thresholds. None of these factors affect OAE measurement

at the outset of an audiologic assessment, normal OAEs are recorded at all test frequencies (OAEs not only are present but amplitudes are within the normal region), then we would expect pure-tone audiometry in the sound booth to result in hearing thresholds around 0 dB HL, and certainly no worse than 15 dB HL. The combination of normal OAE findings and even a mild sensory hearing loss is not expected, and must be investigated until an explanation is found. Possible reasons for discrepancies between OAE and pure-tone findings are reviewed next. The reverse pattern of findings, that is, abnormal OAEs in a patient with entirely normal pure-tone thresholds (hearing levels 15 dB or better), also requires an explanation. In short, the findings for OAEs and the pure-tone audiogram must be scrutinized in combination. If findings for these two independent measures of auditory function are not in agreement, an explanation (technical problem, a possible cause based on patient history, or some type of auditory dysfunction) must be sought.

Troubleshooting in OAE Measurement and Analysis

OAE analysis is entirely dependent on selection of a proper test protocol, adequate test conditions, and taking into account the nonpathologic factors and pathology (e.g., middle-ear and/or cochlear auditory dysfunction) that influence findings. Only with plenty of clinical experience can the clinician develop the ability to rapidly recognize, and effectively solve, problems in OAE measurement. However, a logical, systematic, and consistent approach to troubleshooting will speed up the learning curve and assist the audiologist in addressing technical or clinical challenges encountered for the first time. A step-by-step approach for trouble shooting in OAE measurement is summarized in Table 6–6. There are, of course, myriad potential problems or combinations of problems and, for each, often more than one possible solution. The guidelines in the table are, therefore, offered simply as a starting point. Some of the problems are rather easily solved or, at least, minimized. Excessive measurement noise is a good example. With few exceptions (perhaps a wiggly infant in a noisy intensive care unit), noise almost always can be diminished with one or more of the strategies listed in Table 6–6. Excessive vernix caseous within the ear canal, once identified, can often be circumvented or removed enough to permit valid OAE measurement. Excessive cerumen, debris, or foreign objects in the ear canal may create a more challenging problem. Like any other audiologic procedure, measurement OAEs requires technical skill developed with varied clinical experience.

Table 6–6. Possible Solutions to Common Problems in the Measurement of Otoacoustic Emissions (OAEs)

Problem	Possible Explanation (s)	Possible Solution (s)
No OAE Activity	• Inadequate stimulus	• Verify stimulus matches target intensity level • Verify correct probe for device • Verify probe is calibrated • Verify probe fit within external ear canal
	• External ear canal blockage	• Conduct otoscopic inspection • Rule out vernix caseous (in newborn infants) • Rule out excessive cerumen or foreign body • Check probe ports for debris and blockage
	• Middle ear dysfunction	• Perform tympanometry to verify or rule out • Refer for medical management
	Cochlear auditory dysfunction	• Perform pure tone audiometry • Obtain detailed patient history (e.g., tinnitus, ototoxic drugs, noise exposure) • Record another type of OAE
Unexpected OAE Activity	Standing waves in ear canal	• Reduce stimulus intensity level • Replace probe and repeat measurement • Assess reliability at all test frequencies • Record transient evoked OAEs (vs. DPOAEs) • Verify stimulus intensity level
High Noise Levels	Excessive ambient noise	• Close door to test room • Separate patient from noise sources • Improve probe fit within external ear canal • Orient test ear away from noise sources • Improve probe fit within external ear canal • Increase signal averaging • Limit stimulus frequencies to >2000 Hz
	• Excessive physiologic noise	• Attempt to reduce patient movement • Verify patient is not chewing • For infants, record OAEs after feeding • Record OAEs during sedation after ABR

7

OAEs and Cochlear Pathophysiology: Children

Cross-Check Principle Revisited and Revised

Over 30 years ago, James Jerger and Deborah Hayes described a fundamental concept underlying pediatric audiologic assessment (Jerger & Hayes, 1976). Using the term "cross-check principle" in referring to a pediatric test battery approach, the authors provided clinical evidence that supported the assertion that "no [audiologic test] result should be accepted until it is confirmed by an independent measure" (Jerger & Hayes, 1976, p. 620). The original cross-check principle was based on the combined clinical application of three procedures: behavioral audiometry, aural immittance measures, and the auditory brainstem response (ABR). Today, the cross-check principle has been revised and improved with the addition of a fourth objective auditory measure—otoacoustic emissions. Table 7–1 summarizes strengths and limitations of the four components of the pediatric test battery, including the rather unique advantages gained by incorporating OAEs into the test battery.

The valuable contribution of OAEs to the pediatric test battery is clearly supported by the multiple diverse advantages listed in Table 7–1. Furthermore, rationale for the clinical application of OAEs is exemplified by the heightened diagnostic power of the cross-check principle resulting from the inclusion of OAEs. Recently, Baldwin, Gajewski, and Widen (2010) examined in a large group of infants ($N = 993$) the specific relation of OAE findings (transient-evoked and distortion product) to tympanometry and minimal response levels (MRLs) as determined with visual reinforcement audiometry. In the words of the authors, "We found that there is good agreement between MRLs and OAEs as we would expect. When the two measures do not agree, tympanometry is likely to provide an explanation at least half of the time. When MRL reliability is questionable, OAEs may be important to the ultimate test interpretation. In the larger clinical picture, the value of the cross-check principle has once again been substantiated" (Baldwin et al., 2010, p. 195).

Virtually all applications of OAEs reviewed in this chapter, and the next chapter, are justified by this rationale. OAEs not only add important information to the diagnostic process, but some of the information provided by OAE findings is not available from any other auditory procedures.

Table 7–1. A Revised "Cross-Check Principle" Test Battery for Pediatric Audiologic Diagnostic Assessment that Includes Otoacoustic Emissions (OAEs).

Procedure	Advantages	Disadvantages
Behavioral Audiometry	• The only true hearing test • Pure-tone audiometry is frequency specific • A measure of communication (speech audiometry) • Assesses cortical auditory function	• Limited value in infants <6 months • Influenced by development and maturation • Influenced by neurological (motor) status • Not site specific for auditory dysfunction • Not ear specific in infants • Limited information on difficult-to-test children
Aural Immittance Measures	• No behavioral response required • Sensitive index of middle ear status • Ear-specific information • Frequency specific information • Assesses retrocochlear function • Assesses brainstem function • Can record in sleep/sedation • Does not require sound-treated room	• Not a true test of hearing • Information limited by middle ear dysfunction • Does not assess higher (cortical) function
Auditory Brainstem Response	• No behavioral response required • Appropriate for all ages • Objective assessment of hearing sensitivity • Ear-specific information • Frequency-specific information • Can record in sleep/sedation • Assesses retrocochlear function • Assesses brainstem function • Does not require sound-treated room	• Not a true test of hearing • Does not assess higher (cortical) function • Generated only by onset-neurons (click stimulus) • May be absent with normal hearing sensitivity • Does not assess communication
Otoacoustic Emissions	• No behavioral response required • Appropriate for all ages • Preneural auditory response • Ear-specific information • Frequency-specific information • Does not require sound-treated room	• Not a true test of hearing • Does not assess neural pathways • Invalidated by middle ear dysfunction • Limited value in estimating hearing level

*Tympanometry and acoustic reflexes (ipsilateral and contralateral).

As a result, OAE measurement can significantly affect accurate and timely diagnosis of auditory dysfunction and, also, can contribute to prompt, appropriate, and effective intervention.

Chapter Organization

A detailed review of each and every clinical application of OAEs in children is far beyond the scope of this chapter, or even this book. Since the late 1980s, literally thousands of papers have been published describing the application of OAEs in many dozens of different etiologies. Of course, OAE findings for hundreds of thousands of patients evaluated by clinical audiologists and other hearing care professionals are never written up, or submitted and accepted for publication in peer-reviewed journals. The purpose of this chapter is to provide the clinical audiologist, or the student of audiology, with a handy summary of expected findings for OAE measurement in selected disorders, diseases, and pathologies that account for the majority of entities encountered clinically. We also review mechanisms for each major category of auditory dysfunction, with special reference to the connection between the mechanism and expected OAE findings. Challenges plus potential problems and solutions in OAE measurement are cited within each category of auditory disorder. Finally, representative published studies, including a succinct explanation of typical findings, are noted in tabular form for each diagnostic category or group of pathologies or disorders.

The reader will probably notice immediately the conspicuous absence in the following review of a long-standing and important clinical application of OAEs—newborn hearing screening. The application of OAEs in the identification and diagnosis of hearing loss is covered thoroughly instead in a companion Plural Publishing Core Clinical Concept Series book entitled *Objective Assessment of Hearing* (Hall & Swanepoel, 2010). Readers responsible for hearing screening programs for infants are also encouraged to turn to *Objective Assessment of Hearing* for a comprehensive and complementary review of information on this valuable and widespread application of OAEs. In this chapter, however, we briefly review the application of OAEs in early detection of cochlear hearing loss in preschool and school-age children.

MIDDLE EAR DISEASES AND DISORDERS

Introduction

Detection and valid measurement of OAEs is critically dependent on the status of the middle ear. The stimulus (stimuli) used to evoke either transient or distortion product OAEs is delivered via a miniature speaker(s), usually within a probe assembly and a sound wave is propagated into the external ear canal toward the tympanic membrane. Technical aspects of sound stimulation, ear canal acoustics, and transmission of energy in a forward direction through the middle ear system were discussed in Chapter 2. Energy generated by OAE sources within the cochlea in response to stimulation is propagated via reverse (outward) middle ear transmission to the external ear canal. Transfer functions for forward and reverse transmission are the product of complex interactions among contributions from the ear canal, the middle ear system (tympanic membrane, ossicular chain, middle ear tendons, and muscles), and the cochlea. Measurable variables in this process include acoustic pressure and volume velocity at each of these anatomic regions (the tympanic membrane, middle ear, base of cochlea). The level and phase of acoustic pressure transfer function measured in the external ear canal is dependent on both forward and reverse (i.e., "round trip") transmission, and varies significantly as a function of frequency (e.g., Keefe, 2007; Keefe et al., 2003a, 2003b; Kemp, 1980; Puria, 2003; Shera & Zweig, 1992; Shera & Zweig-Keefe, 2007, 1992).

Soon after the discovery of OAEs, published papers emphasized the marked negative impact of assorted middle ear pathologies on OAEs (e.g., Bonfils, Bertrand, & Uziel, 1988; Kemp, Ryan, & Bray, 1990; Probst, Lonsbury-Martin, & Martin, 1991). In the early years of clinical application of OAEs, conventional thinking argued against an attempt to record responses from ears with known or suspected middle ear dysfunction. The importance of the middle ear to OAE measurement was generally recognized, but most clinicians and clinical researchers assumed logically that OAE measurement and analysis would be adversely influenced and essentially contaminated by middle ear abnormalities. Indeed, according to the prevailing opinion at the time, documented middle ear dysfunction often precluded valid measurement of OAEs. In short, middle ear dysfunction was viewed to be a complication and a problem to be avoided whenever possible in the clinical application of OAEs, particularly in patient populations at risk for abnormal middle ear function (e.g., young children).

The influence of the middle ear on OAE measurement has, of course, not changed over the years. However, our understanding of the impact of middle ear condition on OAEs has evolved significantly. Careful documentation of middle ear status with tympanometry, other admittance measurements, or wideband reflectance techniques is still critical for accurate analysis and meaningful interpretation of abnormal OAE findings. In a newborn infant population, information derived from middle ear measures, such as acoustic admittance and energy reflectance, improves the prediction of hearing status with OAEs (Keefe et al., 2003b). Furthermore, statistical analyses of relations among middle ear measures and OAEs has quantified the variance in OAE findings due to middle ear status and assists in determining when the absence of OAEs can be attributed to middle ear dysfunction (Keefe et al., 2003a) In recent years, we have witnessed two major shifts in thinking about the middle ear and OAEs. As suggested above in the brief discussion of forward and reverse middle ear transfer functions, OAEs are now utilized as a tool for basic investigation of normal and abnormal middle ear transmission, and for testing noninvasively in human subjects system models for middle ear transmission. Secondly, as evidenced by the rapidly growing literature reviewed in this section, OAEs are becoming recognized as a sensitive index of middle ear status that can provide clinically valuable information in patients with assorted middle ear abnormalities.

Negative Middle Ear Pressure

Negative middle ear pressure, usually secondary to Eustachian tube dysfunction, is a very common disorder in the clinical setting, particularly with children. Although not a direct measure of middle ear pressure, tympanometry can easily and quickly detect positive or negative deviations from normal pressure. Artificial manipulation of external ear canal pressure is used to infer the influence of pressure deviations on OAEs. Studies consistently show maximum OAE amplitude with ear canal pressure of 0 daPa, and then decreased amplitude for higher or lower pressures. The exact effect of external ear canal pressure on OAEs, however, varies considerably depending on stimulus frequency and type of OAE (transient-evoked versus distortion product). TEOAE amplitude decreases about 2.5 dB at +100 or at −100 daPa of pressure, with greater changes in amplitude at more extreme pressures, but only for stimulus frequencies below approximately 2000 to 2500 Hz (e.g., Naeve et al., 1992; Trine et al., 1993). For stimulus frequencies below 2000 Hz, DPOAE amplitude also is diminished with deviations of pressure in the external ear canal. Amplitude changes are on the order of 8 dB at the +100 and −100 daPa points, but the actual effects vary depending on stimulus properties (e.g., intensity) and the shape of the DPOAE input-output function (e.g., Osterhammel et al., 1993; Plinkert et al., 1994). For higher stimulus frequencies, ear canal pressure has little or no effect on OAE amplitudes. This finding has clinical implications, of course, for the application of OAEs in hearing screening and diagnosis of auditory dysfunction. That is, if the exclusive goal of screening is the detection of cochlear auditory dysfunction, then it makes sense to limit stimuli to a high-frequency region to avoid the negative influence of middle ear dysfunction. On the other hand, sensitivity of OAE measurement to peripheral auditory dysfunction in general (middle ear and cochlear abnormalities) is increased when OAEs are elicited with a wider range of stimulus frequencies (including lower frequencies).

It is important to note that artificial alterations in external ear pressure, and the secondary influence on middle ear transmission, are not equivalent to middle ear disorders even at apparently comparable pressures (e.g., −100 daPa). Also, creation of positive pressure within the external ear canal moves the tympanic membrane inward and, therefore, is the equivalent to negative pressure within the middle ear space. There really is no way to experimentally manipulate middle ear pressure in order to study the effect pressure changes on OAEs. Increasing the pressure in the ear canal may serve as a surrogate for studying some types of negative middle ear pressure (Sun & Shaver, 2009). Clinical findings confirm that TEOAEs are generally not detected from ears with abnormal negative pressures exceeding −200 daPa (Owens et al., 1993).

Given the documented negative impact of middle ear pathology on detection of OAEs, and the high prevalence of negative middle ear pressure clinically, it is important to consider compensation for negative pressure in OAE measurement. There is evidence that recording TEOAEs at peak tympanogram pressure (compensated for the amount of negative pressure) increases amplitude (about 2 dB) and enhances detection (Hof et al., 2005). Instrumentation with the capacity for recording OAEs after compensation for negative pressure is discussed further in Chapter 10, New Directions in Research and Clinical Application.

Ventilation Tubes (Grommets)

OAEs can be recorded in patients with ventilation tubes (also called grommets and tympanostomy tubes) in the tympanic membrane. OAE measurement is not medically contraindicated and poses no risk to the patient with ventilation tubes. The characteristic pattern of OAE findings associated with patent (unoccluded) ventilation tubes is a reduction of amplitude, especially for lower stimulus frequencies (e.g., Amedee, 1995; Owens et al., 1993; Richardson et al., 1996; Tilanus et al., 1995). DPOAE findings in a patient with bilateral ventilation tubes are shown in Figure 7–1. Pure-tone audiometry showed a very mild low-frequency conductive hearing deficit. DPOAE measurement was conducted as part of a diagnostic assessment of auditory processing. In Figure 7–2, a ventilation tube is clearly visible in each tympanic membrane.

Rather than focusing on the possibility that OAEs are absent in patients with ventilation tubes, it is wise clinically to apply OAE findings as an objective measure of the integrity and function of ventilation tubes (e.g., Chang, Jang, & Rhee, 1998; Cullington, Kumar, & Flood, 1998). Normal OAE amplitudes across a wide frequency region provide rather conclusive evidence that: (1) the ventilation tubes are patent and effective in resolving middle ear effusion, (2) middle ear function is generally intact, and (3) significant cochlear (outer hair cell) dysfunction can be ruled out. Of course, as an objective measure, OAEs can be recorded to evaluate auditory function even in young children, months or even years before pure-tone audiometric assessment is possible.

The majority of patients (over 80%) have measurable postoperative TEOAEs immediately following insertion of grommet type tympanostomy tubes (Charlier & Debruyne, 2004; Fritsch, Wynne, & Diefendorf, 2002), whereas normal OAE findings are less likely for T-type tubes (Fritsch et al., 2002). OAE findings before and after grommet insertion are closely correlated with the outcome of pure-tone audiometry, and with the likelihood of detecting a conductive hearing loss due to middle ear effusion (Saleem et al., 2007). Griffiths and colleagues (2007) found that the proportion of children with detectable TEOAEs increased from 20% immediately postoperatively to 82% at a 6-week postoperative clinic visit. Similarly, Zhang et al. (2004) reported increased DPOAE amplitude immediately after tympanostomy tube insertion for middle ear effusion and further increase in amplitude 2 weeks later. OAEs also are present in the majority of children with middle ear effusion immediately following management with myringotomy, without insertion of tympanostomy tubes (Niedzielska & Katska, 2002).

In a related and rather clever study, Nelson et al. (2009) examined the auditory status of 21 healthy adult normal hearing volunteers with otoacoustic emissions, tympanometry, and pure-tone audiometry during ear canal suctioning. During suctioning, the peak intensity level reached 111 dB in the external ear canal near the tympanic membrane. Nonetheless, the authors found no significant changes in DPOAEs after suctioning.

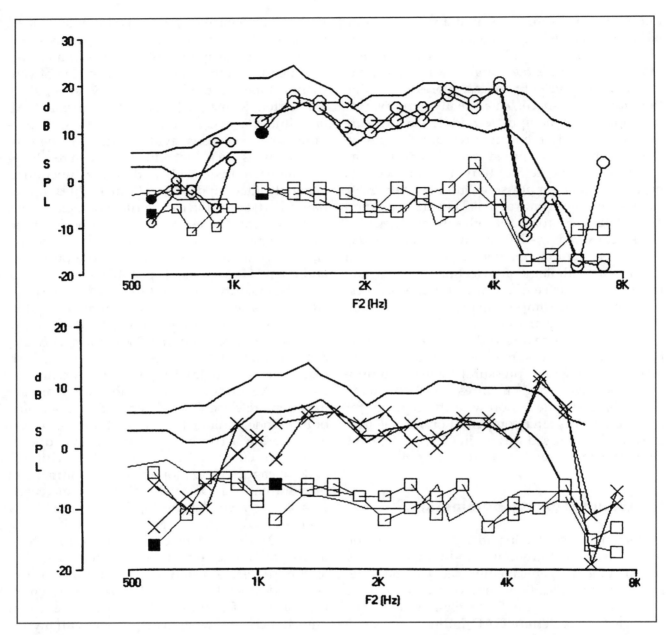

FIGURE 7–1. Distortion product otoacoustic emissions (DPOAEs) recorded from a 5-year-old girl with patent ventilation tubes bilaterally and mild low-frequency conductive hearing loss.

Middle Ear Effusion and Otitis Media

Absence of detectable transient and distortion product OAEs is a typical finding in children with otitis media or middle ear effusion, especially for low stimulus frequencies (e.g. Choi et al., 1999; De Felice et al., 2008; Koivunen et al., 2000; Mendez-Ramirez & Altamirano-Gonzalez, 2006; Owens et al., 1993; Yeo et al., 2002; Zhao et al., 2003). The likelihood that OAEs will be absent and, when present, their amplitude will be reduced, is directly correlated with stimulus frequency, the volume of middle ear effusion, and with the extent of tympanometric abnormality (e.g., Kei et al., 2007; Yeo et al., 2002; Zhao et al., 2003). In populations at high risk for middle ear disorders,

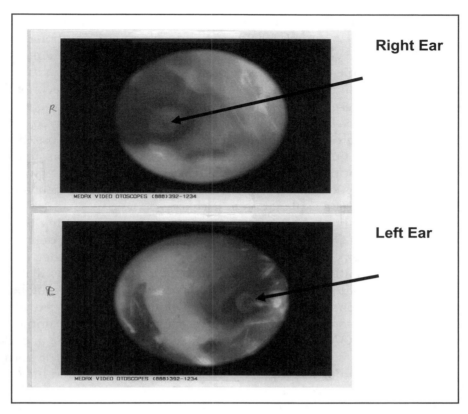

FIGURE 7–2. Video-otoscopy images for a 5-year-girl showing patent ventilation tubes bilaterally. The child had a mild low-frequency conductive hearing
loss. DPOAE findings are shown in Figure 7–1.

such as Down syndrome, DPOAE abnormalities
are not uncommon (Hess et al., 2006; Kumar et al.,
2008). Recent studies provide evidence suggesting
that the absence of DPOAEs in middle ear effusion
is related to the thickness of mucous fluid and the
amount of total protein (Park et al., 2007). At the
other end of the otitis media spectrum, decreased
DPOAE levels (by an average of 3.5 dB in the 2000
to 4000 Hz region) can be correlated with a history of middle ear disease, and the severity of past
infection, even in normal-hearing children (Job &
Nottet, 2002). Yilmaz and colleagues (2006) also
have found TEOAE and DPOAE abnormalities to
be more likely in adult normal hearers with a history of otitis media. Yeoman et al. (2003) describe
high prognostic value for TEOAEs and for DPOAE
input-output functions in predicting response to
medical treatment for middle ear effusion. Estimation of thresholds with DPOAE input-output
functions may distinguish between conductive

(middle ear) and cochlear hearing loss in newborn
infants. Akdogan and Ozkan (2006) also reported
that DPOAEs were helpful in confirming middle
ear status during medical management of otitis media. Failure of OAE screening of newborn
infants also may be related to the presence of
confirmed middle ear effusion (Boone, Bower, &
Martin, 2005). Finally, a study in young Aboriginal
children in western Australia confirmed that those
who failed TEOAE screening at age 1 to 2 months
were 2.6 times more likely to later develop otitis
media than the children who passed the screening
(Lehmann et al., 2008). The authors recommend
routine OAE screening in health service settings
to identify children at risk for otitis media.

In distinct contrast to the strong link between
middle ear effusion and abnormal or absent
OAEs just described, there are published papers
and occasional clinical anecdotal reports of normal OAEs in patients with audiometric signs

of middle ear effusion, particularly within the high-frequency region (Koike & Wetmore, 1999; Konradsson et al., 1999; Lonsbury-Martin et al., 1994; Tas et al., 1994). An example of such a disconnection among audiometric signs is a finding of normal OAEs (not just the presence of detectable OAEs) in a patient with a type B (flat) tympanogram, a finding usually associated with middle ear fluid. Discrepancies between OAEs and other audiologic test procedures in apparent middle ear disorders highlight the "crosscheck role" of OAEs in a pediatric test battery. The explanation for the discrepancy rarely is readily apparent. The discrepancy, that is, the failure of OAEs to "crosscheck" the results of tympanometry, demands further clinical investigation in an attempt to reach a correct diagnosis. Additional audiologic procedures, such as high probe frequency tympanometry, wide band reflectance, pure-tone audiometry, or even another type of OAE (TEOAE or DPOAE) might yield valuable clues. And, a thorough otologic examination would be in order to properly delineate middle ear status. Follow-up diagnostic audiologic assessment, with or without medical intervention, would almost always be warranted.

SCREENING PRESCHOOL AND SCHOOL-AGE CHILDREN FOR COCHLEAR DYSFUNCTION

Rationale for Preschool and School-Age Screening

Considering the voluminous clinical literature describing test performance for OAEs in newborn hearing screening, there are remarkably few published studies of focusing on sensitivity and specificity of OAEs in the identification of cochlear hearing impairment in preschool and school-age children. Of course, detection of hearing loss in children in the age range of 3 months to 5 years is just as critical as it is for newborn infants. Ongoing screening for hearing loss after the neonatal period is important for at least four reasons. First and foremost, hearing loss in young children can adversely affect speech and language acquisition, social development, emotional and psychosocial status, and early preacademic skills, including reading readiness. Children in impoverished settings and of lower socioeconomic status are at relatively greater risk for the negative influences of hearing loss, and delayed identification and intervention for hearing loss. Second, preschool hearing screening permits detection of young children with hearing loss who, for whatever reason, were not screened as neonates before hospital discharge.

Third, screening during the preschool years will detect progressive or delayed-onset cochlear hearing loss that was either not present or not detectable during infancy. Recent research confirms that the prevalence of hearing loss may almost double from the neonatal period to the early school years. For example, Fortnum, Summerfield, Marshall, Davis, and Bamford, (2001) reported that prevalence of permanent hearing loss within 17,160 children increased from 1.07% at age 3 years to 2.05% at age 9 to 16 years. The authors stated, "Relative to current yields of universal neonatal hearing screening in the United Kingdom, which are close to 1/1000 live births, 50 to 90% more children are diagnosed with permanent childhood hearing impairment by the age of 9 years" (Fortnum et al., 2001, p. 536). This statistic alone justifies a systematic approach for hearing screening in the preschool population.

Finally, ongoing screening efforts are required to detect conductive hearing loss secondary to middle ear disease. Otitis media is a well-recognized and common etiology for acquired hearing loss in preschool children. Chronic conductive hearing loss warrants early identification, close monitoring, and, if not responsive to medical therapy, audiologic intervention.

Well-respected professional organizations recommend periodic hearing screening of preschool and school-age children, including the American Academy of Pediatrics (APA, 2000), the American Speech-Language and Hearing Association (ASHA), and the National Institutes of Health (NIH). Current American Academy of Pediatrics guidelines call for hearing screening at 4, 5, 6, 8, and 10 years of age.

Protocol for Preschool and School-Age Screening

There is no "gold standard" for evaluating hearing status in newborn infants. OAE test performance in neonates usually is validated initially and rather indirectly by screening outcome for ABR, with subsequent verification of hearing status at 6 months or later by pure-tone audiometry. In contrast, the "gold standard" for hearing screening in preschool and school age children is outcome for pure-tone hearing screening usually for test frequencies of 1000, 2000, and 4000 Hz at an intensity level of 20 dB HL. (e.g., American Academy of Audiology [AAA] Position Statement, 1997b; American Speech Language Hearing Association [ASHA], 1997, 2002).

The limitations of pure-tone hearing screening in the preschool years are well appreciated, especially by audiologists and others who have attempted this challenging task. Behavioral audiometry techniques do not reliably estimate auditory status in children less than 1 year of age, even when performed by highly skilled audiologists. Pure-tone hearing screening of young children is confounded by a variety of factors, including cognitive status of the child, noise in the test setting, and skill and experience of the tester. Often, the responsibility for preschool and school-age hearing screening in a school setting or physician's office falls to nursing personnel who lack adequate training, expertise or experience. Under the best of circumstances, pure-tone hearing screening of preschool children is time consuming. Unfortunately, as clearly documented by published studies (e.g., Berg et al., 2008; Halloran et al., 2005; Sideris & Glattke, 2006), an unacceptably high proportion of preschool children, often the majority, cannot be reliably screened with pure-tone techniques for the reasons cited above, plus the inability of some young children to follow directions or sustain attention to the task. According to Halloran and colleagues (2005), the number of 3-year-old children unable to be screened is up to 33 times greater than number of the older (school-age) children who cannot be tested. In short, pure-tone techniques are not feasible for hearing screening of a preschool pediatric popu-lation. Pure-tone screening of older preschool and school-age children by nonaudiologic personnel in typical test environments is characterized by the dual problems of low sensitivity and specificity (e.g., Halloran et al., 2009). A more valid screening approach is required.

An objective measure of auditory function, OAEs are quick, technically rather simple and, perhaps most importantly, not influenced by the troublesome listener variables. If OAEs are good enough for detection of hearing loss in newborn infants that might interfere with speech and language acquisition, then it seems logical that they can also be used to detect similar hearing loss in preschool children. Still, it is reasonable to ask, "How do OAEs measure up against the gold standard?" The answer to that question is not readily apparent from published studies. The OAE screening protocol (e.g., stimulus intensity level and frequencies) and criteria used to define pass versus refer outcomes obviously are very important factors in determining the sensitivity and specificity of OAEs in preschool screening.

Published findings argue against reliance on a simple SNR criterion for a pass or refer outcome, as often done with newborn hearing screening. In their study of a large series of 6-year-old children, Lyons et al. (2004) clearly demonstrated that the use of a fixed DP-noise floor difference, such as ≥5 dB for one or more test frequencies, as the Pass criterion, will result in too many false negative outcomes. These researchers found that DPOAE screening alone missed from 32 to 38% of children who failed pure-tone hearing screening and who also had abnormal tympanometry. Adding tympanometry to the DPOAE screening protocol resulted in a substantial improvement in the hit rate for objective hearing screening.

DPOAE hearing screening test performance, at least sensitivity to hearing loss (as defined by pure-tone screening), will be improved by more rigorous criteria for a pass outcome. An argument might be made for using OAEs (at least DPOAEs) in preschool screenings with more rigorous criteria (e.g., a SNR of >6 dB plus an absolute DP amplitude of >0 dB SPL). The extensive statistically based research of Michael Gorga and colleagues at Boys Town (admittedly with adult subjects) supports the performance of these criteria in detecting almost all persons with pure-tone thresholds more than 20 dB HL for DPOAE f2 frequencies of 2000 Hz and higher. This point is illustrated by the distributions for normal hearing versus hearing-impaired persons as a function of DPOAE amplitude, shown in Figure 7–3. DPOAE amplitude is shown on the x-axis. Distributions of Pass and Refer outcomes are shown in relation to the criterion for a Pass outcome. Two criteria must be met for a Pass outcome: (1) A SNR of ≥6 dB (confirming the presence of a DP) and (2) minimal DP amplitude of 0 dB (approximately the lower limit of normal DP amplitude). The use of these combined criteria will detect almost 100% of ears with pure-tone hearing thresholds of more than 20 dB HL.

Referral rates for preschool hearing screening with DPOAEs may be higher than the rates that are acceptable for newborn hearing screening (indicated as "false positive" outcomes in Figure 7–3). However, in contrast to the multiple problems associated with a high false positive rate in newborn hearing screening, the cost of a higher referral rate is minimal in preschool and school-age populations. It simply means that pure-tone audi-

ometry must be performed to confirm or rule out a hearing loss. Pure-tone screening usually can be conducted on site with no additional cost or other concerns. Using a clinical DPOAE screening device, the second author regularly applies these two criteria successfully over the frequency region of 2000 to 5000 Hz in screening children age 6 months through 4 years who are enrolled in Head Start Programs. DPOAE screening rarely requires more than 2 minutes per child. Children yielding a Refer on DPOAE screening immediately undergo otoscopic inspection and tympanometry.

The National Center for Hearing Assessment and Management (NCHAM) has developed detailed evidence-based guidelines for early identification of hearing loss of infants and toddlers during well-child visits and in educational settings (e.g., Head Start Programs). Written educational materials are supplemented with a multimedia instructional package and hearing screening record forms. This information can be readily accessed or requested at the NCHAM Web site (http://www.infanthearing.org).

Steps in the NCHAM screening guidelines are:

■ Visual inspection, usually by a nurse

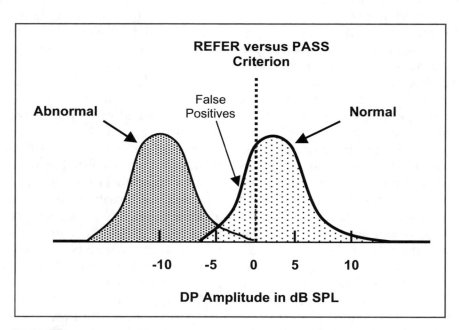

FIGURE 7–3. Distributions of persons with normal hearing sensitivity (≤20 dB HL) versus sensory hearing impairment (>20 dB HL) screened with DPOAEs.

■ Otoacoustic emissions (OAEs), usually by a nurse

■ Tympanometry performed by audiologist or physician if a child fails OAE screening

NCHAM also delineates categories for the outcome of hearing screening, including:

■ Cannot test
■ Child uncooperative
■ Excessive environmental noise
■ Pass
■ Refer (Repeat screening next day. Then, after 3 refer outcomes, refer to health care provider)

Literature on Preschool and School-Age Screening with OAEs

Successful application of OAE screening techniques in preschool children is documented by clinical evidence reported in recently published papers (e.g., Berg et al., 2008; Dille, Glattke, & Earl, 2007; Eiserman et al., 2008; Halloran et al., 2010; Ho et al., 2002; Hunter et al., 2007; Lyons, Kei, & Driscoll, 2004; Psillis et al., 2007; Yin et al., 2009). Selected information from these studies on OAEs in preschool and school-age screening is summarized in Table 7–2. At least six general conclusions can be drawn from the literature on

Table 7–2. Summary of Information from Published Studies of Hearing Screening with OAE Techniques in Preschool and Young School-Age Children

Study (Year)	Type of OAE	Other Tests	N	Age (yrs)	Population	Refer Rate	Comments
Lyons et al. (2004)	DPOAE	Tympanometry	1003	4–8	Early school	NA*	Hit rate up to 0.90
Sideris and Glattke (2006)	TEOAE	PTA Tympanometry	200	2–6	Preschool	21%	
Berg et al. (2006)	DPOAE	Tympanometry PTA	4003	2–9	Mixed	13.6%	Plus 6.5% CNT
Dille et al. (2007)	TEOAE DPOAE		33	0.5–4	Preschool	52% 48%	TEOAE DPOAE
Hunter et al. (2007)	DPOAE	Tympanometry	421	0–2	AI*	30%	Otitis media common
Psillis et al. (2007)	DPOAE	Tympanometry	76	1–5	Preschool	NA	
Eiserman et al. (2008)	DPOAE		4519	≤3	HS*	18%	Plus 8% CNT
Georgalas et al. (2008)	TEOAE	Tympanometry	196	6–12	School age	32%	Correlate with tymp
Hild et al. (2008)	DPOE	Tympanometry PTA	512	10–69	SO*	24%	
Yin et al. (2009)	TEOAE	PTA	744	2–6		5.5%	Sensitivity of 100%

*PTA = pure tone audiometry; NA = Not available; HS = Head Start; CNT = Could not test; SO = Special Olympics; AI = American Indian.

hearing screening of preschool and young school-age children with either TEOAEs or DPOAEs:

1. Test time is generally brief (less than 1 minute per ear)
2. Screening is simple for nonaudiologists,
3. Screening is efficient,
4. Compliance (acceptance) by children is high,
5. Test performance is acceptable across the age range, with referral rates reported from 5.5% up to almost 50%, and
6. DPOAE test performance is enhanced by tympanometry.

We'll take this opportunity to emphasize once again the relatively modest consequences of a higher referral rate for preschool and school-age hearing screening in comparison to the varied and serious problems associated with a high referral rate in newborn hearing screening. Preschool and school-age children who yield a refer outcome with DPOAE screening simply undergo pure-tone hearing screening, either immediately or at a follow-up test session. The many negative consequences of an overly high referral rate in newborn hearing screening are well documented in the literature (e.g., JCIH, 2007).

The work of Lyons et al. (2004). lends further support to the detection of most children with hearing loss using criteria for an OAE pass versus refer outcome that are more rigorous than a simple 6 dB SNR. The Lyons et al. research also confirms the importance of including tympanometry with OAEs to detect middle ear dysfunction, and the possibility that children who pass pure-tone screening can have cochlear dysfunction that is detected with DPOAEs. There also is literature on the reliance of OAE screening of hearing in children participating in the Special Olympics, confirming the value of OAEs in detection of either middle ear or sensory hearing impairment.

GENETIC ETIOLOGIES

Much hearing loss can be traced to genetic origins. Early identification and diagnosis of geneti-cally related hearing loss can lead not only to early intervention to improve communication but also to earlier and more accurate definition of the specific genetic disorder. OAEs offer a variety of advantages for diagnosis of genetic hearing loss and genetic disorders, ranging from the ability to objectively assess auditory function of newborn infants to sensitivity to either conductive or sensory hearing impairment to the capability to determine whether the genetic disorder involves outer hair cells. Literature describing OAE findings in selected genetic disorders is summarized in Table 7–3. As noted in the next section, a literature review can be conducted very quickly via the Internet and scientific search engines.

Review of the literature, and the findings summarized in Table 7–3 permit the following general conclusions about the roles of OAEs in genetic disorders. OAE measurement permits early and site-specific detection of cochlear dysfunction of genetic origin. OAE abnormalities can be documented in newborn infants years before behavioral audiometry is feasible. Furthermore, OAEs can detect "subclinical" auditory dysfunction before it is apparent with pure-tone audiometry. Early evidence of OAE abnormalities may predict which family members are likely to subsequently develop clinically important hearing loss. OAEs also allow for regular, objective, ear- and frequency-specific monitoring of auditory dysfunction in persons who are at risk for progressive hearing loss, including infants and young children. Distinctive patterns of OAE abnormalities may be the auditory signature of families with genetically related cochlear dysfunction. In combination with other auditory procedures (e.g., the ABR), OAEs permit differentiation of cochlear versus retrocochlear or central nervous system auditory dysfunction, either of which may be found in some genetic disorders.

Auditory neuropathy is a component of or associated with multiple and diverse diseases and disorders. Published evidence clearly points to a genetic explanation for a subset of patients with auditory neuropathy. Auditory neuropathy is now well appreciated as a diagnostic category of hearing loss, albeit a rather heterogeneous collection of disorders. Therefore, we will devote the next section to auditory neuropathy, rather than including the disorder within the genetic etiologies.

Table 7–3. Selected Literature Describing Otoacoustic Emissions (OAEs) Findings in Genetic Disorders

Disorder	OAE Type	Findings	Study (Year)
GJB Gene Mutation	TEOAE	OAEs contributed to identification and diagnosis of nonsyndromic hearing impairment in young children	Shi et al. (2005)
		OAEs identify at birth some but not all children with hearing impairment in GJB2 mutations (survey study).	Norris et al. (2006)
	TEOAE	Some newborn infants screened with 35delG mutation had abnormal OAE hearing screening results.	Bathelier et al. (2004)
Connexin 26 Mutations	DPOAE	Outer hair cell dysfunction confirmed with OAEs. Low amplitudes for OAEs in carriers versus noncarriers	Engel-Yeger et al. (2003)
DFNA13/COL11`A2	TEOAE	OAEs confirmed cochlear site of auditory dysfunction in Dutch family with low- and mid-frequency hearing loss.	DeLeenheer et al. (2004)
A1555G mutation (Mitochondrial 12S rRNA Gene)	TEOAE	OAEs were absent in deaf A1555G carriers confirming a cochlear site of dysfunction	Bravo et al. (2006); Matsunaga et al. (2005); Xing et al. (2000)
35delG Mutation A827G Mutation		TEOAE abnormalities in infants with these mutations	Nivoloni et al. (2010)
DFNA 11 (VIIA Mutation)	TEOAE DPOAE	OAEs contributed to confirmation of progressive cochlear auditory dysfunction	Tamagawa et al. (2002)
T7511C Mitochondrial Mutation	TEOAE DPOAE	OAEs contributed to differentiation of cochlear versus retrocochlear auditory dysfunction in Japanese family with nonsyndromic sensorineural hearing loss.	Ishikawa et al. (2002)
Autosomal Dominant	DPOAE	Most (83%) of family members showed distinctive similar abnormal pattern for DPOAEs	Oaken et al. (2002)
Nonsyndromal Hearing Loss	DPOAE	OAE measurement resulted in early detection of low-frequency progressive sensory hearing loss in four generations of a Tennessee family mapping to 4p16.3 gene	Lesperance et al. (1995)
Mitochondrial Myopathy	TEOAE	Most patients had absent TEOAEs when pure-tone hearing sensitivity was normal confirming subclinical cochlear dysfunction.	Korres et al. (2002)
Unspecified Genetic Disorder	TEOAE DPOAE	OAE abnormalities or absence was found for the majority of Chinese family members with genetic progressive hearing loss.	Yu et al. (2000)
	DPOAE	Reduction in OAE amplitudes and absence of OAEs for high-frequency stimulation can identify subclinical cochlear dysfunction.	Ke et al. (2000)

115

AUDITORY NEUROPATHY SPECTRUM DISORDER (ANSD)

Background on ANSD

Otoacoustic emissions play an important role in the detection of auditory neuropathy spectrum disorder (ANSD), and in defining the site of dysfunction during the subsequent diagnostic process. Indeed, the explosion of articles in the mid-1990s describing what was then referred to as auditory neuropathy, and more recently, relabeled as auditory neuropathy spectrum syndrome, was a direct result of the increased and more widespread availability of clinical devices for OAE measurement. Since then, hundreds of publications confirm the clinical value of OAEs in various subgroups of this heterogeneous clinical entity. A complete review of the literature on auditory neuropathy is not appropriate for this chapter, or this book. A recent monograph entitled "Guidelines for Identification and Management of Infants and Young Children with Auditory Neuropathy Spectrum Disorders" (Denver Children's Hospital, 2008) offers a comprehensive and up-to-date source of information. With contributions from a panel of experts, the Guidelines address such important topics as auditory neuroscience underlying ANSD, protocols for identification and diagnosis with behavioral and objective auditory measures, and various management options. The reader also is encouraged to read firsthand some of the recent review articles on ANSD (e.g., Berlin, Morlet, & Hood, 2003; Gibson & Sanli, 2007; Vlastarakos et al., 2008), the lengthy portions of textbooks devoted to the topic (e.g., Hall, 2000, pp. 439–460; Hall, 2007, pp. 138–154), or an entire book on ANSD (Sininger & Starr, 2001). Another Plural Publishing Core Clinical Concepts Series book entitled *Objective Assessment of Hearing* (Hall & Swanepoel, 2010) also includes a very recent and detailed discussion of auditory neuropathy covering identification, diagnosis, and management.

According to recommendations made by Rapin and Gravel (2003, 2006), inner hair cell disorder does not fall within the definition of auditory neuropathy (i.e., acoustic nerve involvement is a requirement). Based on accumulated research findings, and clinical experience, a simple pattern of auditory findings (e.g., normal OAEs, absent ABR, and very poor speech perception) may be too often and too quickly attributed to auditory neuropathy when, instead, the explanation is a true sensory (cochlear) hearing impairment. ANSD also may be mistakenly diagnosed in children with cochlear nerve deficiency, as the two distinct entities share similar patterns of auditory findings (Adunka et al., 2006; Buchman et al., 2006). Interestingly, Buchman and colleagues (2006) reported that only one of nine children with cochlear nerve deficiency (i.e., small or absent 8th cranial nerve on MRI) had detectable DPOAEs. We echo the plea of Rapin and Gravel (2003) for more diagnostic specificity in the definition of "auditory neuropathy," based on physiologic and pathologic evidence of neural disorder.

Summary of Literature on OAE Findings

ANSD is associated with a rather diverse collection of disorders, ranging from neurologic diseases to genetic origins. Among children with hearing loss, auditory neuropathy or a pattern of auditory findings suggesting auditory neuropathy may account for as many as 8 to 10% (e.g., Foerst et al., 2006; Kirkim et al., 2008; Ngo et al., 2006; Rodríguez Domínguez et al., 2007; Xoinis et al., 2007). Several investigators have conducted widespread measurement of OAEs in populations of children with known hearing loss, such as those attending schools for deaf and hearing-impaired children (e.g., Cheng et al., 2005; Duman et al., 2008; Foerst et al., 2006; Lotfi & Mehrkian, 2007; Tang et al., 2004; Wang et al., 2007). Collectively, the results of these studies confirm the possibility of recording normal OAEs in up to 5% of children with severe to profound hearing loss.

OAE were present, and sometimes even normal, for a rather remarkable proportion of the children (up to 10%). The most likely explanation for the finding of outer hair cell integrity in serious hearing loss is inner hair cell abnormality, rather than neural disorder. A series of independent papers highlights the association of an auditory neuropathy type pattern of findings with otoferlin (OTOF) mutations (DFNB9 locus), other autosomal dominant inherited forms of hearing impairment (e.g., Dutch FFNA8/12), connexin 26 mutations, and Waardenburg syndrome. Each of these aberrations probably constitutes a genetic form of inner hair cell or "endocochlear" hearing loss (e.g., Cheng et al., 2005; Forli et al., 2006; Jutras et al., 2003; Kothe et al., 2004; Loundon et al., 2005; Plantinga et al., 2007; Ramsebner et al., 2007; Santarelli et al., 2008; Starr et al., 2001, 2004; Tekin, Akcayoz, & Incesulu, 2005; Xing et al., 2007), although a synaptic defect between the inner hair cells and the afferent auditory fibers cannot always be ruled out. OAEs in genetically based hearing loss help to localize the sensory deficit to the inner rather than the outer hair cells.

The literature indicates that OAEs can contribute importantly not only to identification of ANSD and definition of the site of auditory dysfunction, but also to differentiation among the possible mechanisms for various etiologies of ANSD. With the development of protocols, statistical algorithms, and instrumentation for clinical measurement of the suppression of OAEs by ipsilateral and contralateral sound, we anticipate that OAEs will provide uniquely valuable information on the status of the efferent auditory system (medial olivocochlear pathways) in auditory neuropathy (e.g., Hood et al., 2003). The measurement of OAEs with suppression paradigms is reviewed in Chapter 10.

OAEs are particularly useful clinically for identification and diagnosis of ANSD because they are readily available, widely applied, and easily recorded without sedation even from newborn infants. Overwhelming evidence from formal investigations and case reports conducted internationally and published in journals worldwide confirms the feasibility and value of OAEs in the diagnosis of bilateral auditory neuropathy (e.g., Akman et al., 2004; Berlin et al., 2005, 2010; Emara & Gabr, 2010; Kirkim et al., 2005; Kumar & Jayaram, 2006; Li et al., 2005; López-Díaz-de-León et al., 2003; Marlin et al., 2010; Ni et al., 2000; Sano et al., 2005; Shehata-Dieler et al., 2007; Truy et al., 2005; Wang et al., 2002, 2003) and also unilateral auditory neuropathy (e.g., Kothe et al., 2006; Madden et al., 2002; Podwall et al., 2002;

Salvinelli et al., 2004; Wang et al., 2007). There also are reports of apparently reversible "auditory neuropathy" in infants, as categorized by a finding of consistently normal OAEs and the reappearance of an ABR (e.g., Psarommatis et al., 2006).

Clearly, the diagnosis of ANSD is not made only with the information obtained from OAE measurement. On the contrary, diagnosis of auditory neuropathy requires a very comprehensive test battery including many electroacoustic, electrophysiologic, and behavioral auditory measures. Again, we recommend the monograph "Guidelines for Identification and Management of Infants and Young Children with Auditory Neuropathy Spectrum Disorders" (Denver Children's Hospital, 2008) for an in-depth review of the diagnostic criteria for ANSD.

There are at least three main limitations to the application of OAEs in ANSD. First, the site-specificity of OAEs is a liability, as well as an asset. The isolated finding of normal OAE findings and presumed outer hair cell integrity does not contribute to the identification or diagnosis of ANSD. OAE measurement must be combined with neural auditory measures (e.g., the ABR) to have clinical value. A second disadvantage of OAEs is their dependence on functional integrity of the middle ear system. Middle ear disorders, common in young children, often preclude valid measurement of OAEs and render them useless in the diagnosis of auditory neuropathy.

And, finally, for reasons that remain unknown, OAEs may disappear over time in patients with ANSD, perhaps as many as one-third, even those with absolutely normal hearing sensitivity (Deltenre et al., 1999; Starr et al., 2000). Interestingly, the cochlear microphonic component of electrocochleography (ECochG) usually persists in these patients. The absence of OAEs in persons who are at risk for auditory neuropathy does not help in ruling out a neural disorder, or in confirming a sensory disorder. That is, even though the combination of normal OAEs and absent ABR raises the distinct possibility of auditory neuropathy, the combination of absent OAEs and an absent ABR does not confirm a sensory loss or rule out auditory neuropathy. ECochG is very useful in differentiating sensory versus neural auditory dysfunction (see Hall, 2007, for details). Taken together, findings from OAE measurement, ECochG, and ABR measurement offer a powerful approach for evaluating sensory and neural function in suspected auditory neuropathy.

OTOTOXICITY

Introduction

If a clinical procedure for monitoring the possible effects of medications on cochlear function were to be designed, with the knowledge we have about ototoxicity, the procedure would probably end up looking just like OAEs. OAEs are remarkably well suited for early detection of damage or dysfunction of the inner ear for the following reasons:

- OAEs are highly sensitive to cochlear (outer hair cell) dysfunction
- Most ototoxic drugs first damage outer hair cells
- OAEs permit earlier detection of cochlear auditory dysfunction than the conventional audiogram (500 to 8000 Hz)
- OAEs are objective, and can be performed on very young and sick patients
- Test time is brief, usually only 1 or 2 minutes
- OAE measurement yields a high degree of frequency selectivity (detail), including up to 8 or 10 frequencies between each audiometric octave frequency
- Information can be obtained for high frequencies up to at least 8000 to 10,000 Hz, and sometimes even higher (see Chapter 2 for complications related to calibrating signals at these frequencies). Because of the extended high frequency limit, DPOAEs are superior to TEOAEs for ototoxicity monitoring.

Research evidence supporting the application of OAEs for monitoring the possible effects of medications on cochlear function in children was first reported over 10 years ago, including the antineoplastic (chemotherapy) agent cisplatin (e.g., Allen et al., 1998; Madasu et al., 1997; Yardley et al., 1998) and antibiotics for cystic fibrosis and infectious diseases (François et al., 1997; Katbamma et al., 1998; Mulheran & Degg, 1997; see Hall, 2000,

for review). Since then, we have seen a steady increase in the number publications confirming the clinical value of OAEs for early detection of inner ear dysfunction. Ototoxicity is one of most common risk factors for infant hearing loss. Early identification of children receiving potentially ototoxic drugs is recommended by the Joint Committee on Infant Hearing (2007). Not surprisingly, the mounting evidence linking changes in OAEs to drug-induced cochlear changes has led to more widespread use of OAEs for this purpose, and to clinical practice guidelines and recommendations for relying on OAEs in monitoring patients (e.g., ASHA, 1993; American Academy of Audiology, Ototoxicity Monitoring: Position Statement and Clinical Guidelines, 2010; National Cancer Institute [NCI] Common Terminology Criteria for Adverse Events [CTCAE], 2009).

A quotation from the American Academy of Audiology Position Statement and Clinical Practices Guideline on Ototoxicity Monitoring (2009) provides a good summary of the current peer-reviewed opinion on strengths and limitations of OAEs in detection of drug-induced auditory dysfunction. As stated in the Guidelines: "OAE measurement in children is a particularly attractive approach for ototoxicity monitoring, namely, as an efficient objective test." Coradini et al. (2007), in fact, demonstrated the importance of incorporating OAEs, particularly DPOAE testing in children and adolescents receiving cisplatin. Their results showed good agreement between high-frequency audiometry (HFA) and DPOAE findings. On the other hand, as noted above, Knight et al. (2007) reported that HFA usually detected ototoxic change earlier than DPOAEs, contrary to expectations. Further, HFA is less affected by otitis media than OAEs. Otitis media is common in children and in immunosuppressed chemotherapy patients, in general, as well as in patients receiving head and neck radiation, and in patients with infections undergoing treatment with aminoglycoside antibiotics. In any event, using a test battery approach increases the chances of obtaining reliable ototoxicity monitoring data over time" (AAA, 2009, p. 8). The Guidelines also define the rationale for reliance on DPOAEs rather than TEOAEs for ototoxicity monitoring, namely, the sensitivity of DPOAE to cochlear dysfunction in a higher frequency region and the presence of DPOAE that can be monitored in patients with greater sensory hearing loss.

One important limitation of OAE testing, as mentioned above, is that the results are significantly affected by middle ear pathology such as otitis media (Allen et al., 1998). That is, OAEs are difficult to record reliably, if detectable at all, in the presence of otitis media (Owens et al., 1992). And, as previously noted, the patient populations receiving ototoxic medications have increased susceptibility to otitis media, which interferes with OAE ototoxicity monitoring. For this reason alone, tympanometry (ideally multifrequency tympanometry) should routinely be evaluated when OAE testing is included as part of the test battery. Hence, OAE measurement probably should not be the sole method of ototoxicity monitoring, because interruptions in monitoring may occur whenever otitis media is present.

Although either transient or distortion product OAEs can used to detect ototoxicity, DPOAEs are applied more often because they can be recorded using high-frequency stimuli, thus providing information on the status of outer hair cell function toward the base of the cochlea. Recall from discussions in Chapters 5 and 6 that with TEOAE measurement it is not, for technical reasons, possible to detect a cochlear response from high-frequency (>5000 Hz) regions of the cochlea. DPOAEs, on the other hand, theoretically can be recorded for stimulation up to the high-frequency limits of the human cochlea (20,000 Hz). Clinical instrumentation now is available for recording DPOAEs with stimuli as high as 16,000 Hz. In fact, given the clear distinction in generation of TEOAEs versus DPOAEs (reviewed in Chapters 2 and 3), we probably should consider monitoring for ototoxicity with both techniques to maximize sensitivity to varied drug-induced pathophysiologic changes in the cochlea. Space does not permit a detailed review of the extensive and rapidly growing literature on OAEs and ototoxicity. Instead, we identify potentially ototoxic medications, briefly note suspected mechanisms for cochlear damage and dysfunction, and then summarize accumulated evidence in support of the clinical application of OAE in monitoring for ototoxicity.

Rationale for Monitoring for Ototoxicity

The obvious overall reason for monitoring for ototoxicity is detection of cochlear dysfunction. Clearly, if a child's health problem was minor and there was no medical contraindication to discontinuing the drug, then therapy would be discontinued at the first clinical evidence of ototoxicity. However, children whose medical management includes potentially ototoxic drugs invariably have serious, often life-threatening health problems, such as infection, excessive fluid (edema) in a vital anatomic region, or neoplasms (tumors) that cannot be treated surgically. Medical personnel responsible for management of children with such diseases would appear to face a difficult dilemma. So, what is done with information obtained from monitoring for ototoxicity, and how does the information affect management of a child?

Optimally, monitoring for ototoxicity prevents permanent sensory hearing loss and, consequently, preserves or permits the development of normal communication (speech and language acquisition). Monitoring cochlear function with OAEs makes it possible, at least in theory, to meet this ambitious objective. In the ideal scenario, monitoring of auditory status with OAEs leads to the detection of outer hair cell dysfunction at a very early stage. Preferably, the change in OAEs is at most a modest decrease in amplitude (perhaps only 4 or 5 dB) for the distortion products resulting from stimulation for only a limited number of the highest frequencies. It might be instructive to illustrate the application of DPOAEs in monitoring for ototoxic-induced auditory dysfunction a review of the literature on the topic.

Figure 7–4 depicts the relation between serial measurement of pure-tone hearing thresholds and DP amplitudes for a 10-year-old boy with an inoperable brainstem glioma who was managed medically with chemotherapy including cisplatin. A change in amplitude of 6 dB from one test date to another is greater than normal DPOAE test-retest variability (discussed in Chapter 6). Rather, the decreased DP amplitude can be viewed as evidence of a change, presumably drug-related, in cochlear (outer hair cell) function. Yet, according to the Brock system, hearing loss in ototoxicity

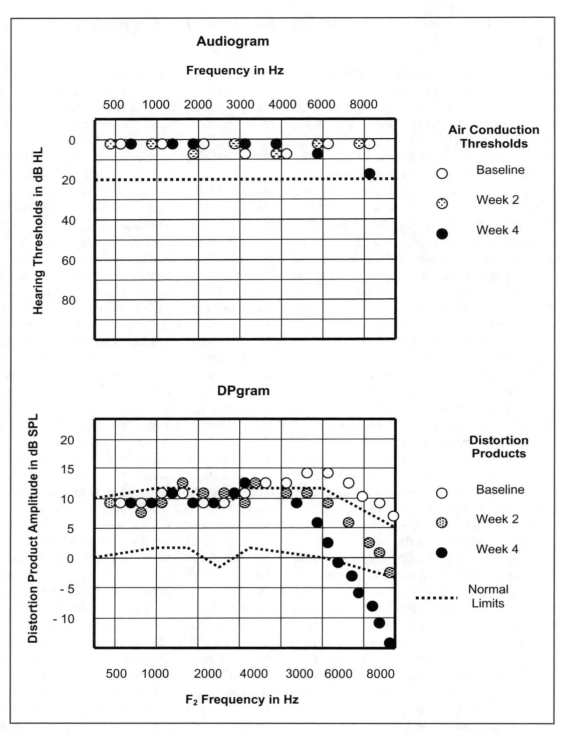

FIGURE 7–4. Pure-tone hearing thresholds and distortion product otoacoustic emissions (DPOAEs) in a 7-year-old boy with a brainstem tumor undergoing medical management with chemotherapy that included cisplatin. Findings are described in the text.

(see description below), the patient would be categorized as a Grade 0 (no deficit).

As seen in Figure 7–4, changes in DPOAE amplitudes occur well before evidence of any deficit in behavioral measures of hearing (e.g., pure-tone threshold). Early detection of cochlear dysfunction, as reflected by the decrease in DP amplitudes, should be reported immediately to the physician (s) responsible for medical management. A decision is made to modify drug therapy so as to effectively treat the underlying medical problem (e.g., infection or tumor) while simultaneously preserving auditory system integrity. Alterations in medical management could include some combination of reducing dosage of the drug, increasing time intervals between administrations of the drug (decreasing cumulative dosages), or using a substitute drug within minimal or no risk for ototoxicity for the patient. Medical status is then closely monitored to verify that there is no progression of the underlying problem with parallel ototoxicity monitoring to rule out delayed onset or progressive drug-induced cochlear dysfunction until it is clear that auditory status is stable. With early detection of clinically subtle changes in cochlear function, and effective modification of the medical therapy, it is sometimes possible to prevent or arrest a hearing loss as documented by the conventional audiogram.

A second and more common scenario leads to a less positive outcome. Baseline audiologic assessment, including DPOAE measurement and perhaps high-frequency audiometry, confirms normal auditory function. Medical therapy begins, with close monitoring of auditory function (once per week) with DPOAEs, including stimulation for frequencies up to 10,000 Hz. Weeks into the course of management with a potentially ototoxic drug, DPOAE amplitudes are decreased for the highest frequencies. The physician in charge of medical management is informed of the significant change in auditory function but decides, based on the patient's overall health status and the nature of the underlying health crisis, to continue with the drug therapy. Monitoring continues to document progression of changes in DPOAEs and, in addition, pure-tone hearing thresholds indicate a hearing loss. Several months later, DPOAEs are not detected for frequencies above 1000 Hz and

there is a hearing loss exceeding 25 dB HL for pure-tone frequencies above 2000 Hz. As monitoring confirmed developing auditory dysfunction, the managing physician was constantly updated but, unfortunately, the patient's medical status remains precarious and drug therapy cannot be altered.

How does monitoring of auditory status affect patient management for this scenario? Early detection of auditory dysfunction prompts communication with the patient, family members, and physician about audiologic intervention options. The patient and family are counseled regularly regarding hearing and hearing loss. Depending on the age of the patient, and the educational setting, teachers and other school personnel may be included in the counseling sessions. Options for audiologic intervention (e.g., hearing aids and FM technology) are reviewed, and implemented as indicated. Monitoring of auditory function is ongoing to ensure that intervention is most appropriate, and to ensure that the patient benefits as much as possible from intervention for the hearing loss, and communication with family members, medical personnel, and others is effective during this very difficult time. If the drug-induced hearing loss worsens yet, simultaneously, the child's underlying health problem actually improves with drug therapy; the physician may re-evaluate the medical plan. In some cases, discontinuing medical therapy halts the progression of hearing loss, while ongoing audiologic intervention permits effective communication.

A third scenario, involving therapeutic protection from cochlear toxicity, is now almost a reality. Extensive experimental investigation, and early clinical trials, confirm that cochlear damage can be minimized or even prevented medically with the use antioxidants and other scavengers for the dangerous free radicals that are involved in ototoxicity. The management challenge is to disrupt the cochlear toxicity by administration of protective agents (e.g., glutathione, D-methionine) without minimizing the effectives of the medical therapy for the underlying medical problem (e.g., tumor or infection). It is likely that the timely use of protective agents will result in the best of both worlds for some patients, that is, highly effective medical therapy for a life-threatening disease with preservation of normal auditory functioning.

Overview of Ototoxic Medications

Dozens of therapeutic drugs are potentially damaging to the auditory and vestibular systems. The actual risk for damage and dysfunction varies substantially among the drugs, among individual patients, and as a function of a wide array of factors. Ototoxicity, monitoring for ototoxicity with OAEs, and protective mechanisms for preventing or minimizing ototoxicity is currently a "hot research topic." Hundreds of published papers describe clinical studies focusing on the effects on cochlear function of specific drugs, or combinations of drugs, in different patient populations. Space constraints here permit only a cursory review of the rapidly growing literature. Again, we recommend for a more detailed discussion on ototoxicity monitoring, including information on the pharmacology of potentially ototoxic drugs, we highly recommend the American Academy of Audiology Position Statement and Clinical Practices Guideline on Ototoxicity Monitoring (2009).

The more common categories of drugs with documented potential for interfering with cochlear physiology include:

- Aminoglycoside antibiotics (e.g., gentamicin, dihydrostreptomycin, kanamycin, amikacin, tobramycin)
- Other antibiotics (e.g., vancomycin)
- Antineoplastic (chemotherapeutic) drugs (e.g., cisplatin, carboplatin, oxiloplatin, nedaplatin, ZD0473, BBR3464, satraplatin, and DFMO, but not vincristine or vinblastine sulfate)
- Loop diuretics (e.g., furosemide [Lasix], bumetanide [Bumex]), and ethacrynic acid [Edecrin])
- Nonsteroidal anti-inflammatory drugs, including salicylates (aspirin)
- Quinine derived drugs (e.g., Larium)
- Environmental chemicals (e.g., mercury, lead, arsenic, solvents)

Incidence, and even configuration, of auditory dysfunction and hearing loss varies among potentially ototoxic drugs. As a rule, auditory dysfunction is permanent (not reversible). It is first detected at the highest frequencies and then proceeds to lower frequencies. If a monitoring protocol includes OAE measurement for stimulation at high frequencies (e.g., >8000 Hz), well above those important for speech perception, early detection of auditory dysfunction can help to prevent hearing loss that interferes with communication. However, with some drugs (e.g., cisplatin), auditory dysfunction initially is more pronounced for a specific frequency region (e.g., 1000 to 2000 Hz, 3000 to 4000 Hz, or 6000 Hz) and less affected for the higher frequencies. Also, a variety of factors influence the likelihood, and the extent, of ototoxicity, including dose per treatment, mode of administration (inhalant, tablet, IV), schedule of administration (e.g., once daily or multiple dialing dosing), cumulative dose, concurrent or previous exposure to the same or other ototoxic drugs, renal function, prior hearing loss, noise exposure, chronological age in children, and individual differences in susceptibility. When these factors are equivalent for adults and children, ototoxicity tends to be more extensive in children. For some drugs, such as cisplatin, hearing loss may occur almost immediately after the initial treatment. Although bilateral hearing loss is typical, unilateral or asymmetric cochlear dysfunction sometimes occurs.

It is important when monitoring for ototoxicity with OAEs to keep in mind that changes in cochlear function can be delayed for some time after the drug is administered. Follow-up screening for ototoxicity should be scheduled (e.g., 3 months later) even if OAE findings initially are within normal limits. Also, ototoxic changes in cochlear function may occur or progress after the medication is discontinued, even after the drug is no longer present in the blood. Progressive ongoing effects of some drugs, particularly when they are administered in combination, may continue for days, weeks, and even months beyond the final dose.

Mechanisms of Ototoxicity and Protection

Aminoglycoside Antibiotics

A detailed review of the mechanisms for the ototoxic effects of different medications is beyond the scope of this book. The reader is referred to a recent textbook devoted entirely to the topic (Campbell, 2006). A literature search will also reveal recent review articles on the topic (e.g., Contopoulos-Ioannidis et al., 2005; Jacob et al., 2006).

Hearing loss due to aminoglycoside antibiotics usually is permanent, but reversible deficits are possible. High frequencies generally are affected initially. Outer hair cells are most susceptible to ototoxicity, but inner hair cell damage occurs with prolonged and/or higher doses. As noted above, auditory dysfunction may be delayed and/or progressive over the course of months after the drug is no longer detectable in blood plasma. Synergistic interactions in ototoxicity can occur for aminoglycoside antibiotics and furosemide, along with noise exposure and renal failure. Cochlear damage with aminoglycosides is secondary to toxic metabolites and to blockage of calcium and ion channels. Production of free radicals contributes to the abnormal biochemical processes underlying cochlear dysfunction. Outer hair cell physiology is affected initially, followed by structural damage to outer hair cells and supporting cells. Cochlear changes first appear in the base and then proceed toward the apical regions. There also is some evidence that the first row of outer hair cells is involved before the second and then the third rows. Protective mechanisms for aminoglycoside ototoxicity are based largely on the proven benefits of vitamins and other micronutrients that serve as scavengers for neutralization of the toxic metabolites.

Antineoplastic Drugs

Antineoplastic drugs are applied for medical management of tumors affecting different organs, including tumors in the nervous system, lung, liver and urogenital system (ovarian, testicular, cervical, bladder tumors). Examples of antineoplastic drugs are cisplatin, carboplatin, vincristine sulfate, and a relatively new drug α-difluoromethylornithine (DFMO). Among these antineoplastic agents, cisplatin is encountered most often clinically. A common clinical reason for administering cisplatin in children is management of solid brain tumors (e.g., brainstem gliomas), neuroblastomas, and hepatoblastoma (liver cancer). Cochlear toxicity with cisplatin is cumu-

lative and dose dependent. A related second-generation platinum drug, carboplatin, is relatively less toxic to the cochlea when administered in isolation. Recent research suggests that hearing loss is rare, occurring in less than 5% of patients (Dean et al., 2008), or DPOAEs are even normal (Dhooge et al., 2006) following carboplatin administration whereas more than 70% of patients treated with cisplatin and carboplatin in combination had hearing loss of Brock Scale grade 1 or higher.

Loop Diuretics

These diuretics, including ferosomide (Lasix) and less often bumetanide, act on the epithelial cells of the loop of Henle in the kidney. With children, loop diuretics are used for management of pulmonary edema and edema secondary to kidney malfunction. Furosemide interferes with stria vascular physiology reducing cochlear energy supply (endocochlear potential) and outer hair cell motility. The ototoxicity of aminoglycoside antibiotics varies substantially among persons, and may be strengthened by concurrent administration of furosemide. The auditory dysfunction or hearing loss caused by furosemide usually is permanent and found in the high- or midfrequency region of the audiogram.

Other Drugs

Quinine-derived drugs are used for treatment of malaria and, less often, leg cramps. Although not common in many developed countries, serious quinine-induced hearing loss is encountered frequently in countries with high prevalence of malaria (e.g., a number of countries on the African continent).

Evidence for Monitoring Drug-Induced Cochlear Dysfunction with OAEs

Summary of Literature

The evidence for monitoring for ototoxicity with OAEs is substantial and compelling. Earliest documentation that OAEs were a sensitive measure of drug-induced cochlear was based on experiments with various small animal models (e.g., gerbil and rabbit). The animal literature is not reviewed here, in part because conclusions drawn from the findings of experiment studies do not necessarily apply to humans. For example, early experimental evidence from research with chinchilla clearly showed an exclusive effect of carboplatin on inner versus outer hair cells (e.g., Hofstetter et al., 1997; Jock et al., 1996), but later clinical research and experience confirmed that carboplatin alone occasionally was associated with outer hair cell damage (e.g., Dean et al., 2008). Findings for selected clinical papers on monitoring for ototoxicity with OAEs in children are summarized in Table 7–4. The references cited in the table represent a fraction of the rather substantial literature on the topic. The reader is encouraged to perform a computer-based literature search to gain a true appreciation of the wealth of information relating OAE findings and ototoxicity.

At least eight consistent themes emerge from the constantly expanding literature on OAEs and ototoxicity: (1) Monitoring with OAEs for drug-induced auditory dysfunction in children is efficient and clinically feasible, even in infants and other children who, because of the effects of their illness, are difficult to test with behavioral techniques. Monitoring with OAEs requires minimal patient cooperation, brief test time, and relatively simple technical skills. (2) Distortion product OAEs provide an adequate degree of frequency resolution for precise detection of cochlear dysfunction; (3) For selected drugs (e.g., cisplatin), the majority of children show significant reductions in DPOAE amplitude; (4) changes in OAEs progress and may persist with ongoing treatment with potentially ototoxic medications; (5) DPOAE findings reveal substantial individual susceptibility (variability) in cochlear toxicity; (6) Greater changes in OAEs are associated with younger age and cumulative dose for some drugs (e.g., cisplatin); (7) Deficits in cochlear function are detected earlier with DPOAEs than with conventional pure tone audiometry; and (8) High-frequency audiometry (HFA), also known as extra-high-frequency audiometry (EFA), is a useful alternative to DPOAE recording for early detection of drug-induced cochlear dysfunction.

The application of OAEs as a monitoring strategy in adults is covered in Chapter 8. The reader is encouraged to periodically generate an updated literature review by performing an Internet-based search (e.g., http://www.nlm.nih.gov) for newly published papers, using key words such as "otoacoustic emissions" and "ototoxicity" or the names of specific drugs. There also are several recent papers describing meta-analyses of existing data on ototoxicity (e.g., Contopoulos-Ioannidis et al., 2005).

Protocol for Monitoring for Ototoxicity

Features of the strategy for monitoring for drug-induced cochlear toxicity with OAEs were mentioned and illustrated earlier with the case in Figure 7–4. When applied to monitoring for ototoxic changes in cochlear function, DPOAEs are recorded and replicated for 5 to 8 frequencies per octave up to the highest test frequency possible with the test equipment (preferably a high frequency of 8000 to 10,000 Hz) using an appropriately sensitive stimulus intensity paradigm (e.g., L_1/L_2 of 65/55 dB SPL). Examples of DPOAE test protocols, include one for monitoring for ototoxic-induced cochlear dysfunction, were shown in Chapter 6.

Young children often produce OAE amplitudes that are larger than the amplitudes for normal-hearing adult subjects, especially for the highest stimulus frequencies. Delaying a report of concern regarding auditory status until OAE amplitudes fall outside of an adult normal region, or worse yet, until there is no detectable OAE response, is inappropriate. When monitoring for possible cochlear toxicity, simply reporting OAE findings as "present" or "absent" is unacceptable. Failing to take note of the early decreases in DPOAE amplitude effectively detracts from the sensitivity of OAEs to cochlear toxicity and may well lead to delayed medical decisions, and poorer patient outcome.

A basic protocol for evaluation and monitoring for ototoxicity includes:

■ Baseline audiologic assessment before the first administration of the drug including pure-tone audiometry (conventional and high frequency) if feasible, tympanometry, and distortion product OAEs, or both TEOAEs and DPOAEs

Table 7–4. Selected Literature Describing Monitoring for Drug-Induced Cochlear Damage (Ototoxicity) with Otoacoustic Emissions

Medication	OAE Type	Findings	Study (Year)
Aminoglycoside Antibiotics			
Gentamicin	DPOAE TEOAE	Both types of OAEs showed significant decreases in amplitude from baseline through 14-day treatment of children with cystic fibrosis, whereas pure-tone thresholds remained normal.	Stavroulaki et al. (2002)
Tobramycin	DPOAE	Detection of ototoxicity in children with cystic fibrosis (CF)	Mulheran and Degg (1997)
	DPOAE	Otoacoustic emissions recorded from children with CF detected cochlear dysfunction, and were particularly useful in children age 4 years and younger.	Martins et al. (2010)
Amikacin	DPOAE	No change found in newborn infants with recommended dose up to 40 days after hospital discharge	Ruggieri-Marone and Schochat (2007)
	DPOAE	Otoacoustic emissions recorded from children with CF and treated with amikacin detected cochlear dysfunction	Martins et al. (2010)
Other Antibiotics			
Vancomycin	DPOAE	No change found in newborn infants with recommended doses	Ruggieri-Marone and Schochat (2007)
Chemotherapeutic Agents			
Cisplatin	TEOAE DPOAE	Early clinical study of cisplatin-induced ototoxicity OAE and pure-tone threshold changes in children treated with cisplatin, but not with carboplatin treatment with cisplatin	Zorowka et al. (1993) Dooge et al. (2006)
Carboplatin	DPOAE	No change in OAEs and pure-tone thresholds with carboplatin treatment	Dooge et al. (2006)
Loop Diuretics			
Quinine-Derived Drugs			
Quinine	TEOAE	Early study of quinine effects in normal volunteers	Karlsson et al. (1991)
Other Drugs and Substances			
Aspirin	DPOAE	Effects of aspirin show considerable individual variability	Brown et al. (199e)
PCB (polychlorinated biphenyls)	TEOAE DPOAE	Correlation in OAEs, especially TEOAEs, with PCB concentrations in 574 12-year-old children	Trnovec et al. (2010)

■ Monitoring every week or every two weeks during administration of the drug,

■ Monitoring at 1 week, 1 month, and 3 months after the drug is discontinued

Criteria for a change in auditory status include a decrease in 15 dB for the pure-tone air conduction threshold at one frequency or 10 dB for two frequencies (in comparison to baseline findings) and/or a decrease in DPOAE amplitude exceeding test-retest reliability (e.g., >2 dB). Behavioral results should be replicated within 24 hours to assure validity of a change in hearing status. The National Cancer Institute (NCI) Common Terminology Criteria for Adverse Events (CTCAE, 2009) Ototoxicity Grades can be used to define the degree of drug-induced hearing loss. Specifically:

Grade 1: An average pure-tone hearing threshold shift or hearing loss of 15 to 25 dB in comparison to baseline findings for two or more adjacent frequencies in one or both ears

Grade 2: An average pure-tone hearing threshold shift or hearing loss of >25 to 90 dB in comparison to baseline findings for two or more adjacent frequencies in one or both ears

Grade 3: The degree of hearing loss warrants intervention, including amplification and speech/language services

Grade 4: Meets criteria for a cochlear implant and also speech/language services

Another system for describing the degree of hearing loss in studies of ototoxicity is the Brock Scale (Brock et al., 1991). Brock grades of hearing loss are defined as follows:

Grade 0: Pure-tone hearing thresholds <40 dB at all audiometric frequencies

Grade 1: Pure-tone hearing thresholds ≥40 dB at only 8000 Hz

Grade 2: Pure-tone hearing thresholds ≥40 dB for frequencies within the range of 4000 to 8000 Hz

Grade 3: Pure-tone hearing thresholds ≥40 dB for frequencies within the range of 2000 to 8000 Hz

Grade 4: Pure-tone hearing thresholds ≥40 dB for frequencies within the range of 1000 to 8000 Hz

Audiologic protocols may be modified depending on the specific ototoxic drug, factors known to increase risk of ototoxicity, and the documentation of auditory dysfunction.

OTHER DISORDERS AND DISEASES

Introduction

Inclusion of OAEs within the test battery is now standard of care for pediatric diagnostic audiologic assessment (e.g., JCIH, 2007). Given the widespread clinical application of OAEs, it is not surprising that published papers describe OAE findings in a wide variety of diseases and disorders. In this final section of this chapter, we briefly review this literature for selected patient populations. Findings are summarized in Table 7–4 for these and other patient populations. In the Internet era, one can almost instantly locate published and, sometimes, unpublished reports of OAE studies for virtually any disease or disorder. A good starting point in a literature search is to utilize the Web site for the National Library of Medicine in Bethesda, Maryland, USA (http://www.nlm.nih.gov), selecting the Medline field for Health Researchers, and then entering key words (e.g., "otoacoustic emissions" and "the disease or disorder"). The search will produce the abstracts of articles including the key words somewhere within the title or text. Articles that seem to be relevant based on the abstract can then be requested directly via e-mail from the author designated for correspondence for detailed review.

Pseudohypacusis

Pseudohypacusis also is sometimes referred to as nonorganic or functional hearing loss. As an objective auditory measure, OAEs offer a logical option for defining auditory status in persons who will not, or cannot, volunteer valid behavioral responses to sound. It is good clinical practice to perform OAE measurement early in an audiologic assessment or at least before the patient is taken into a sound booth for pure-tone and speech audiometry. If OAEs are normal and, at the beginning of behavioral assessment, the patient shows even a mild deficit in pure-tone hearing thresholds, the troubling conflict in findings must be resolved. The discrepancy warrants serious efforts to rule out technical explanations (e.g., problems with earphone or earphone placement), nonpathologic subject factors (e.g., patient didn't understand the task), or pathologic subject factors (e.g., isolated inner hair cell dysfunction or neural dysfunction, including auditory neuropathy spectrum disorder).

The rationale for recording OAEs in the assessment of pseudohypacusis, and some limitations of OAEs, are summarized in Table 7–5. Although the advantages of OAE measurement in the assessment of pseudohypacusis are compelling, the constraints also must be pointed out. OAEs have no diagnostic value in patients with abnormalities of the middle ear, even mild dysfunction. Also, OAEs may be absent or markedly abnormal in persons with true cochlear (outer hair cell) dysfunction and superimposed functional hearing loss ("functional overlay"). OAEs are a valuable component of the diagnostic test battery, but they are by no means a test of hearing. OAE measurement must always be supplemented with other independent auditory measures.

To maximize clinical efficiency and minimize test time and general frustration, it makes sense to discover invalid behavioral findings early in the diagnostic process. Clinical experience confirms that transient or distortion product OAEs can contribute to quicker, easier, and more confident diagnosis of pseudohypacusis (e.g., Saravanappa, Mepham, & Bowdler, 2005). OAEs are particularly helpful in the evaluation of persons seeking monetary compensation or financial gain for hearing loss. Patients with a history of noise exposure or head injury may very well have some degree of true organic auditory dysfunction involving the cochlea. In such cases, the absence of OAEs secondary to cochlear dysfunction doesn't rule out the possibility of pseudohypacusis, or an exaggerated hearing loss. Further diagnostic assessment with other audiologic procedures (e.g., ABR or

Table 7–5. Summary of the Rationale for the Application of OAEs in the Evaluation of Pseudohypacusis

Strengths

- Do not require a behavioral response from the patient
- Results are not influenced by motivation
- Results are not influenced by cognitive status
- Results generally are not influenced by state of arousal
- Measurements can be made with patient sedated or anesthetized
- Results are not influenced by the patient's native language
- Patient is not required to follow detailed verbal instructions
- Results not influenced by motor status
- High degree of sensitivity to peripheral (outer hair cell) auditory dysfunction
- Site-specific information on auditory dysfunction
- Valid measures are possible from infants and young children
- Reasonable test time

Weaknesses

- Do not measure "hearing"
- Abnormal finding does not invariably indicate hearing loss
- Single measure generally provides limited information on hearing status
- No information on speech perception or understanding

ASSR) may be necessary for frequency-specific estimation of hearing thresholds.

Pediatric populations at risk for pseudohypacusis include children who have suffered emotional trauma and/or abuse. Among school-age children, there also is a tendency (for unknown reasons) to encounter malingering more often with adolescent girls. Children who are feigning a hearing loss, or who have a psychological basis to their hearing loss, may have OAEs with amplitudes well within normal limits across a frequency region of 500 to 8000 Hz. Whenever there is a discrepancy between behavioral threshold measures and OAEs, the possibility of neural auditory dysfunction must always be considered, and immediately ruled out with appropriate techniques (e.g., ABR). The combination of normal OAE *and* ABR findings, yet depressed pure-tone and/or speech thresholds, argues strongly for a diagnosis of pseudohypacusis. Once the diagnosis of pseudohypacusis is made with objective measures like OAEs, informing the patient about the evidence of normal cochlear function (in age-appropriate language of course) may lead to appropriate cooperation during a reassessment, and valid behavioral estimations of thresholds (Balatsouras et al., 2003).

Should OAEs be routinely recorded, from each new patient, as part of the diagnostic audiologic test battery? For the pediatric patient population, the answer to this fair question is unequivocally "yes." Standard of care for hearing assessment of children, as defined by the Joint Committee on Infant Hearing (2007), clearly requires the inclusion of OAEs in the test protocol. A clinical and financial argument also can be made for beginning adult hearing assessments with OAE, at least for new patients who have not undergone previous assessment. The peer-reviewed literature, summarized below, clearly confirms the sensitivity of OAEs to even subtle outer hair cell dysfunction.

Clinical research confirms that transient or distortion product OAEs can contribute to quicker, easier, and more confident diagnosis of pseudohypacusis in children. Saravanappa and colleagues in London (Saravanappa, Mepham, & Bowdler, 2005) reported encountering 10 to 20 children with the diagnosis of pseudohypacusis among the 3000 to 3500 children they evaluate each year. The authors describe transient evoked OAE (TEOAE) findings for 31 children between the ages of 5 and 16 years. OAEs contributed to the diagnosis in all cases, and independently confirmed pseudohypacusis in 12 children. Interestingly, Saravanappa et al. (2005) state that, "In most cases . . . explaining to children and parents regarding the results of otoacoustic emissions resulted in improvement in hearing and disappearance of the condition . . . otoacoustic emissions helped in treating pseudohypacusis" (p. 1236).

The problem of mismanagement with misdiagnosis of pseudohypacusis also was confirmed

recently in a study reported by Holenweg and Kompis (2010). Over a 6-year period, 40 patients (25 females and 15 females) seen in a Swiss otolaryngology clinic were ultimately diagnosed with pseudohypacusis, including 18 adults and 22 children. For all of the patients, objective test procedures (OAEs and/or ABR) yielded normal or near-normal findings. Interestingly, 1 out of 5 patients (2 children and 7 adults) were fit with hearing aids at a previous clinic before pseudohypacusis was diagnosed.

In a detailed description of auditory findings for 47 children with pseudohypacusis (they use the term nonorganic hearing loss), Morita et al. (2010) found evidence of DPOAEs in all children with normal middle ear function. Six children in the series with ventilation tubes had abnormal DPOAEs. The authors cite advantages of OAEs in the diagnosis of pseudohypacusis, namely they are "noninvasive, easy, and quick to perform." Echoing others who have systematically investigated pseudohypacusis, Morita and colleagues (2010) emphasize the dangers and hazards to the patients associated with late or misdiagnosis of pseudohypacusis including: increased cost of health care, patient (and health care provider) involvement in litigation, steroid treatments, inappropriate medical treatment (e.g., steroids), and inappropriate surgical procedures (e.g., exploratory tympanotomies).

Finally, a study by Ioannis and colleagues (2009) suggests that pseudohypacusis should be considered in children with sudden hearing loss. Forty-eight children seen at the clinic with sudden hearing loss were included in the study. Among this group, 27 had unilateral hearing loss and 29 were female whereas only 19 were boys. The average age of children with pseudohypacusis was 10.4 years. History revealed poor academic performance, family conflicts, or a new baby in the family for most of the children. Following a comprehensive medical and audiologic evaluation, 26 (54%) of the children were diagnosed with pseudohypacusis, whereas the remaining 46% had confirmed organic etiologies for their hearing loss. OAEs were part of the test battery for all subjects. Normal OAEs were recorded for 45 of the 52 ears of pseudohypacusic patients. Middle ear abnormality precluded valid OAE measurement for the other 7 ears (of 5 children). A quote from Ioannis and colleagues (2009) offers a vivid description of the diagnostic, and therapeutic, value of OAEs in this population: "Otoacoustic emissions were used in all children who participated in this study and in some cases their role as 'lie detector' produced a striking and immediate result" (p. 1860).

Sudden Infant Death Syndrome (SIDS)

Ruben et al. (2007) published a preliminary report suggesting a relationship between TEOAE newborn hearing screening findings and risk for SIDS. The authors identified, at a birthing and pediatric hospital with a long history of universal newborn hearing screening, 31 infants who had died of SIDS and then formed a control group of 31 infants matched on important medical characteristics without SIDS. Children who succumbed to SIDS had a statistically significantly reduced TEOAE amplitude to noise floor difference for the right ear in comparison to the control group. There was no difference between groups for TEOAE findings with stimulation of the left ear. As you might suspect, the findings of this study generated a great deal of interest in the general media and, of course, among parents of children who were already victims of SIDS. Concerned by reports of the study in the lay media, mothers who lost children to SIDS called audiologists supervising newborn hearing screening programs asking about OAE findings for other siblings who had undergone newborn hearing screening.

The report on OAEs and SIDS, published in a pediatric journal, raises a number of important issues. One issue is the major difference between a statistically significant finding based on the analysis of data for a relatively large group of subjects versus the predictive value of the finding clinically for a single patient. In other words, can a lower TEOAE SNR for the left versus right ear be used to accurately predict a child with SIDS? Are all of the children who fail hearing screening unilaterally, and for the left ear, likely to later be the victim of SIDS? Have the results of this study been replicated with a larger and entirely different sample of infants? Does the apparent relation between TEOAEs and SIDS also apply to infant

hearing screening with DPOAEs? This question is particularly relevant as the majority of hearing screening programs in many countries (e.g., the United States) rely on DPOAE, rather than TEOAE. Is risk for SIDS evident in the results for infant hearing screening with ABR?

Sickle Cell Disease

Walker, Stuart, and Green (2004) conducted a study of auditory function in 12 normal-hearing African American children aged 6 to 14 years diagnosed with sickle cell disease (SCD). Interestingly, in comparison to an age- and gender-matched control group, children with SCD had significantly larger DPOAE amplitudes. The authors could not offer an explanation for this finding. Children with SCD tend to have smaller ear canal volumes than non-SCD children, apparently due to slower than normal growth patterns. Walker et al. (2010) ruled out the influence of ear canal volume on the findings by demonstrating no significant difference in equivalent volume as calculated from aural immittance measurements. Middle ear pressure differences between the SCD and control group also were eliminated as a factor in the findings.

Williams Syndrome

Williams syndrome (WS) is a developmental disorder characterized by distinctive facial features and personalities, cardiovascular disorders, cognitive deficits, and in about 95% of patients, hyperacusis (marked intolerance to high-intensity sounds). The majority of children with WS have hearing loss, including high-frequency sensorineural and mixed hearing loss. Children with WS have a tendency to develop excessive cerumen and also middle ear disease. Marler et al. (2010) studied auditory function in 81 children and adults diagnosed with WS. In the study, the authors compared DPOAE findings in 14 subjects with WS who had normal hearing sensitivity and middle ear function (average age of 14 years) with the DPOAE findings for an age-matched control group of 14 subjects. In the authors words: "The results from this analysis provide further support for the hypothesis of cochlear outer hair cell dysfunction in individual with WS who have otherwise "normal" hearing" (Marler et al., 2010, p. 259). Marler and colleagues (2010) offer a detailed discussion of possible middle ear and cochlear mechanisms that might explain the findings.

Autism

Communicative disorders, especially language impairment, are a diagnostically essential component of what is now known as autism spectrum disorder. An audiologic assessment to rule out peripheral hearing loss typically is part of the diagnostic process for children with autism spectrum disorder. The hearing assessment usually is scheduled as language impairment is documented in early childhood. Later, in school-age years, more comprehensive assessment of auditory processing may be requested for high functioning children meeting diagnostic criteria for autism spectrum disorder. Diagnostic audiologic assessment is, of course, quite challenging for children with autism spectrum disorder due to their limited verbal language skills, impairments in social interaction, and the behavioral characteristics associated with the disorder. As an objective measure of auditory function, OAE measurement is a logical component of the test battery to rule out hearing loss in young children with autism spectrum disorder. Children with autism often are highly sensitive to tactile stimulation, and defensive or even combative when touched. Proper placement of earphones (including insert earphones) for ear-specific audiologic assessment, including measurement of OAEs, is sometimes problematic with autism spectral disorder.

Recently published clinical reports confirm that children with autism spectrum disorder have normal peripheral auditory function when evaluated with objective measures such as OAEs and the auditory brainstem response (e.g., Gravel et al., 2006; Grewe et al., 1994; Khalfa et al., 2001; Tas et al., 2007; Tharpe et al., 2006). The findings of Tas and colleagues (2007) are representative of the literature. The Turkish researchers recorded TEOAE measures, and the ABR, in an investigation of auditory function in 21 children (70% males) with the diagnosis of autism. Age ranged from 2 to 7 years, with a mean age of 3.8 years. The study also included an age-matched control group ($N = 12$). As recorded with the ILO 88 device, TEOAEs were present in 83% of the children with autism (52 ears) and all subjects in the control group. Presence of TEOAEs was defined by a reproducibility value of $\geq$50%. Tas et al. (2007) state, "In the present study, we have shown that TEOAE and ABR can be used to evaluate hearing in children with autism objectively. Objective assessments of auditory function may be more practical than behavioral testing of children with autism, especially for those who are uncooperative" (p. 77).

OAEs, in combination with an objective measure of auditory threshold (e.g., the ABR) are a feasible first step in the assessment of hearing function in children with autism spectrum disorder. Based on research reports and clinical experience, we would expect normal findings for most children with ASD. In contrast, behavioral responses to sound often are unreliable and/or outside of the normal region. This latter finding emphasizes the value of OAEs as a "cross-check" for pure-tone and speech audiometry in the assessment of autism spectrum disorder.

Auditory Processing Disorder (APD)

Auditory processing disorder (APD) was defined in 2000 (Jerger & Musiek, 2000, p. 467) as an "auditory specific" disorder at any level within the auditory pathways, and not limited to the central nervous system. There are a number of reasons to routinely include OAEs in the test battery for diagnosis of APD. Specifically, OAEs are:

- **Site Specific:** Information is provided specifically on outer hair cell function.
- **Sensitive to Cochlear Dysfunction:** OAEs are highly sensitive to cochlear auditory dysfunction. Abnormal OAEs can be recorded in persons with normal audiograms.
- **Frequency Specific:** OAEs provide information on auditory function at many frequencies, include those between conventional audiometric octave frequencies.

- **Quick and Easy to Record:** Test time even for a diagnostic protocol is usually less than 2 minutes per ear. The measurement technique is relatively easy to learn.
- **Objective:** As a nonbehavioral procedure, OAEs are independent of listener variables that may confound audiologic assessment of children for suspected APD, for example:

  - Age (chronological and developmental)
  - State of arousal
  - Fatigue
  - Motivation
  - Cognition
  - Language

In view of these clinical attributes and advantages, the Bruton Conference Panel recommended inclusion of OAEs in the test battery for APD assessment (Jerger & Musiek, 2000). This recommendation, along with many other guidelines included in the 2000 report was subsequently questioned by a collection of audiologists (Katz et al., 2002). In a retort, however, Jerger & Musiek (2000) with input from other participants at the Bruton Conference, pointed out:

> One important dimension in the differentiation diagnosis of APD is to differentiate APD from speech understanding problems due to malfunction at either the auditory periphery or the low brainstem level. We know that a pure-tone audiogram within "normal limits" does not guarantee normality at the auditory periphery . . . The best, and virtually only, technique we currently have available for excluding this possibility [peripheral disorder] is evoked OAEs. (p. 20)

Clinical research confirms the assumption that OAEs are useful in detecting peripheral auditory function not apparent from the pure-tone audiogram (Hall & Johnston, 2007). In a series of 65 consecutive children evaluated for APD, DPOAE measurement was conducted with a typical diagnostic test protocol (e.g., stimulus intensities $L_2 = 55$ dB, $L_1 = 65$ dB; $f_2/f_1 = 1.22$; 5 frequencies per octave for an f_2 over the range of 500 to 8000 Hz), with amplitude analyzed according to an appropriate normative data. Over three-fourths of the children had normal pure-tone thresholds (<20 dB HL) within the speech frequency region, but 35.5% had abnormal OAEs. A composite audiogram and DPgram for the group of 35 children undergoing APD assessment is displayed in Figure 7–5. Clearly, on the average, DPOAEs reflect cochlear auditory dysfunction that is not evident with pure-tone audiometry, a so-called "subclinical" finding.

What clinical value is derived from the finding of abnormal OAEs in a child undergoing an evaluation for APD? The inclusion of OAEs in the test battery assures that peripheral dysfunction will be detected even in children with no other abnormal audiometric findings. Based on further diagnostic audiologic assessment, more detailed history, or otologic consultation, it usually is possible to determine whether the abnormal OAEs reflect middle ear or inner ear (outer hair cell) dysfunction. In either event, the goal of confirming (or ruling out) peripheral auditory dysfunction in the APD evaluation is achieved. In addition, abnormal peripheral auditory dysfunction as confirmed by OAEs may be related to performance on measures of central auditory processing, such as temporal integration and gap detection and to patient complaints of difficulty hearing in background noise (e.g., Smurzynski & Probst, 1999; Zhao & Stephens, 2006). At the very least, abnormal results of central auditory processing tasks should be interpreted in the context of OAEs.

According to surveys of audiologists' practices (Emanuel, 2002), OAE measurement typically is not among the procedures incorporated into auditory processing test battery. We strongly recommend diagnostic measurement of distortion product OAEs with five or more frequencies per octave across a wide frequency region (e.g., 500 to 8000 Hz) as a mandatory procedure in the assessment of APD. There is some evidence that sophisticated procedures for measurement and analysis of OAEs also have value in the diagnosis of APD. For example, children with APD have abnormal findings on a contralateral suppression paradigm involving transiently evoked OAEs in comparison to a control group (Muchnik et al., 2004; Sanches & Carvallo, 2006). Interestingly, in a study of children with specific language impairment (SLI), Clark et al. (2006) found no abnormalities in contralateral suppression of TEOAEs.

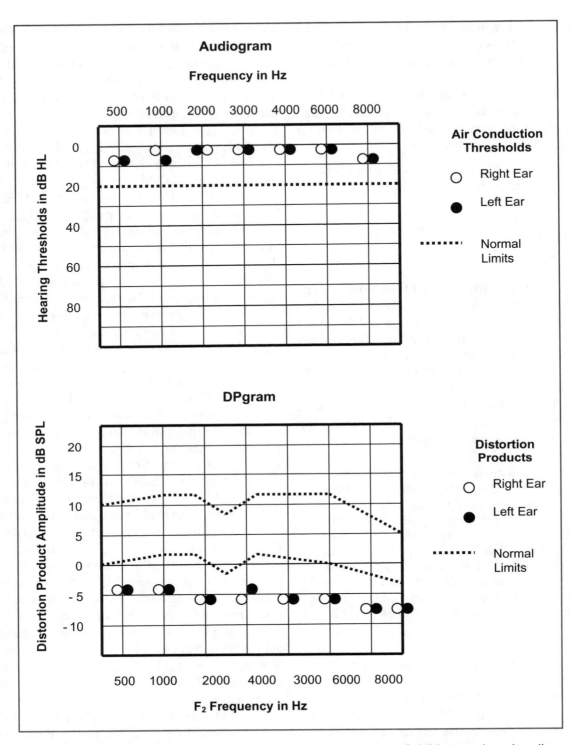

FIGURE 7–5. Composite audiogram and DPgram for a series of children undergoing diagnostic assessment for auditory processing.

The difference in findings for children with APD versus SLI is entirely consistent with the rationale for OAE application in APD outlined above. That is, OAEs permit confident assessment of auditory processing without the confounding influence of other listener variables, such as language. Also, based on these admittedly limited published data, it would appear that measurement of suppression of OAEs by sound (and the medial olivocochlear system) offers an "auditory specific" marker of auditory processing disorders. Further investigation of ipsilateral and contralateral suppression of transient and distortion product OAEs in children with APD, along with alternative OAE analysis strategies (e.g., phase), certainly seems to be warranted.

||| 8 |||

OAEs and Cochlear Pathophysiology: Adults

INTRODUCTION

Otoacoustic emissions are not just for children. The review of applications of otoacoustic emissions in children in the previous chapter left no doubt that the inclusion of OAEs in the pediatric audiologic test battery is now viewed as standard of care, and endorsed by multidisciplinary groups such as the Joint Committee on Infant Hearing (2007). In this chapter, we make the case that OAEs also can contribute importantly to the identification and diagnosis of auditory dysfunction in varied adult patient populations. Compelling evidence from a large and rapidly expanding peer-reviewed literature confirms the role of OAEs in detection of auditory dysfunction and differential diagnosis of a remarkably diverse collection of auditory disorders. There are published reports of OAE findings for almost every imaginable disease. Although occasionally a single paper describes documentation of cochlear dysfunction for a disease, more typically results are reported independently in a series of articles by different investigators for a common disease, such as diabetes or arthritis.

Without doubt, there are more published studies on OAEs in noise- or music-induced hearing loss than for any other etiology for auditory dysfunction. The literature on noise and OAE is itself very diverse, including formal and often large-scale investigations of noise-related auditory dysfunction in industrial workers (some also exposed to chemical substances), military personnel, musicians, telephone users, patients undergoing magnetic resonance imaging (MRI), and persons involved in recreational activities ranging from music to hunting.

Hundreds of papers also have been published on the role of OAEs in the diagnosis of other primarily adult otologic diseases and disorders, among them tinnitus, Ménière's disease, drug-induced cochlear toxicity, and retrocochlear pathologies. In short, there is ample evidence-based rationale and justification for the application of OAEs in selected adult patient populations. OAE findings permit early identification of auditory dysfunction, and therefore the potential for prevention of hearing loss. The unequaled sensitivity of OAEs to cochlear dysfunction is a consistent theme in this chapter. Abnormal OAE findings reveal outer hair cell dysfunction well before any changes are noticeable for pure-tone thresholds. Early detection, of course, usually leads to early intervention. Likewise, early intervention for auditory disorders almost always is associated with better patient outcome. The relatively brief time required to record OAEs clinically, at a modest charge, is often a critical first step in an efficient and accurate diagnostic process and, as a consequence, more timely, appropriate, and effective management. Finally, we inform the reader that clinical investigations in adult populations involving suppression of OAEs with ipsilateral and/or contralateral noise are summarized in Chapter 9.

At the outset of the review of OAE applications in adults, we again encourage readers to periodically generate an updated literature review by performing a internet based search (e.g., http://www.nlm.nih.gov) for newly published papers, using key words such as "otoacoustic emissions" and "noise" or the etiology, disease, or disorder of interest to the reader.

SOUND-INDUCED COCHLEAR DYSFUNCTION

The rationale for applying OAEs in the early detection of sound-induced cochlear dysfunction is best summed up by the old adage "An ounce of prevention is worth a pound of cure." Exposure to high-intensity (greater than 90 dB SPL) sounds can damage the cochlea and cause hearing loss. Of course, the higher the sound intensity, the shorter the "safe" exposure time, and at the extremely high intensity levels (>130 dB), any exposure can be dangerous. The reader may notice that the term "sound" is used here instead of the term "noise." The source or type of sound, and the venue for

sound exposure, are not really important factors in determining whether the cochlea is damaged. As clearly illustrated by the following review, cochlear dysfunction can arise from exposure to a diverse collection of high-intensity sounds, including different types of noise, speech, or music. The noise may be generated from weapons and equipment in military activities (e.g., rifles, tanks, jet aircraft), various types of engines in the private sector (e.g., aircraft, race car, motorcycle), industrial equipment, and even diagnostic devices commonly employed in health care facilities (e.g., MRI scanners). In recent years, literature has appeared on the potential dangers of excessive exposure to cell phone sounds. Finally, even exposure to esthetically appealing sound during recreation, entertainment, or employment (e.g., a symphony orchestra, band of amplified instruments, or a personal listening device like an MP3 player) can certainly pose risk to cochlear integrity and hearing.

Therefore, it is not the type of sound but its intensity and duration that are critical considerations in risk for auditory dysfunction. The common denominator for any sound exposure is its potentially deleterious effect on cochlear function and structure. The reader is referred to a number of sources, including research and review articles in scientific journals, monographs, and textbooks for detailed information on the many ways excessive sound exposure can alter cochlear physiology and anatomy. New theories and facts on biochemical changes in the cochlea, and mechanisms for protecting the ear and, specifically, hair cells from the toxic effects of excessive sound exposure, are published regularly. A literature search will quickly reveal many thousands of peer-reviewed publications, and a wealth of current information on the topic. What follows here is a brief review of a small proportion of papers to highlight selected findings for the major groups of persons at risk for noise-induced auditory dysfunction. Additional papers, but certainly not an exhaustive list, are cited in Table 8–1.

Table 8–1. Selected Literature Describing Findings for Studies of Otoacoustic Emissions in Sound-Induced (Noise or Music) Hearing Loss in Adults

Population	OAE Type	Findings	Study (Year)
Noise			
Industrial	TEOAE	Early study confirming sensitivity of TOAEs in detection of industrial (metal factory) noise-induced cochlear dysfunction	Sliwinska-Kowalska et al. (1999)
	TEOAE	Early systemic study showing that OAEs are more sensitive than pure-tone audiometry in detecting cochlear (outer hair cell) damage	Desai et al. (1999)
	TEOAE DPOAE	OAE amplitudes decreased in workers exposed to occupational noise, with reduction also in efferent suppression of TEOAEs	Sliwinski-Kowalska and Kotylo (2001, 2002)
	TEOAE	Noise exposure associated with reduced TEOAE amplitudes and contralateral acoustic suppression	Xue and Zhong (2002)
	TEOAE	Abnormalities in TEOAEs recorded with linear and nonlinear mode, including the contralateral ears of persons with unilateral noise-induced hearing loss	Moleti et al. (2002)
	TEOAE DPOAE	Changes in both types of OAEs with exposure to work-related noise for a large sample of subjects ($N = 5072$). Music exposure had no effect on OAEs	Engdahl and Tambs (2002)
	DPOAE	Amplitude decreased with noise exposure for oil well drilling workers in China. OAEs more sensitive to noise effects than pure-tone audiometry	Zhang et al. (2000)
	DPOAE	Early review paper describing effect of noise exposure on early detection of noise-induced auditory dysfunction	Avan et al. (2001)
	DPOAE	Decreased amplitudes in the 4000 to 6000 Hz region provided early detection of noise-induced auditory dysfunction	Han et al. (2003)
	DPOAE	Length of employment in construction trades was related to the magnitude of decrease in amplitude and slight changes in pure-tone hearing thresholds.	Seixas et al. (2004)
	DPOAE	As an objective test, OAEs can be used in the assessment of hearing status in persons applying for occupational hearing loss compensation	Chan et al. (2004)
	TEOAE	Time of exposure to noise in workers was related to significant decrements in TEOAE signal-to-noise ratio	Wang et al. (2004)
	TEOAE	Latency of OAEs changed in adult workers following exposure to jet engine noise	Jedrzejczak et al. (2005)
	DPOAE TEOAE	OAEs confirmed cochlear dysfunction in noise exposure and generally agreed with pure tone thresholds. DP-grams showed better correlation with audiograms than TEOAEs	Avian and Bonfils (2005)

continues

Table 8–1. *continued*

Population	OAE Type	Findings	Study (Year)
	DPOAE	DPOAEs documented small but significant changes with noise exposure associated with construction industry apprentices during their first 3 years of work, even for noise evaluating the effectiveness of hearing protection devices.	Seixas et al. (2005)
	DPOAE	Occupational noise was associated with DPOAE abnormalities even when pure-tone hearing thresholds remained within normal limits	Marques and da Costa (2006)
	TEOAE	Reduction in TEOAE amplitude in the 1000 to 4000 Hz region following industrial noise exposure. Pure-tone audiometry showed changes in 3000 to 8000 Hz region.	Barros et al. (2007)
Military	DPOAE	Early study documenting OAE abnormalities secondary to "micro- mechanical-traumatic and biochemical-metabolic" damage to outer hair cells in persons exposed to antitank rocket launchers, submachine gunfire and hand-grenade explosions.	Oeken (1998)
	DPOAE TEOAE	Amplitudes were reduced with military noise exposure were more sensitive than pure-tone audiometry in detecting noise damage. The objectivity of OAEs was an advantage in the assessment of medicolegal cases	Attias et al. (2001)
	TEOAE DPOAE	Exposure to automatic gunfire during shooting produced a reduction in amplitudes for both types of OAEs, mostly for the 1000- to 3000-Hz region. OAEs were more sensitive to changes than pure-tone audiometry.	Konokopa et al. (2001)
	TEOAE	OAE provide a useful approach for monitoring for noise-induced hearing loss in Italian air force personnel. TEOAE more accurate than pure-tone audiometry in detection of cochlear dysfunction.	Lucertini et al. (2001)
	TEOAE	OAE abnormalities are found in the apparently normal ear contralateral to the ear with noise-induced hearing loss in Italian air force personnel, indicating "subclinical" cochlear damage	Lucertini et al. (2002)
	TEOAE DPOAE	Early review article documenting the use of both types of OAEs (especially DPOAEs) in the early detection of noise-induced hearing loss	Marshall et al. (2001)
	TEOAE DPOAE	Changes were recorded with both types of OAEs in persons exposed to high intensity levels of noise with a high correlation in findings for each type. Authors conclude that "it is premature to use OAEs in hearing conservation programs."	Laspsley et al. (2004)
	TEOAE	OAEs provided early indication of cochlear dysfunction in engine drivers and were predictive of subsequent hearing loss	Mukhamedova (2003)
	DPOAE	Lower amplitudes and longer latencies were found in miners with an 8- to 15-year history of noise exposure versus control subjects	Namyslowski et al. (2004)
	DPOAE	Industrial workers showed significant reduction in amplitudes versus control subjects, even beyond changes in the pure-tone audiogram	Balatsouras (2004)

Table 8–1. *continued*

Population	OAE Type	Findings	Study (Year)
	DPOAE	Decrease in amplitudes were correlated with flight time in hours for a group of 105 pilots versus a control group of nonpilots. There was considerable individual variability in susceptibility to noise-induced changes in DPOAEs	Zhang et al. (2004)
	TEOAE	Short-term exposure to rifle sounds in male hunters and candidate policemen produced significant decreases in OAE amplitude, but abnormalities were prevented by ear protection (earmuffs)	Pawlaczyk-Luszczynska et al. (2004)
	DPOAE	Even a single exposure to military jet aircraft engine noise resulted in decreased DPOAE amplitudes, and greater changes resulted in decreased DPOAE amplitudes, and greater changes than those found for pure-tone thresholds	Konopka et al. (2004)
	TEOAE	Military recruits exposed to rifle sounds in training had evidence within minutes to 3 hours after exposure of temporary hearing threshold shift correlated with reductions in TEOAEs	Olzewski et al. (2005)
	TEOAE	Reduction in amplitudes were documented in the 2000- to 4000-Hz region following exposure to impulse noise during military service	Konopka et al. (2005)
	DPOAE	Police officer exposure to impulse noise from shooting was associated with a reduction in OAE amplitude. OAEs could be used to document recovery	Balatsouras et al. (2005)
	DPOAE TEOAE	Military service with exposure to impulse noise from weapons and explosions was associated with decreased amplitude for both types of OAEs within the 1000- to 4000-Hz region.	Konopkas et al. (2006)
	DPOAE	Workers in fiberglass and metal-product manufacturing plants exposed to noise and styrene showed more abnormalities in exposed to noise and styrene showed more abnormalities in DPOAEs and hearing sensitivity than those exposed only to noise	Johnson et al. (2006)
	TEOAE	TEOAE amplitudes were changed following small-arms fire by 15 soldiers. Analysis of the relation between TEOAEs and multiple exposures to noise offers possible evidence of the presence of a protective mechanism for acoustic trauma-related hearing loss.	Dudevany and Furst (2006)
	TEOAE DPOAE	Reduced OAE amplitudes as a function of years of experience in working in a ship engine room for new recruits. Abnormal TEOAE findings predicted noise-induced hearing loss.	Shupak et al. (2007)
	TEOAE	Soldiers exposed to rifle sounds who used hearing protection (earmuffs) showed no change in OAEs confirming adequate attenuation of sound.	Olzewski et al. (2007)
	TEOAE	Amplitudes decreased with exposure to small arms fire of Israel soldiers. TEOAEs contributed to prediction of vulnerability to cochlear dysfunction with noise exposure.	Dudevany and Furst (2007)

continues

Table 8–1. *continued*

Population	OAE Type	Findings	Study (Year)
Recreational	DPOAE	Early study documenting OAE abnormalities secondary to "micro mechanical-traumatic and biochemical-metabolic" damage to outer hair cells in persons exposed to fireworks, rock concerts, and minor explosions (exploding car battery and tire).	Oeken (1998)
	TEOAE	Reduction in OAE amplitude in young adults (medical students) was associated with leisure activities (e.g., visiting discotheque), even with no change in pure-tone thresholds.	Rosanowski et al. (2006)
	DPOAE TEOAE	Significant reduction in TEOAE amplitude, but not DPOAE amplitude, following use of an MP3 player at 97 and 102 dB SPL by young normal-hearing adults.	Keppler et al. (2010)
Telephone	TEOAE	No change in OAE amplitudes after exposure to 10 minutes of mobile telephone use	Ozurtan et al. (2002)
	TEOAE	No change following exposure to high-frequency pulsed electromagnetic fields (EMFs) produced by mobile placed at the mastoid region.	Monnery et al. (2004)
	DPOAE	No significant change in amplitudes with exposure to electromagnetic fields of global system for mobile (cell phone) communication	Janssen et al. (2005)
	DPOAE	No change in DPOAEs, even with inverse fast Fourier transform and time-windowing domain analysis, with 10-minute exposure to electromagnetic fields at maximum power.	Parazzini et al. (2005, 2007)
	TEOAE	No changes in OAEs were noted following exposure to electromagnetic field of 900 to 1800 MHz produced by cellular telephones	Mora et al. (2006)
	TEOAE	No change in temporal or spectral fine structure with exposure to cell phone EMFs for 10 minutes	Paglialonga et al. (2007)
Medical Devices	TEOAE	Decrease in OAE amplitudes in persons exposed to sound levels of 122 to 131 dB during magnetic resonance imaging (MRI).	Randonskij et al. (2002)
	DPOAE	With sound levels of 79.5 to 86.5 dBA (peaks to 120 dB). In MRI scanner there were no changes in DPOAEs when sound-damping head supports and ear protection were used.	Wagner et al. (2003)
	DPOAE	Amplitudes were reduced in the ear contralateral to mastoid drilling immediately after otologic surgery with improvement.	Karatas et al. (2007)
Snoring	TEOAEs	Abnormal OAEs and hearing thresholds found in one ear of "bed partners" of chronic snorers consistent with noise-induced auditory dysfunction	Sardesai et al. (2003)
Noise Protection	DPOAE	Magnesium offered protection for temporary shift of DPOAE I/O and pure-tone threshold shifts in young normal-hearing adults exposed for 10 minutes to 90 dB SL of white noise	Attias et al. (2004)

Rationale for OAE Noise Measurement

The rationale for OAE measurement in persons at risk for sound-induced cochlear dysfunction is multifaceted, compelling, and supported by substantial evidence published in peer-reviewed literature. With OAE measurement, it is possible to: (1) identify sound-induced cochlear dysfunction before it is apparent with pure-tone audiometry; (2) identify persons who are likely to be at risk for further cochlear dysfunction despite normal hearing sensitivity; (3) closely, frequently, and objectively monitor cochlear function in persons at risk for sound-induced abnormalities; (4) monitor the effectiveness of hearing conservation strategies or efforts to protect the ear from hair cell dysfunction and death with micronutrients; (5) counsel individual patients and develop group hearing conservation programs to maximize prevention of permanent sound-induced hearing loss; (6), evaluate the affected interoctave frequency regions within the cochlea with precision; and (7) identify cochlear dysfunction in apparently normal hearing ears contralateral to ears with sound-induced hearing loss. The literature contains reports of evidence in support of each of these clinical applications for OAEs, and probably others as well. A small sample of the hundreds of articles on OAEs and noise are summarized in Table 8–1. The interested reader will easily find many more papers with a formal literature search.

Industrial

Overwhelming evidence in the peer-reviewed literature confirms the sensitivity of OAEs to industrial noise-induced auditory dysfunction. Some articles on OAEs and industrial noise are noted in Table 8–1. Sisto and Italian colleagues (2007) report a representative and comprehensive study of the application of TEOAEs and DPOAEs in 217 young (18 to 35 years) workers exposed to industrial noise. The study "confirms the sensitivity and specificity of OAEs for the detection of hearing loss, with a particular emphasis on mild hearing loss" (p. 399). Care was taken to ensure that hearing loss was due to noise exposure. The authors (Sisto et al., 2007) add that, "good test performance was found using the DPOAE data only,

and inclusion of the information from TEOAEs added no predictive power to the test" (p. 400).

OAEs have value in predicting the likelihood of permanent threshold shift in the noisy industrial setting, as they do in military personnel (see below). Shupak and colleagues (2007) investigated auditory function longitudinally in a group of 135 ship engine room recruits over a period of 2 years, comparing results to a control group of 100 subjects with no noise exposure. Reductions in TEOAE and DPOAE amplitudes were associated with occupational noise exposure. Abnormal TEOAE findings at 1 year of employment "had high sensitivity (86 to 88%) and low specificity (33 to 35%) for prediction of noise-induced hearing loss after 2 years" (Shupak et al., 2007, p. 745). The predictive value of the TEOAEs was minimized, however, by a high false-positive rate. DPOAEs were not well correlated with the pure-tone audiogram, prompting the authors to conclude that the DP-gram should not be applied as an objective index of pure-tone thresholds in noise-induced hearing loss.

A study by Avan and Bonfils (2005) raises the possibility of direct inner hair cell and/or auditory neuron dysfunction, with sparing of the outer hair cells, in a subgroup of 27 persons with a history of industrial noise exposure. According to the authors, "all patients suffered from sudden and permanent hearing losses due to one acoustic trauma in one or both ears, and none had any other likely cause of hearing loss" (p. 70). Animal investigations also have confirmed the possibility of preferential inner hair cell damage following noise exposure. The relevant implication for this discussion, of course, is the insensitivity of OAEs to non-outer-hair cell noise-induced auditory system damage. The same authors also found that DP-grams with 10 frequencies per octave "performed better than transient-evoked OAE spectra" in the detecting cochlear hearing loss as reflected by a highly frequency-specific Békésy technique for measuring hearing thresholds. The study highlights the importance of coupling OAE (preferably DPOAE) measurement with estimation of pure-tone hearing thresholds for the initial documentation of noise-induced auditory dysfunction to avoid "false negative" errors in the detection of hearing loss. However, OAEs can be used to monitor cochlear status in persons who have cochlear dysfunction that is linked to abnormalities in OAEs.

Another practical application of OAEs, with potential value in any hearing conservation program, is objective verification of the effectiveness of hearing protection. Briefly, baseline OAEs document cochlear status before noise exposure. Hearing protection strategies are implemented to prevent noise-induced cochlear dysfunction and hearing loss. The effectiveness of the hearing conservation program, including compliance with hearing protection requirements, is verified with follow up OAE measurements. Bockstael and colleagues (2008) provided strong evidence in support of this application of OAEs. Both TEOAEs and DPOAEs were recorded from 31 adult subjects before gunfire practice. All subjects then used hearing protection devices during a 5-day military practice period. Follow-up OAE measurement after noise exposure showed no statistically significant difference in amplitude values.

Finally, a recent paper by Kujawa and Liberman (2009) showed in mice that DPOAE thresholds could return to normal values after recovery from a temporary threshold shift, but neural degeneration could be severe and continue for prolonged periods of time. In this work, the authors recorded DPOAE and ABR thresholds that returned to normal in a week or two after the noise exposure event. However, suprathreshold metrics of ABR did not return to normal values and histology confirmed a loss of afferent nerve terminals. To make matters worse, the auditory nerve continued to degenerate even when no subsequent changes were noted in the OAE measures or ABR thresholds. The reader is reminded that OAEs evaluate the auditory system up to the outer hair cells and are blind to abnormalities beyond (Kujawa & Liberman, 2009).

Chemical Substances and Noise. Although reports are somewhat conflicting (see Table 8–1), it appears that a combination of exposure to high levels of noise and (via inhalation) styrene may increase the likelihood of cochlear dysfunction. An international collection of investigators (Johnson et al., 2008) recently found that 313 "Workers exposed to noise and styrene had significantly

poorer pure-tone thresholds in the high-frequency range (3 to 8 kHz) than the controls, noise-exposed workers, and those listed in a Swedish age-specific database" (p. 45). Styrene is toxic to the auditory system, even more so than other inhalants (e.g., toluene). Industrial exposure to styrene is common in manufacturer of fiberglass products (e.g., boats). DPOAEs can be helpful in documenting cochlear, specifically outer hair cell, damage and differentiating it from retrocochlear or central nervous system involvement (e.g., speech perception and cortical auditory evoked response abnormalities) related to exposure to styrene. In contrast, Prasher et al. (2005) reported less change in DPOAEs and TEOAEs for a group of mill workers exposed to solvents and intermittent noise than for a group exposed to noise alone.

Military

As we noted at the outset of the discussion, noise-induced auditory dysfunction occurs without regard to the type or source of the noise. The two most important factors or properties of noise in predicting auditory dysfunction are the intensity level and the duration. Without doubt, many persons in active duty military service are exposed repeatedly to very high levels noise and are at great risk for acoustic trauma resulting from brief impact-type noise as well as chronic noise exposure. Any list of military noise sources is long and diverse, including sounds produced by various types of rifles, larger cannons and guns, hand grenades, rocket propelled grenades, jet and conventional aircraft engines, heavy equipment (trucks, tanks), and now, in modern conflicts, high intensity blasts produced by improvised explosive devices (IEDs). Attias and colleagues (2001) were among the first to document in a large study the sensitivity of TEOAEs and DPOAEs to noise-induced cochlear dysfunction. Subjects were 283 subjects with noise exposure, 176 subjects with noise exposure yet normal audiogram findings, and 310 young military recruits with normal audiograms and no history of noise exposure. Patterns of abnormalities as a function of test frequency for OAEs were similar to the configurations of noise-induced hearing loss. Importantly, evidence of "noise-induced emissions loss" was found in persons with normal audiograms. As a screening measure for detection of noise-induced hearing loss, OAEs had high sensitivity (79 to 95%) and high specificity (84 to 87%). The authors cautiously concluded that, "OAEs provide objectivity and greater accuracy, complementing the behavioral audiogram in the diagnosis and monitoring of cochlear status following noise exposure" (Attias et al., 2001, p. 19).

The literature contains dozens of published papers confirming the value of OAEs in early detection and ongoing documentation of cochlear auditory dysfunction secondary to military noise exposure (see Table 8–1). The topic of military noise, noise exposure, and hearing protection certainly warrants a longer discussion than space permits herein. We advise the interested reader to conduct a literature search on the topic and to review a few recent articles. Be prepared to encounter a remarkably large number of peer-reviewed articles reporting OAE findings for studies of persons exposed to a wide assortment of sounds.

From the earliest years of OAE research, a group of researchers at the United States Naval Submarine Medical Research Laboratory in Connecticut have published studies on noise exposure in military personnel (e.g., Marshall & Heller, 1998; Marshall et al, 2001; Miller et al, 2006). An aircraft carrier is one of the consistently noisiest environments in military service. Noise on the flight deck during aircraft launches ranges from 126 to 148 dBA peak levels. Even in the hanger bay below the flight deck, sound levels exceed 120 dBA. There is really no quiet (<85 dBA) location onboard the aircraft carrier. As noted by Miller and colleagues (2006), "Double hearing protection ideally can provide attenuation up to 30 dB, but cannot provide sufficient attenuation to remove the risk of NIHL in the extreme noise levels present on an aircraft carrier. Furthermore, unlike many noise-hazardous industrial environments, there may be no truly quiet time for ears to recover from these shipboard exposures" (p. 281). The authors collected pure-tone threshold and TEOAE and DPOAE data from 338 sailors before and after a 9-month period on a Nimitz-class aircraft carrier including 6 months at sea. The subjects were exposed to noise at different locations on or below the flight deck. In agreement with many other studies of noise exposure in various settings, OAEs consistently showed more evidence of noise-induced auditory dysfunction than pure-tone audiometry. One of the clear findings of the study was the high number of subjects with permanent threshold shift (PTS) with either very low level or absent OAEs. Notably, OAE findings contributed to the prediction of PTS. In the words of Miller et al. (2006), "Diminished OAEs are predictive of subsequent hearing loss if the sailor remains in the noise-hazardous environment" p. 293).

Cell Phones

A remarkable number of investigators have applied OAEs to assess the possible damaging effects of cell phone use on cochlear function. As seen from the summary in Table 8–1, most studies have yielded negative findings for relatively brief periods of cell phone use (e.g., 10 minutes) even when OAEs are analyzed very carefully to detect even subtle changes. To date, there are apparently no published longitudinal investigations of the effect of cell phone use over long but not atypical periods of time (e.g., years of use).

Recreational

Exposure to potentially damaging levels of noise may be associated with a wide variety of recreational and leisure activities, ranging from attendance at sporting events, to hunting and target shooting, to riding vehicles (e.g., motorcycles), to listening to music while exercising. Of course, even pleasurable and fun activities may pose a serious risk to cochlear integrity. Furthermore, neither the specific recreational activity nor the specific role a person has in the recreational activity accurately predict the risk for auditory dysfunction. Consider, for example, a professional automobile race, such as a NASCAR or a Formula One event. Drivers, pit crews, race officials, and spectators are exposed to high levels of noise. Motorcycle riding is another common example of risk for recreational noise exposure. It typically is assumed that risk for rider noise exposure during motorcycle riding is greater for motorcycles that produce higher noise levels. In fact, the two most important factors contributing to noise levels during motorcycle riding are speed and the intensity level of wind noise reaching the ears (Pierce & Hall, 2007). Motorcycle noise levels may put bystanders at risk for cochlear dysfunction, but they are not the only risk factor for a single rider. Regardless of the type of recreational activity or a person's level of involvement (e.g., participant or spectator), risk of cochlear dysfunction and possible hearing loss is determined by conventional factors (i.e., sound intensity levels and duration of exposure).

Torre and Howell (2008) published one of the modest number of papers on the topic of noise exposure during aerobic exercise classes. DPOAEs were recorded before and after 50-minute exercise classes during which young adult (19 to 41 years) subjects (48 women and 2 men) listened to music. The mean noise level of the music was 87.1 dBA, but minute-by-minute peak noise levels of 90.5 and 99.7 dBA were documented with dosimeters. Following aerobic exercise, DPOAE amplitudes were lower but a significant decrease was noted only for a single frequency (6000 Hz). Although not conclusive, the findings of this study suggest the possibility of noise-induced auditory dysfunction associated with exposure to high intensity levels during exercise classes.

Music

The potentially damaging effects of music on auditory function have been appreciated for many years. The reader is referred to several comprehensive reviews of the risk to hearing of music (Lindhardt, 2008; Petrescu, 2008; Størmer & Stenklev, 2007) and, specifically, music and OAEs (Hall, 2000). Risk is not related to genre of music but, rather, to sound levels and duration of exposure. Risk of music-induced auditory dysfunction is not limited to performers. In addition to musicians, depending on the venue, stage workers, security personnel, ushers, bartenders, and servers are at risk due to repeated exposures to high intensity sound. Even disc jockeys are at risk for music-induced cochlear damage. Of course, patrons of musical performances also are at risk, albeit to a lesser extent. Symphony musicians are certainly not exempt from possible noise-induced hearing loss. In fact, for several reasons symphony musicians are conceivably at greater risk than other musical genres. Symphony musicians often make a living from their musical skills, that is, music is a night and day job. Consequently, symphony musicians often are exposed to musical sound when performing, rehearsing with the orchestra, practicing independently, and serving as music teachers to students of various ages. The symphony musician typically is situated in a consistent location within the orchestra, with no option for increasing the distance from other performers who generate high sound levels with their instruments (e.g., the brass or percussion sections). And, it would be considered improper, in bad taste, and perhaps against policy or regulations for a professional symphony musician to wear earmuffs or, perhaps, any visible form of hearing protection. Ironically, this profession depends entirely on music for livelihood yet the music may also damage their most important instrument—their hearing—and thereby end their career. In contrast, performers in other music genres (e.g., rock musicians) may be able exercise a variety of these options to reduce risk for sound-induced hearing loss.

The following review of a small sample of recently published papers describing OAE findings following exposure to music will highlight diverse forms of risk. Santos and Brazilian colleagues (2007) documented auditory function with pure-tone audiometry, TEOAEs, and DPOAEs in 30 disc jockeys (DJs) before and after sound exposure during their typical work in nightclubs. Sound levels in the nightclubs ranged from 93.2 to 109.7 dBA. There was a significant decrease in amplitude for TEOAEs and DPOAEs following music exposure, as well as evidence of temporary pure-tone threshold shift.

Classical musicians clearly are at risk for music-induced cochlear dysfunction, and permanent hearing loss (for a recent review, see O'Brien, Wilson & Bradley, 2008). Documentation of decreases in OAE amplitudes, and temporary or permanent changes in pure-tone thresholds, has been reported following rehearsals and performances for theater orchestra musicians (Nataletti et al., 2007), symphony musicians (Jansen et al., 2009), and professional singers (Hamdan et al., 2008). One common theme in the literature describing auditory status in classical music performers is the substantially greater sensitivity of OAEs versus pure-tone thresholds to cochlear dysfunction. For example, Hamdan et al. (2008) concluded that, "Subtle cochlear dysfunction can be detected with TEOAE measurement in a subset of normal-hearing professional singers" (p. 360). Cochlear dysfunction and hearing loss are, of course, a well-known occupational hazard for rock and roll musicians. The remarkable sensitivity of both TEOAEs and DPOAEs to cochlear dysfunction in musicians representing a wide spectrum of genres (e.g., rock and roll, country, alternative rock, punk rock, etc.) is thoroughly documented (see Hall, 2000; Maia & Russo, 2008).

The introduction of MP3 and personal audio player technology has led to tremendous interest in the lay press regarding the potentially damaging effects of exposure to music. At first glance, the latest generation of personal music devices would appear to increase the risk for sound-induced auditory dysfunction. Physical features of MP3 type devices that contribute to their widespread popularity also may be factors in their heightened potential for damaging the ear. That is, modern devices are small and lightweight permitting their use almost everywhere and anytime (e.g., while studying, exercising, driving vehicles, and even sleeping). With the option for downloading music the user conceivably can listen to new songs constantly (24 hours 7 days per week) for years on end, perhaps over the course of a lifetime. Finally, comfortable miniature high fidelity in-the-ear transducers (earphones) encourage the user to listen to music for long periods of time at, potentially, high effective (at the tympanic membrane) intensity levels without disturbing others. There is absolutely no question that early MP3 devices were capable of producing sound levels in the ear canal in excess of 125 dB SPL. In response to public and governmental concerns about the dangers to hearing, some manufacturers of MP3 devices have voluntarily or by law begun to limit output levels. Nonetheless, excessive exposure to music and sound-induced auditory dysfunction remains a distinct possibility for any MP3 user and, depending on the duration and level of sound exposure, a certainty.

Some of the rapidly growing literature on MP3 devices, and other personal audio players, and hearing is cited in Table 8–1. A recent article by Montoya and colleagues in Spain (Montoya et al., 2008) illustrates findings from retrospective studies of persons with a history of MP3 device sound exposure. TEOAE and DPOAE findings for 40 ears of a group of young (19 to 29 years) adults with a history of using MP3 devices were compared to those of a large age-matched control group (N = 232 ears) with no exposure to MP3 devices. Gender was balanced for each group. OAE amplitudes were decreased and thresholds increased in the group of MP3 users. The extent of OAE abnormalities was related to the years and hours per week of MP3 use, even for subjects with normal hearing sensitivity by pure-tone audiometry. Montoya et al. (2008) concluded that, "Cochlear impairment caused by MP3 player noise exposure may be detectable by analyzing TEOAEs and DPOAEs before the impairment becomes clinically apparent" (p. 718). Of course, studies of recreational sound exposure and auditory function must include a carefully selected group of control subjects. LePrell et al. (2010) report deficits in hearing sensitivity and DPOAEs in college-age students who self-report normal hearing. Also, a widely cited large-scale cross-sectional analysis of audiometric data from the period of 1988–1994 versus 2005–2006 yielded evidence of an increase in the prevalence of hearing loss in adolescents aged 12 to 19 years in the United States (Shargorodsky et al., 2010). In other words, cochlear abnormalities in adolescent and young adult populations appear to be more common than previously suspected, even for persons without a history of exposure to music from a personal audio player.

Protective Agents and Mechanisms

One of the most exciting auditory research discoveries in the past decade was the finding in animal studies that certain substances could protect the inner ear from noise-induced damage. Clinical trials have left no doubt of the possibility for prevention or reduction of noise-induced hearing loss by timely oral administration of micronutrients, including selected vitamins and magnesium. Current research is focused on discovering the combination and amount of micronutrients that will produce the optimal outcome (e.g., LePrell et al.,

2007). In recent and ongoing clinical investigations, OAEs play an important, really primary, role as an early and sensitive index of cochlear status. The significance of this research effort is hard to exaggerate. Exposure to high intensity levels of noise is the most common cause of acquired sensory hearing loss. Prevention of noise-induced hearing loss would be a major step toward eliminating hearing loss in general, at least in adults, and would benefit millions of persons around the world.

As an example of research on protective mechanisms for noise-related auditory dysfunction, Attias and Israeli colleagues (2004) describe the application of DPOAEs in monitoring cochlear function in a study of the effect of magnesium intake on noise-induced temporary threshold shift. Subjects were 20 normal hearing (pure-tone thresholds <20 dB HL) men between the ages of 16 and 37 years enrolled in a double-blind clinical trial. Neither the subjects nor the investigators were aware of the details of the substance administered during the study. Baseline audiometric and DPOAE I/O measurements were completed initially before any noise exposure or ingestion of any therapeutic agent. Subjects then evaluated audiologically again after they took either a placebo or 122 mg Mg (magnesium) for 10 days. Subjects were exposed to a 90 dB SL (sensation level) white noise for 10 minutes. The authors reported a correlation between blood magnesium levels and both pure-tone and DPOAE threshold shift reduction.

TINNITUS

Tinnitus, a phantom auditory perception, is a major health problem for a surprisingly large number of people. Tinnitus generally *is* considered debilitating when it seriously affects quality of life, such as disruption of sleep. Tinnitus is a symptom, not a disease or pathology. One of the first objectives in the formal assessment of a patient with tinnitus is to discover whether it is a symptom of a disease or pathology that requires medical management. For the majority of persons with bothersome tinnitus, well over 80%, there is no active disease process. Rather, the tinnitus is related to cochlear dysfunction associated with

either aging and/or noise exposure. The accepted mechanism of tinnitus involves cochlear abnormalities and, specifically, hair cell dysfunction (for review see Baguley, 2002; Eggermont, 2007). Damage to outer and/or inner hair cells associated with a variety of etiologies can lead to disruption in normal resting activity of afferent auditory (8th cranial) nerve fibers, deviations from the normal level of activity for neurons in regions of the auditory brainstem, and reorganization of the function of higher centers within the auditory system. The consistent reports of abnormal OAEs in persons with tinnitus certainly support the premise that tinnitus originates from cochlear hair cell damage (e.g., Hall, 2000; Sztuka et al., 2006).

OAEs play an important role in the objective documentation of hair cell dysfunction as the physiologic origin of tinnitus. Clinical experience and research dating back over 15 years clearly indicates that OAE abnormalities are a common, almost invariable finding, in persons with debilitating tinnitus (Ceranic et al., 1995; Shiomi et al., 1997; Hall, 2000). Reductions in DPOAE amplitude may also occur in temporary noise-induced tinnitus (Job, Raynal, & Kossowski, 2007). There are, however, occasional and rather infrequent reports of normal OAEs in a sizable proportion of tinnitus patients, including DPOAEs (Gouveris, Maurer, & Mann, 2005; Onishi, Fukuda, & Suzuki, 2004) and TEOAEs (Montoya et al., 2007). At this juncture, it is important to emphasize the distinction between "normal OAEs" (i.e., amplitudes consistently within an appropriate normal region) versus "present OAEs" (i.e., OAE amplitudes that are more than 6 dB above the noise floor but still below normal limits). Entirely normal OAE findings are rare in persons with tinnitus, whereas a finding of present but abnormal OAEs in tinnitus is not uncommon.

Continuing the theme that has been emphasized already in this review of the literature, OAE abnormalities almost always are documented even for persons with hearing thresholds that are within a clinically normal range. Evidence reported in the literature, and summarized in this section, supports the inclusion of OAE measurement as a routine component in the clinical test battery for tinnitus assessment. As a component in the diagnostic test battery, OAEs provide a quick, sensitive, and objective confirmation of cochlear dysfunction in persons with tinnitus.

When reported simply and confidently to the patient, information from OAE measurement validates an auditory abnormality and suggests a logical physiologic explanation for the perception of tinnitus (Hall, 2000, 2004). Ironically, the person with bothersome tinnitus who has been incorrectly told by a health care provider that, "There is nothing wrong with you," "You have normal hearing," and/or "The tinnitus sound is just in your head" is invariably comforted by the news that OAE findings indicate cochlear dysfunction. OAE findings confirm what the patient suspected—there is something wrong with the ear (or ears). Indeed, in the all-important counseling process for patients with bothersome tinnitus, OAE findings have real therapeutic value.

A correspondence exists between the perceived pitch of tinnitus and OAE test frequencies yielding abnormally reduced amplitudes or no detectable OAE activity (Bartnik et al., 2004; Hall, 2000; Satar, Kapkin, & Ozkaptan, 2003), although the relationship is not always precise (Pospiech et al., 2003). For the purpose of documenting cochlear dysfunction in the frequency region of perceived tinnitus, DPOAEs are preferable to TEOAEs. The constrained upper frequency range for TEOAEs is a serious limitation to their routine application in the assessment of tinnitus patients. TEOAEs cannot be consistently recorded above 5000 Hz, even in persons with completely normal cochlear functioning. However, TEOAEs may offer a clinical tool for assessing cochlear fine structure in persons with tinnitus. For example, Paglialonga and colleagues (2010) reported "minor dysfunction of the cochlear active mechanisms" in 23 normal-hearing patients with tinnitus in comparison to a control group.

With a sufficient number of stimulus frequencies per octave (e.g., ≥5), it is possible with DPOAEs to identify regions of outer hair cell dysfunction up to frequencies of 8000 Hz or higher. Although most patients match their tinnitus pitch to frequencies in the region of 2000 to 4000 Hz, it is not uncommon for tinnitus to be perceived for higher frequencies. For example, in a study of 20 normal-hearing persons with tinnitus, Satar, Kapkin, and Ozkaptan (2003) reported a mean perceived tinnitus frequency of 6300 Hz for the left ear and 6200 Hz for the right ear. Other formal studies involving tinnitus populations confirm the likelihood of OAE abnormalities for frequencies above 3000 Hz (e.g., Pospiech et al., 2003). On the other hand, a controlled study of contralateral suppression of TEOAEs in 97 patients with tinnitus revealed no clear evidence of medial olivocochlear efferent system dysfunction (Geven et al., 2010).

The repeatedly reported finding of abnormal OAE findings in persons with tinnitus and normal hearing sensitivity (e.g., Hall, 2000; Granjeiro et al., 2008; Mao et al., 2005; Nottet et al, 2006; Ozimek et al., 2006; Satar, Kapkin, & Ozkaptan, 2003; Shiomi et al,, 1997; Sindhusake et al., 2003) has multiple clinical implications. As already noted, OAE abnormalities provide objective confirmation to the audiologist, otolaryngologist, and patient of a logical physiologic explanation for the subjective (perceived) tinnitus (i.e., cochlear dysfunction). The high degree of frequency specificity of OAEs allows for rather precise description of the cochlear origin of the tinnitus. OAE abnormalities often are found well within typical audiometric octave frequencies. OAE findings also may contribute to information about the laterality of tinnitus (Satar, Kapkin, & Ozkaptan, 2003). It is not uncommon for a patient to report the perception of unilateral tinnitus in the presence of bilateral cochlear dysfunction and, sometimes, hearing loss. Presumably, when tinnitus is more pronounced in one ear than the other, the tinnitus is perceived as unilateral. Bilaterally abnormal OAE findings suggest the possibility of bilateral tinnitus. Whether tinnitus is unilateral or bilateral may affect decisions regarding the most appropriate management strategy for the patient.

There are few studies of OAE changes associated with treatment of tinnitus. Successful tinnitus management strategies largely focus on altering the representation of the tinnitus in the central nervous system, rather than the cochlear origin of tinnitus. The reason for this bias in tinnitus management is probably quite apparent. Plasticity of the central nervous system presents a greater opportunity for treatment-induced changes, whereas there is little chance of positive alterations in peripheral (e.g., hair cell or auditory nerve) function with even intensive therapy. A study reported by Kalcioglu and Turkish colleagues (2005) is an exception to this statement.

Thirty patients with tinnitus were treated with intravenous lidocaine. Immediately after lidocaine infusion, four of the patients reported suppression of their tinnitus. Measurement of DPOAE input/output functions showed significant changes in the lidocaine responders, but also in the nonresponders. Unfortunately, with lidocaine, relief from tinnitus is temporary. Subjects reported return of the perception of tinnitus within weeks of the treatment.

For many years, acupuncture has been proposed for treatment of tinnitus, although results of formal clinical trials are not promising. De Azevedo and colleagues (2007) report an intriguing investigation of the effect of acupuncture on TEOAEs in a group of 38 patients with tinnitus. According to the authors, there was a significant difference in OAE amplitude for subjects receiving "needle acupuncture . . . applied to the temporoparietal point corresponding to the vestibulocochlear area" versus a control group that were given acupuncture in another "unrecognized" cranial location.

In another interesting study reported by Job and colleagues in France (2004), reductions in DPOAE amplitude were found in young normal hearing and psychologically normal subjects following target practice with weapons. Subjects with tinnitus who were most anxious and tense showed the largest decreases in DPOAE amplitudes.

Malingering (Pseudohypacusis)

As summarized in Table 8–2, papers occasionally appear in the literature supporting the clinical value of OAEs in the diagnosis of adults with pseudohypacusis. The literature on pseudohypacusis in children, reviewed in Chapter 7, is also not extensive. Presumably, papers in the literature underestimate the actual clinical application of OAEs as an objective tool for the audiologic evaluation of patients with various nonorganic or functional hearing impairments, including pseudohypacusis, malingering, and psychological disorders. Our clinical experience confirms the usefulness of OAEs in identifying patients at risk for nonorganic hearing loss. That is, recording OAEs at the outset of a complete audiologic assessment invariably contributes to early detection of possible nonorganic hearing loss. At the least, OAE findings prevent the erroneous diagnosis of a sensory hearing loss.

The following scenario is supported by anecdotal reports of clinical audiologists (e.g., Hall, 2000). A 40-year-old male patient is evaluated with the conventional, yet outdated, test battery consisting only of air- and bone-conduction pure-tone audiometry and simple speech audiometry (e.g., word recognition performance). Results show a bilateral relatively flat configuration hearing loss in the range of 45 to 50 dB HL. Given the patient's hearing complaints and the audiologic findings, high-end digital hearing aids were recommended. However, the patient obtained no verifiable benefit from binaural amplification. Months later, he still complains of problems with speech understanding in all settings, particularly while working as a clerk in a busy auto parts store. Despite multiple return visits to the clinic for adjustments and modifications, the patient claims that the hearing aids are useless. The patient requests a letter documenting that, because of the hearing loss, he should receive disability benefits. At one of many follow-up visits to the clinic, a different audiologist decides to perform otoacoustic emissions to better assess cochlear function. DPOAE findings are entirely normal throughout an f_2 stimulus region from 500 to 8000 Hz. Based on these findings, comprehensive auditory evoked response assessment is performed, confirming auditory thresholds well within the normal region bilaterally.

Recording OAEs at the beginning of first clinic visit for this patient would have immediately raised a high degree of suspicion about a nonorganic hearing loss. A few minutes of time invested in OAE measurement would have saved considerable clinic time and expense consumed with the unsuccessful hearing aid fitting. The finding of normal OAEs in combination with an apparent sensory hearing loss demands further assessment with electroacoustic measures (e.g., acoustic reflexes) or with electrophysiologic techniques (e.g., tone-burst ABR or ASSR) to verify auditory thresholds. In possible pseudohypacusis, OAEs can reduce test time and increase diagnostic accuracy. Ultimately, the information provided by OAEs minimizes the likelihood of mismanagement.

Table 8–2. Selected Literature Describing Findings for Studies of Otoacoustic Emissions in Adults With Various Diseases and Disorders

Disease or Disorder	OAE Type	Findings	Study (Year)
Diabetes	DPOAE	Average amplitudes were reduced significantly in adults with insulin-dependent diabetes mellitus; ABR latencies were also delayed	Lisakowska et al. (2001, 2002)
	DPOAE	Lower DPOAE amplitude and latency in adult patients compared to a control group.	Park et al. (2001)
	TEOAE	Lower TEOAE amplitude or absent responses in adults with type I diabetes, plus >5 dB hearing sensitivity deficit in the 1000 Hz region.	Ottaviani et al. (2002)
	TEOAE	Absent responses were in most adults with type II diabetes and associated with sensory hearing loss.	Rózanska-Kudelska et al. (2002)
	DPOAE	Decreased DPOAE amplitudes at 4000 Hz in non-insulin-dependent diabetes (NIDDM) even in normal hearing sensitivity.	Erdem et al. (2003)
	DPOAE	Deficits in OAEs, pure-tone audiometry, and speech reception thresholds for elderly patients with type II diabetes versus age-matched control subjects. Deficits were greater for the right versus left ears.	Frisina et al. (2006)
	DPOAE	Abnormalities of DPOAE amplitude and contralateral suppression in adult patients with diabetes mellitus.	Wang and Zhong (1998)
Arthritis	DPOAE	No difference in findings for patients with rheumatoid arthritis (RA) versus gender- and age-matched control group.	Halligan et al. (2006)
	TEOAE	Decreased reproducibility and amplitude in rheumatoid m arthritis, with significant correlation between disease duration and diminished amplitude. Conductive hearing loss (air-bone gap) decreased following stapedectomy, but TEOAEs showed no improvement.	Salvinelli et al. (2006)
	TEOAE DPOAE	Normal OAEs in rheumatoid arthritis patients, even in patients with hearing loss	Bayazit et al. (2007)
	TEOAE	Absence of TEOAEs in 18% of RA patients with normal hearing sensitivity, and serve as early marker of asymptomatic hearing loss.	Murdin et al. (2008)
Behçet disease	TEOAE	Reduced amplitudes even in patients with normal pure-tone hearing thresholds	Muluk and Birol (2007)
Pseudohypacusis	DPOAE	No statistically significant difference in the proportion of normal versus pseudohypacusis subjects with normal DPOAE findings.	Wang (2008)
	DPOAE	OAEs contributed to the detection of pseudohypacusis In 66% of patients with inconsistent behavioral audiometry findings who complained of post-traumatic hearing loss.	Li et al. (2006)

continues

Table 8–2. *continued*

Disease or Disorder	OAE Type	Findings	Study (Year)
Pseudohypacusis *continued*	TEOAE	Inclusion of tympanometry and TEOAEs in a test battery resulted in significantly improved pure-tone thresholds for a group of 72 patients with suspected pseudohypacusis.	Balatsouris et al. (2003)
	TEOAE DPOAE	Normal OAEs in patients diagnosed with malingering and functional hearing loss.	Xu et al. (2005)
Vestibular neuritis	TEOAE	TEOAE findings showed no evidence of subclinical cochlear (outer hair cell) involvement in 28 patients with vestibular neuritis.	Xu et al. (2006)
Miscellaneous Disorders			
Chronic alcoholism	TEOAE	Absence of TEOAEs with sensory hearing loss in 30 chronic alcoholics	Niedzielska et al. (2001)
Renal failure	DPOAE	Decreased amplitudes in 31 patients with chronic renal failure undergoing dialysis versus a control group.	Gierek et al. (2002)
HDR syndrome	TEOAE	Hypoparathyroidism, deafness, and renal dysplasia (HDR syndrome) secondary to haploinsufficiency of the zinc finger transcription factor GATA3 in 2 patients was associated with absent OAEs.	Van Looij et al. (2006)
Alport syndrome	TEOAE	OAEs present in 40% and absent in 60% of patients with hearing loss.	Viveiros et al. (2006)
Urinary lithiasis	TEOAE	Patients with urinary lithiasis treated with electro-hydraulic extracorporeal shock wave lithotripsy (ESWL) showed no change in OAEs versus control group. Patient group wore "earheadings."	Muluk et al. (2006)
Sjögren syndrome	DPOAE	Abnormal pure-tone and DPOAE thresholds in 22 female patients with this "cell-mediated immune disorder."	Hatzopoulos et al. (2002)
Autoimmune conditions	TEOAE DPOAE	Absence of both types of OAEs in patients immunomediated inner ear disease," associated with rising or flat configured hearing loss.	Quaranta et al. (2002)
	DPOAE	Documentation of return of cochlear function in a patient with systemic lupus erythematosus and antiphospholipid syndrome with suspected "transient ischemic attack in stria vascular region of right cochlea.	Digiovanni and Nair (2006)
Fabry disease	TEOAE	The majority (54%) of 22 male patients had abnormal cochlear function as evidenced by pure-tone audiometry and OAEs.	Germain et al. (2002)
	TEOAE	Abnormal responses reflected cochlear hearing loss and inner ear damage (unilateral or bilateral) in 8 to 14 male and female patients with Fabry disease.	Conti and Sergi (2003)
Hypertriglyceride	DPOAE	Significant decrease in amplitude in persons with hypertriglyceridemia versus control group, even though hearing sensitivity was normal.	Pan et al. (2000)

Table 8–2. *continued*

Disease or Disorder	OAE Type	Findings	Study (Year)
Hypertriglyceride/ Hyper-cholesterolemia	TEOAE	Negative correlation between cholesterol serum level and TEOAE amplitude in patients with hyperlipidemia versus a control group.	Namyslowski et al. (2003)
Iron deficiency anemia	DPOAE	No difference in amplitudes or pure-tone thresholds in patients with iron deficiency anemia versus control subjects.	Cetin et al. (2004)
Kearns-Sayre syndrome	TEOAE	OAE abnormalities in patients with chronic progressive external ophthalmoplegia due to mitochondrial DNA deletion or mtDNA tRNA A3243G point mutation even when patients were asymptomatic (no hearing loss).	Kornblum et al. (2005)
Williams syndrome	DPOAE	DPOAE results confirmed cochlear dysfunction in patients with normal hearing sensitivity.	Marler et al. (2010)
Sleep apnea	TEOAE DPOAE	Patients with "obstructive sleep apnea/hyponea syndrome" showed decreased OAE amplitude. Amplitudes increased significantly following management with uvulopalatopharyngoplasty.	She et al. (2004)
Myasthenia gravis	TEOAE DPOAE	Reduced amplitude for both types of OAEs in comparison to a control group.	Hamed et al. (2006)
Muscular dystrophy	TEOAE	Patients with facioscapulohumeral muscular dystrophy had reduced reproducibility and dystrophy had reduced reproducibility and amplitudes versus a control group.	Balasouras et al. (2007)
Chronic obstructive pulmonary disease	TEOAE	Reduced amplitudes correlated with presumably secondary to hypoxia	el-Kady et al. (2006)
Spinal anesthesia	TEOAE DPOAE	Decreased amplitudes immediately after surgery with spinal anesthesia with full recovery by 15 days postoperatively.	Karatas et al. (2006)
Thyroid hypofunction	TEOAE	Abnormal OAEs in patients with normal pure-tone hearing thresholds.	Khechinaschvili et al. (2007)
Fibromyalgia	DPOAE TEOAE	No difference between experimental group and control group for either type of OAE	Yilmaz et al. (2005)
Cardiovascular disease (CVD)	DPOAE	Women with history of myocardial infarction were twice as likely to have abnormalities as those with no history. There was no association between CVD and OAE findings for men.	Torre et al. (2005)
Progestin	DPOAE	Women treated with estrogen and progestin had lower amplitudes than women treated with estrogen or a control group.	Guimaraes et al. (2006)

continues

Table 8–2. *continued*

Disease or Disorder	OAE Type	Findings	Study (Year)
Ankylosing spondylitis	TEOAE	Lower reproducibility and amplitude in 32 patients with ankylosing spondylitis versus a control group	Erbek et al. (2006)
	DPOAE	Significantly lower amplitudes in patients with versus a control group.	
Idiopathic sudden hearing loss	TEOAE	Presence of OAEs soon after onset of hearing loss predict recovery from ISHL.	Lalaki et al. (2001)
	DPOAE	DPOAE recovery was delayed in comparison to return of hearing thresholds suggesting residual cochlear dysfunction.	Zhang et al. (1999)
	TEOAE	Abnormalities in OAEs, but no correlation of findings with type of sudden deafness.	Mozota Núñez et al. (2003)
	TEOAE DPOAE	OAEs confirm inner ear status in low-frequency sudden hearing loss, but have no value in predicting the efficacy of osmotic therapy.	Canale et al. (2005)
	TEOAE DPOAE	Small OAEs immediately after onset of sudden hearing loss are associated with poor outcome, whereas large OAEs are rarely associated with poor hearing outcome.	Hoth (2005)
	DPOAE	DPOAE amplitude soon after onset of hearing loss significant prognostic indicator of recovery rate for the return of hearing.	Chao and Chen (2006)
	DPOAE	Presence of OAEs soon after onset of sudden hearing loss predicted good hearing improvement.	Ishida et al. (2008)
Diplaucusis	TEOAE DPOAE	Reduction in amplitude for both types of OAEs. Frequency shifts in fine structure was associated with changes in diplaucusis.	Knight (2004)
Bacterial meningitis	DPOAE	Subtle "subclinical" cochlear dysfunction reflected by OAEs even in patient with normal hearing sensitivity by pure-tone audiometry.	Mulheran et al. (2004)
Chronic hydrocephalus	DPOAE	Phase of OAEs with postural changes differs for normal subjects versus patients with chronic hydrocephalus treated with shunts.	Chomicki et al (2007)
Otosclerosis	TEOAE	Presence of OAEs rules out diagnosis of otosclerosis	Probst (2007)
Stapedectomy	DPOAE	No OAEs before stapedectomy but 67.3% of 34 patients had OAEs after stapedectomy	Gierek et al. (2004)
	DPOAE	No DPOAEs preoperatively in patients with otosclerosis, For most patients, there were no DPOAEs at 5 days and 1-month postoperatively.	Filipo et al. (2007)
	TEOAE DPOAE	OAEs were absent preoperatively and did not return immediately, 3, or 6-months postoperatively.	Herzog et al. (2001)

Table 8–2. *continued*

Disease or Disorder	OAE Type	Findings	Study (Year)
Superior canal dehiscence syndrome (SCDS)	TEOAE	Standard TEOAE analysis showed no abnormalities, off-line duration analysis (moving time window analysis) showed significant shortening.	Thabet (2010)
Aberrant internal carotid artery		Abnormal OAE findings in a 52-year-old woman with aberrant internal carotid artery and conductive hearing loss.	Cho and Cho (2005)
Duane retraction syndrome	DPOAE	No significant difference in DPOAE amplitude between DRS versus control groups.	Sevik et al. (2008)
Cigarette smoking	DPOAE	Significantly lower amplitudes and increased I/O DP thresholds in high frequencies in 12 smokers versus 12 nonsmokers.	Negley et al. (2007)
Nicotine	DPOAE	No effect of nicotine on OAEs.	Harkrider et al. (2001)
Systematic high blood pressure	DPOAE	Absence of OAEs for 4000- to 8000-Hz region	Esparza et al. (2007)

Analysis of OAE findings permits early identification of the major types of adult auditory problems commonly encountered in the clinic, including noise-induced and ototoxic-drug-induced auditory dysfunction, sudden onset hearing loss, and related disorders (e.g., tinnitus). In such cases, information from OAEs will have an immediate and critical impact on medical and audiologic management and even prevention of hearing loss. As just noted, normal OAEs documented at the outset of the assessment can minimize the likelihood of time-consuming behavioral testing that, in retrospect, yielded invalid findings. The minute or two required to record normal OAEs also may preclude the need for other rather lengthy test procedures (e.g., bone conduction audiometry). Finally, and perhaps most importantly, OAEs contribute to more efficient, accurate, and precise diagnosis of hearing loss. Misdiagnosis of auditory dysfunction can, of course, lead directly to mismanagement of hearing loss. Information from simple OAE measurement may prevent inappropriate, unnecessary, expensive, and potentially harmful treatment options, such as amplification, for children who really have normal hearing thresholds.

The problem of mismanagement with misdiagnosis of pseudohypacusis was confirmed recently in a data-based study reported by Holenweg and Kompis (2010). Over a 6-year period, 40 patients (25 females and 15 females) seen in a Swiss otolaryngology clinic were ultimately diagnosed with pseudohypacusis, including 18 adults and 22 children. For all of the patients, objective test procedures (OAEs and/or ABR) yielded normal or near-normal findings. Interestingly, one-out-of-five patients (2 children and 7 adults) were fit with hearing aids at a previous clinic before pseudohypacusis was finally diagnosed.

A modest number of papers in the literature support the clinical value of OAEs in the diagnosis of adults with pseudohypacusis (Balatsouras et al., 2003; Wang, 2003). Balatsouras and Greek colleagues (2003) demonstrated in a formal study of 72 adult patients with suspected pseudohypacusis the benefits of including objective measures (tympanometry and OAEs) in the test battery. When TEOAEs and tympanometry were performed along with pure-tone audiometry, the average PTA was 32.7 dB HL. When the test battery consisted only of behavioral measures (the control condition), patients with suspected pseudohypacusis yielded a PTA of 52.9 dB HL. The authors concluded that "the inclusion of TEOAEs and tympanometry in an audiologic protocol in the evaluation of patients suspicious of pseudohypacusis resulted in a significantly greater threshold improvement on repeat modified pure-tone audiometry when compared to the improvement observed for a control group in which these tests were not performed (Balatsouras et al., 2003, p. 518). Once the diagnosis of pseudohypacusis is made with objective measures like OAEs, informing the patient about the evidence of normal cochlear function (in age-appropriate language, of course) may lead to appropriate cooperation during a reassessment, and valid behavioral estimations of thresholds. (Balatsouras et al., 2003).

Wang (2003) reported no statistically significant difference in the proportion of normal hearing subjects versus pseudohypacusis subjects with normal DPOAE findings. According to Li et al. (2006), OAEs contributed to the detection of pseudohypacusis in 66% of patients who yielded inconsistent behavioral audiometry findings and who complained of post-traumatic hearing loss. Oishi and colleagues (2009) offer perhaps the most compelling recent evidence-based argument for objective measures in the diagnosis of pseudohypacusis, particularly OAEs. In a report of 6 cases of adults (all females) with sudden pseudohypacusis, the authors conclude, "Although the presence of distinctive clinical features (age, gender, and past history) is important for suspecting psychogenic hearing loss, objective audiologic tests such as otoacoustic emissions are essential for diagnosing some cases" (p. 279), adding that: ". . . OAEs are easy to perform and are time-saving. Therefore, OAEs can be very valuable tests for distinguishing PHL [psychogenic hearing loss] from ISSHL [idiopathic sudden sensorineural hearing loss]" (p. 282).

Idiopathic Sudden Sensorineural Hearing Loss

Since the early 1990s, a number of studies have consistently confirmed the value of OAEs in documenting cochlear auditory dysfunction in sudden idiopathic hearing loss (see Table 8–2). The word "idiopathic" in the terminology for this type of hearing loss implies that the mechanism or actual cause is not known. According to findings for limited studies to date (e.g., Schweinfurth et al., 1997), OAEs may contribute to diagnosis of the underlying mechanism for sudden hearing loss (e.g., vascular, viral, or some other pathophysiologic process). However, at the least, OAEs permit quick differentiation of sudden sensory hearing loss from rapid onset or recognition of neural auditory dysfunction and from an abrupt appearance of nonorganic hearing loss, such as conversion deafness (Hoth & Bonnhoff, 1993; Lalaki et al., 2001; Sakashita et al., 1991; Schweinfurth et al., 1997; Truy et al., 1993; Yilmaz et al., 2007). There are conflicting reports on the application of OAEs in prognosis of outcome for patients with sudden hearing loss. The presence of OAEs, and specifically a finding of normal OAEs, soon after the onset of the hearing loss generally is associated with a good outcome in a relatively short time period (e.g., Chao & Chen, 2006; Hoth, 2005; Ishida et al., 2008; Lalaki et al., 2001; Mozota Núñez et al., 2003). Almost inevitably, OAEs are not detectable in patients with poor recovery from sudden hearing impairment or "deafness" (Ishida et al., 2008). This statement assumes that retrocochlear auditory dysfunction is ruled out conclusively with auditory evoked responses and/or neuroradiologic imaging. The combination of normal OAEs and abnormal pure-tone hearing thresholds is also compatible, of course, with hearing loss secondary to inner versus outer hair cell dysfunction. Conversely, a finding of no detectable OAE activity, or small amplitude and abnormal OAEs within days after the onset of hearing loss portends poor outcome. Finally, recovery from sudden hearing loss probably is not complete until OAEs also return. Pure-tone audiometry usually shows recovery of hearing thresholds before the reappearance of normal OAEs. However, if OAEs have not returned to normal, then recovery probably s incomplete. If medical management for sudden idiopathic hearing loss (e.g., steroids) is discontinued before OAEs have normalized, there is the possibility that hearing loss will return (e.g., Hall, 2000). OAEs can help to predict outcome, and even rate of recovery, for sudden hearing loss. OAEs also can serve as an objective index for determining when medical management can be discontinued.

Ménière's Disease

Objective, noninvasive, and ear-specific identification of two clear patterns of OAEs in Ménière's disease represents one of the most dramatic illustrations of the value of OAEs in differentiating pathophysiologic processes in ear disease. Of course, OAEs also are used in combination with medical history, clinical findings, and other auditory (e.g., ECochG) and vestibular (ENG/VNG, VEMP) procedures for diagnosis of Ménière's disease (e.g., Chen & Young, 2006). The conventional theories on mechanisms of Ménière's disease have assumed a disruption in the mechanical properties of the cochlea, such as dampened basilar membrane mobility and disconnection of outer hair cells from the tectorial membrane, secondary to endolymphatic hydrops (increased pressure of cochlear pressure). As anticipated, for most patients with Ménière's disease, OAEs are abnormal or absent consistent with outer hair cell dysfunction (e.g., de Kleine et al., 2002; Hall, 2000). Both TEOAE and DPOAE amplitudes are reduced with very mild hearing loss, and OAEs are not detected for greater degrees of hearing loss (e.g., de Kleine et al., (2002). OAE abnormalities in patients with Ménière's disease are similar to OAE abnormalities in other forms of sensory hearing loss. Importantly, a proportion of patients with unilateral Ménière's disease as diagnosed with recommended methods, show abnormal OAEs on the apparently uninvolved ear (e.g, de Kleine et al., 2002; Hirschfelder et al., 2005), suggesting bilateral pathology with subclinical findings on one ear.

Approximately 20 to 30% of patients with the diagnosis of Ménière's disease and associated hearing loss do not have OAE abnormalities. In fact, patients in this subgroup of Ménière's disease have robust and normal OAEs, even in the presence of mild to moderate apparently sensory hearing loss (e.g., Bonfils et al., 1988; Bartoli et al, 1992; Cianfrone et al, 2000; Fetterman, 2001; Hall, 2000; Harris & Probst, 1992; Hof-Duin & Wit, 2007; Ohlms et al, 1991; Sakashita et al., 1991; van Huffelen et al., 1998;). The two highly divergent patterns of OAE findings in Ménière's disease suggest more than one pathophysiologic process may underlie the disorder. Cochlear hair cell involvement is common in the majority of patients with Ménière's disease. Endolympathatic hydrops also can be associated with biochemical abnormalities and disruption of neural innervation of hair cells within the cochlea (e.g., Nadol, 1989), and preservation of the integrity of outer hair cell function and the presence of OAEs. Whether or not differentiation of the two types of Ménière's disease contributes to decisions regarding management, and improved outcome, is not yet known.

Glycerol, a diuretic, is a drug sometimes used in the diagnosis of Ménière's disease. Improvement in auditory status with glycerol-induced dehydration (a positive glycerol test finding) is consistent with endolymphatic hydrops. Change in OAEs following administration of glycerol in patients diagnosed with Ménière's disease include increased amplitudes with, decreases for DPOAE thresholds extracted from I/O functions (Jablonka et al., 2003; Levina, 2004; Magliulo et al., 2004; Sakashita et al., 2001; 2004) and, for TEOAEs, increased reproducibility (e.g., Li & Zhong, 1999; Zeng & Zhong, 2006). Furthermore, studies with both types of OAEs confirm that they are a more sensitive and simple indicator of improved auditory status than pure tone audiometry. As an objective test, OAEs have a major advantage over pure-tone or speech audiometry for this application, as patients are often feeling very poorly (e.g., headache, nausea, lethargy) after glycerol administration when the audiologic assessments are carried out. Some apparently fun-loving patients describe their state after glycerol as resembling a very bad hangover after a night of excessive drinking.

Drug-Induced Hearing Loss (Ototoxicity)

The clinical challenge in medical management with potentially ototoxic drugs of any patient with a disease is to maximize the therapeutic benefits of the medication to preserve and prolong life, while minimizing the negative impact of the same medication on hearing and, in turn, quality of life. The rationale for monitoring auditory status during chemotherapy for neoplasms or other medical therapy (e.g., antibiotic treatment for infection, is the same for adults as for children; see Chapter 7).

The goal is early detection of drug-related cochlear dysfunction. Without close and direct monitoring of auditory function, it is not possible to predict accurately which patients will develop ototoxicity-related hearing loss versus those whose hearing will be preserved. A variety of factors affect individual susceptibility to drug-induced auditory dysfunction. The typical physical examination by a physician, and patient report, are simply inadequate for early detection of auditory dysfunction. Without formal and frequent monitoring of auditory function, hearing loss is likely to be suspected only when the patient begins to experience difficulties with speech understanding in communicative situations.

In contrast to the rather extensive literature on monitoring cochlear function for possible ototoxicity in children (reviewed in Chapter 7), relatively few papers describe their use with adults. There are a few practical explanations for why ototoxicity monitoring with OAEs is less common and critical clinically for adults. For most adults, behavioral pure-tone audiometry is a feasible option for screening and diagnosis of auditory dysfunction. Clinical studies have confirmed the important role of extra- or extended high-frequency pure-tone threshold measurements for early detection of hearing loss in adults (e.g., Fausti et al., 1994; Ress et al., 1999). There is some evidence suggesting that DPOAE amplitudes are repeatable for the frequency range of 8000 to 16,000 Hz (Dreisbach et al., 2006). Repeatability in DPOAE findings would, of course, be an important requisite for clinical monitoring of ototoxicity. Although valid pure-tone audiometry may not be possible in very ill adults, the difficulties associated with behavioral audiometry in infants and young children are certainly less common in an adult population. Also, even before therapy with potentially ototoxic medications is initiated, adults may lack detectable OAEs, particularly for higher test frequencies, because of pre-existing and non-drug-related sensory hearing loss (e.g., Ress et al., 1999). Noise-induced hearing loss and hearing loss with aging (presbycusis) are two common examples. Nonetheless, as summarized in Table 8–3, there are a modest number of papers describing the application of OAEs in patients with various disorders undergoing medical therapy with different potentially ototoxic drugs.

In an extensive and recent multisite study, Reavis and colleagues (2008) report DPOAE findings for 53 subjects (51 men and 2 women) treated medically with cisplatinum ($n = 36$), carboplatin ($n = 10$), or antibiotic drugs (aminoglycosides amikacin or gentamicin or another antibiotic (glycopeptide) vancomycin ($n = 7$) who were enrolled from three Veterans Administration Medical Centers in the United States. Age ranged from 46 to 82 years (average age 59 years). Importantly, the authors conducted a preliminary study with four normal subjects of test-retest reliability for DPOAE amplitude to determine how much decrease should be considered clinically significant. Based on the outcome of this pilot project, with the instrumentation used in the study (Intelligent Hearing Systems Smart OAE device), a decrease in DPOAE amplitude of ≥4 dB SPL at two or more adjacent f2 frequencies was selected as the criteria for a significant change. For these criteria, the anticipated false positive rate (the likelihood that a change would be incorrectly identified) was approximately 5%. Curiously, DPOAE data were reported only for 53 patients (90 ears) who demonstrated shifts in behavioral hearing thresholds for at least one ear rather than all patients treated with the above-noted drugs. Among this group, DPOAEs were detected in 91% of the ears and could therefore be monitored during the study. Among the patients (82 ears) with baseline DPOAEs, 78% showed a significant (as defined above) decrease (referred to by the authors as "a DPOAE hit") in amplitude during the course of drug therapy. No differences in DPOAE changes were found as a function of the type of drug taken by the patients. Interestingly, DPOAE changes preceded behavioral changes in hearing in 32.8% of the ears, the changes for the two techniques coincided also in 32.8% of the ears, and DPOAE changes actually appeared after the decrease in pure-tone thresholds in 34.4% of the ears (by an average of 63.9 days). It is important to keep in mind that behavioral audiometry in the study consisted of extended high-frequency threshold measurement using 1/6-octave frequency increments.

Table 8–3. Selected Literature Describing Monitoring for Drug-Induced Cochlear Damage (Ototoxicity) with Otoacoustic Emissions, Arranged According to the Medications Used Therapeutically

Medication	OAE Type	Findings	Study (Year)
Aminoglycoside Antibiotics			
Gentamicin	DPOAE	Intratympanic injection of gentamicin for treatment of Ménière's disease according to accepted therapeutic guidelines (27 mg/ml each week) produced no change	Perez et al. (2005)
Tobramycin	DPOAE	Initially, amplitude is reduced before changes are observed in pure-tone hearing thresholds. Most adult patients with cystic fibrosis who are treated with tobramycin have absent OAEs, thus limiting the effectiveness of monitoring ototoxic changes with OAEs.	Hall et al. (2004)
Chemotherapeutic Agents			
Cisplatin	DPOAE	Changes in DPOAE amplitudes detected cochlear toxicity with cisplatin of adults for testicular cancer earlier than pure-tone audiometry. Lower frequencies 65 dB SPL, precluding monitoring with OAEs.	Biro et al. (2006)
	TEOAE	Abnormal findings were found in 53% of 32 patients with testicular cancer treated with cisplatin.	Strumberg et al. (2002)
	DPOAE	Few subjects in the age range of 42 to 80 years (average of 62 years) had detectable OAEs at baseline assessment for a stimulus intensity level of 65 dB SPL, precluding monitoring with OAEs.	Ress et al. (1999)
	DPOAE	DPOAE findings alone indicate whether hearing has changed with cisplatin treatment, especially for the highest frequencies.	Reavis et al. (2010)
DFMO	DPOAE	No change in pure-tone audiometry or DPOAE amplitudes for adults treated with low dose or oral DFMO for colorectal polyps.	Doyle et al. (2001)
Quinine Derived Drugs			
Quinine	TEOAE	Early study of quinine effects in normal volunteers.	Karlsson et al. (1991)
Other Drugs, Substances and Treatments			
Aspirin	DPOAE	Effects of aspirin show considerable individual variability.	Brown et al. (1993)
Salycilate (aspirin)	TEOAE	Disappearance of OAEs on first day of aspirin overdosage, with reappearance on day 5 after aspirin was discontinued.	Wecker and Laubert (2004)

continues

Table 8–3. *continued*

Medication	OAE Type	Findings	Study (Year)
Dexamethasone	TEOAE	Intratympanic injection was associated with a decrease in reproducibility but not the signal-to-noise ratio of TEOAEs. There was no drug effect.	Yilmaz et al. (2005)
Antithyroid drug	DPOAE	Propylthiouracil (PTU) caused bilateral sensory hearing loss with abnormality of DPOAEs.	Sano et al. (2004)
Deferoxamine	DPOAE	Ototoxic effects of deferoxamine therapy in patients with lifelong transfusion-dependent anemia.	Delhaye et al. (2008)
Radiotherapy	TEOAE	OAEs were normal or abnormal in a series of patients with nasopharyngeal carcinoma undergoing radiotherapy depending on whether cochlear or retrocochlear pathways were involved.	Guo et al. (2005)
	DPOAE	Statistically significant decreases in DPOAE amplitude in patients with radiotherapy for head and neck cancers were noted immediately with treatment, and preceded changes in pure-tone thresholds.	Yilmaz et al. (2008)
Styrene	TEOAE	Occupational chronic exposure to styrene (40 to 50 ppm) In a boat building plant had little effect on TEOAEs. Improvement was noted during work holidays.	Triebig et al. (2008)
	TEOAE	Workers in a fiberglass-reinforced plastic boat building factory exposed to styrene for an average of 7.5 years showed no consistent or significant changes in pure-tone thresholds or TEOAEs.	Hoffman et al. (2006)
Aluminum	DPOAE	Cochlear dysfunction in hemodialysis patients was enhanced by serum aluminum levels.	Chu et al. (2007)

*DFMO = difluoromethylornithine.

Although OAEs are most commonly applied to monitor auditory function in chemotherapy or antibiotic therapy, recent studies confirm other perhaps equally valuable applications. A study reported by Delehaye and colleagues in Italy (2008) illustrates the use of OAEs in early detection of auditory dysfunction in patients with transfusion-dependent anemias whose medical management includes deferoxamine. Repeated transfusions results in excessive iron that, in turn, can lead to life-threatening and/or -shortening cardiac toxicity. Iron balance is maintained with deferoxamine (chelation therapy), a drug with documented cochlear ototoxicity. Significant reductions in DPOAE amplitudes for frequencies above 3000 Hz were detected before changes were noted in pure-tone audiometry. The authors (Delehaye et al., 2008) conclude that: "As [an] ototoxicity screening tool DP-gram was extremely sensitive and superior to pure-tone audiometry" (p. 198). We believe, because of several advantages of DPOAEs (noninvasive, objective, rapid, easy to use, sensitive), this testing should be implemented in the audiologic follow-up of the patients under chelating treatment" (p. 201). There also is a report of the use of DPOAEs for early detection of cochlear dysfunction associated with accumulation of aluminum in patients with renal failure

(Chu et al., 2006). Toxins are known to build up in the blood of patients with chronic renal failure or end-stage renal disease undergoing hemodialysis.

The themes in the literature on ototoxicity are quite obvious, and common to most substances causing auditory dysfunction. OAEs, and particularly DPOAEs, offer an objective, quick, noninvasive, and highly sensitivity index of even subtle early changes in outer hair cell function secondary to the toxic effects of the substances. Early detection of cochlear dysfunction opens the possibility of prevention of hearing loss that interferes with communication.

Retrocochlear Pathology

Acoustic Tumors. Reports describing the exploration of OAEs in the diagnosis of auditory dysfunction in persons with acoustic tumors (really vestibular schwannomas) date back to the earliest years of clinical research of the new technique (e.g., Bonfils & Uziel, 1988; Robinette & Bauch, 1991). Some early clinical investigators speculated based on data collected in the operating room that OAEs would play a role in monitoring intraoperatively cochlear function during surgeries putting the auditory system at risk (e.g., Cane et al., 1992; Telischi et al., 1995). However, intraoperative monitoring of cochlear function with OAEs has not become a widely accepted clinical service. Several factors probably account for uncommon application of OAEs in monitoring cochlear function, including the measurement challenges associated with high noise levels and a desire to present to the ear ipsilateral during surgery click stimuli used to record ECochG and ABR. Both ECochG and ABR are valuable intraoperative measures of cochlear and retrocochlear function (e.g., Hall, 2007).

There is plenty of evidence confirming that inclusion of otoacoustic emissions in the diagnostic test battery, along with procedures such as the ABR, enhances the precise differentiation of sensory versus neural hearing loss (e.g., Komazec, Lemajic, & Vlaski, 2004; Quaranta et al., 2001; Sone et al., 2007). OAEs also offer the possibility of contributing to the diagnosis of cochlear dysfunction coexisting with or secondary to space-occupying retrocochlear pathology. Assuming the absence of sensory loss of other causes, normal cochlear function generally is associated with smaller tumors, due to preservation of intact blood flow to the cochlea. However, tumor size is not always linked to the presence or absence of OAEs unless hearing loss also present (e.g., Gouveris, Victor, & Mann, 2007). Early and partial compromise of vascular function within the internal auditory canal typically leads to cochlear dysfunction and is reflected by abnormalities in OAEs, even if it is not always evident in pure-tone audiometry (e.g., Gouveris et al., 2007). Abnormal OAE findings in patients with acoustic tumors (vestibular schwannomas) imply cochlear involvement, even if the tumor is small (e.g., Baguley, Jones, & Moffat, 2003). Paradoxically, OAE amplitudes may actually be increased when recorded from ears ipsilateral to acoustic tumors, even in the presence of a mild pure-tone hearing loss, in comparison to the opposite ear (Gouveris & Mann, 2004). Although a mechanism is not proven, it would seem reasonable to raise the possibility of efferent system involvement to explain the unusually large OAE amplitudes. This theory is supported by evidence of reduced contralateral suppression of TEOAEs in patients with cerebellopontine angle tumors (Aniol-Borkowska et al., 2005).

Because of their sensitivity to subtle cochlear dysfunction, OAEs also may play an important role in decisions regarding management of patients with acoustic tumors (e.g., Kim et al., 2006). With normal OAEs at most or all test frequencies and, by inference, intact cochlear function, there is greater likelihood of hearing preservation with surgical management of acoustic tumors (e.g., Kim et al., 2006). That is, surgical decompression of the auditory nerve may remove or reduce neural auditory dysfunction while leaving cochlear function intact. Clinical studies show that hearing recovery with surgery, an actual improvement rather than simply preservation of hearing, is possible if OAEs are normal preoperatively (Inoue, Ogawa, & Kanzaki, 2003). On the other hand, if serious cochlear dysfunction exists preoperatively secondary to compromise of the blood flow via the internal auditory artery, then hearing abnormalities will not be reversed by tumor removal and surgical manipulations during removal of tumor compressing the artery may lead to further disruption of a consistent supply of blood to the cochlea.

Cochlear dysfunction secondary to tumor effects, as evidenced by OAE (specifically DPOAE) abnormalities, is more likely for acoustic tumors than for other nonacoustic tumors within the cerebellopontine angle (Mobley et al., 2002). On the other hand, extension of acoustic tumors into and filling the internal auditory canal (IAC) are no more likely to cause cochlear abnormalities, again as detected with DPOAEs, than tumors that only partially enter the IAC Odabasi et al., 2003).

OAEs can be used to evaluate effects of surgical and nonsurgical management of acoustic neurinomas (vestibular schwannomas) on auditory function. Gamma knife radio-surgery is an alternative to conventional surgical therapy. Although initially considered a safer alternative, Ottaviani et al. (2002) reported worsening of TEOAEs with gamma knife radiosurgery versus pretreatment status for 9 of 12 patients even though the ABR remained unchanged. The authors hypothesized that auditory dysfunction was due to radiation effects on the cochlea. A few investigators have attempted to monitor cochlear function intraoperatively with OAEs (Morawski et al, 2004; Zaouche et al., 2005). Morawski et al. (2004) continuously recorded DPOAEs from 20 patients during surgical removal of cerebellopontine angle tumors. Because of the relatively high ambient noise levels in the operating room, OAE measurement usually is limited to a mid- to high-frequency region (2000 Hz and above). Changes in DPOAEs were noted intraoperatively, beginning with the highest frequencies (close to 6000 Hz). In some cases, reductions in DPOAE amplitude were only temporary, and full recovery was documented. In other patients; however, DPOAEs disappeared permanently. The loss of DPOAEs was associated with several surgical events, including microcoagulation of small blood vessels, surgical debulking of the tumor, and operative manipulations that stretched or compressed structures within the IAC.

Auditory Neuropathy Spectrum Disorder (ANSD). The extensive, and still rapidly growing, literature describing application of OAEs in detection of auditory neuropathy spectrum disorder (ANSD) in children, including adolescents, was reviewed in Chapter 7. Although onset of ANSD in adulthood is highly unusual, a rather sizable number of published papers report OAE findings in the identification of ANSD apparently "acquired" during adolescence (e.g., Duman et al., 2008; Zang et al., 2002, 2008). In general, the principles derived from the pediatric literature on OAEs in ANSD pertain also with young adults. There is a fundamental lesson to be learned from this accumulated clinical experience. Very simply, OAEs should routinely be included in the basic test battery whenever a new patient undergoes audiologic assessment. The presence of retrocochlear dysfunction, including forms of ANSD, will immediately be suspected during the assessment if normal OAEs are recorded in a person with any degree of hearing loss by pure-tone audiometry. And, absence of acoustic reflexes and extraordinarily poor word recognition scores will raise additional concerns about possible ANSD. Then, diagnosis of ANSD can be confirmed with findings from ECochG and ABR measurement, along with a thorough neurologic assessment.

Diabetes

Within the period of 1998 to about 2003, a handful of reports published independently by researchers from around the world described DPOAE abnormalities in adults with the diagnosis of diabetes (refer back to Table 8–2 miscellaneous). The characteristic finding was a statistically significant reduction in the amplitude of DPOAEs in comparison to control subjects, even in the presence of normal hearing sensitivity. Lisowaska and colleagues in Poland (Lisowaska et al., 2001., 2002) hypothesized that the finding " . . . indicates the existence of an alteration in cochlear micromechanics in diabetic patients" and "the impaired functional properties of the outer hair cells are probably caused by early metabolic complications in diabetes (among other things nonenzymatic glycation related to hyperactivity of free oxygen radicals . . . " (Lisowska et al., 2001, p. 199). Interestingly, the investigation of cochlear function in diabetes using OAEs has not continued since the early studies.

Arthritis

Conflicting OAE findings recently reported for rheumatoid arthritis range from normal even in patients with hearing loss to abnormal with a correlation between OAE amplitude and duration of the disease (see Table 8–2). The results of one study (Murdin et al., 2008) suggest that OAEs may serve as an early marker of cochlear dysfunction in persons with rheumatoid arthritis yet normal hearing sensitivity. Differences in the diagnosis criteria for rheumatoid arthritis and in the criteria for enrollment as subjects among studies may account for some of the discrepancies in OAE findings. Nonetheless, there is tantalizing evidence in support of diagnostic value for OAEs in rheumatoid arthritis. Given that OAEs are noninvasive, objective, and widely accessible and that rheumatoid arthritis is a very common chronic disease, additional research certainly is warranted.

OTHER DISORDERS

The literature contains dozens of papers describing OAE findings for a rather remarkable array of diseases and disorders, as summarized earlier in Table 8–2. Clearly, for many of the disorders, more research is warranted before the clinical value of OAEs is established. Review of the literature yielded only a single published paper or only a few papers, but some promise of clinical usefulness, for a variety of diseases and disorders (e.g., systemic high blood pressure, chronic hydrocephalus, bacterial meningitis, diplacusis, ankylosing spondylitis, cardiovascular disease, thyroid hypofunction, chronic obstructive pulmonary disease, myasthenia gravis, sleep apnea, Fabry disease). One good example of an intriguing and potentially fruitful line of research is a study on the effects of cigarette smoking on cochlear function as defined by DPOAEs. Using a clinical device (Mimosa Acoustics CubeDis, Eden Prairie, MN), Negley and colleagues (2007) recorded DPgrams at $L_1 = 70$ and $L_2 = 70$ dB SPL and at $L_1 = 65$ and $L_2 = 50$ dB SPL, and then input/output (I/O) functions for DPOAE for f_2 frequencies of 2000, 3000, and 4000 Hz with intensity descending in 10 dB steps from 80 to 20 dB SPL. Subjects were 12 young (20 to 30 years) healthy adults (5 male and 7 female) who had a history of smoking for 5 to 8 years and a control group of 12 age- and gender-matched nonsmokers. Although all of the subjects in both groups had normal hearing sensitivity (<25 dB) throughout the conventional audiometric frequency region (250 to 8000 Hz) and for high frequencies up to 20,000 Hz, pure-tone thresholds were significantly poorer for smokers. Also, DPOAE amplitudes were reduced (on the order of 3 to 9 dB for moderate stimulus intensity levels), and I/O thresholds increased in the smokers. The authors (Negley, 2007), in agreement with other investigators of auditory function and smoking, speculate that a mechanisms involving vascular insufficiency to the cochlea secondary to factors such as atherosclerotic narrowing and occlusion of blood vessels, and vasospasm induced by nicotine. This study illustrates the sensitivity of OAEs to rather subtle cochlear dysfunction. Considering the widespread abuse of tobacco, the findings suggest that OAEs may play a role in addressing a major global public health problem.

For other diseases (e.g., Duane retraction syndrome, iron deficiency syndrome, fibromyalgia), limited data suggest that OAEs have little or no value in identification or diagnosis.

III 9 III

Efferent Auditory System and Otoacoustic Emissions

INTRODUCTION

One area of clinical work that is poised to become an important element of the diagnostic arsenal is the evaluation of the olivocochlear (OC) efferent system via OAEs. You were introduced to structure and function of the auditory efferent system, especially the olivocochlear pathways, in Chapter 1. Here, we will discuss the important influence of the OC efferent system on OAEs in more detail, and strategies for clinical measurement of the OC efferent system. Activation of the OC efferent system modulates OAEs in a complex manner that is being actively explored by scientists around the world. However, in general, activation of the OC efferent system inhibits OAEs, as evidenced by a reduction in their amplitude. This inhibition of OAE activity is consistent with the effect of the OC efferent system on the basilar membrane and the auditory nerve. The functional role of the auditory efferent system is not fully understood at this time. We begin with a brief discussion of the functional anatomy of the OC efferent system. Later, we review current thinking on the function of the OC efferent system and its influence on different types of OAEs.

Functional Anatomy

The descending (efferent) auditory pathways originate in the cortex and terminate in the cochlea. In mammals, the olivocochlear bundle (OCB) forms the final segment of the auditory efferent system (Guinan, 2006; Guinan, War, & Norris, 1983; Rasmussen, 1946; Warr, 1975; Warr & Guinan, 1979). A schematic representation of this final stage of the efferent system is presented in Figure 9–1. Lateral (L) and medial (M) OC fibers originate at the lateral and medial OC nuclei and form the OCB that target each cochlea. Note that the OCB entering each cochlea contains LOC and MOC fibers from both sides of the brainstem. From the perspective of the cochlea on a given side of the head, the OCB can then be thought to contain both *crossed* and *uncrossed* bundles.

The LOC and MOC fibers are distinct. The MOC fibers are thick, myelinated, and terminate at the base of OHC. In contrast, the LOC fibers are thin, unmyelinated, and terminate on the dendrites of the auditory nerve fiber. In animal studies, direct recordings from MOC fibers have been made following electrical stimulation. MOC fibers can also be acoustically stimulated. In contrast, there are no reports of direct recording of LOC fiber activity following direct (electrical) stimulation, and it is not clear if they can be stimulated acoustically.

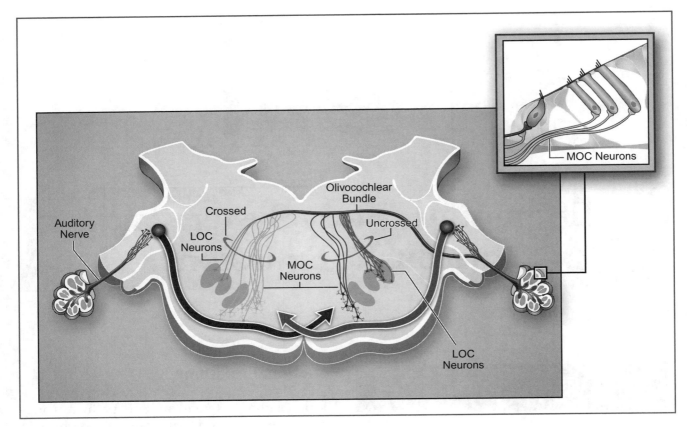

FIGURE 9–1. A stylized representation of the olivocochlear (*OC*) pathway based on Guinan (2006). The main panel shows a transverse section of the brainstem along with both cochleae and the afferent neurons of the cochlear nerve to the cochlear nucleus, continuing across the brainstem. Efferent connections going to only one cochlea are shown for clarity. Lateral olivocochlear (*LOC*) neurons in blue arise from the S-shaped nucleus on either side of the brainstem. Medial olivocochlear (*MOC*) neurons arise more medially. Crossed and uncrossed groups of LOC and MOC neurons form the olivocochlear bundle (*OCB*) that enters the cochlea. The specific efferent innervation inside the cochlea is displayed in the inset. MOC neurons terminate directly on the OHC, whereas LOC neurons terminate on the dendrites of the afferent neurons of the auditory nerve.

Our knowledge of the MOC system is much more complete than that of the LOC system.

To understand the relationship between crossed (C) and uncrossed (U) OCB, let us conduct a thought experiment where we are interested in measuring the modulation of OAEs in the right ear of a patient. As is customary, the OCB will be stimulated acoustically and the change in OAE characteristics will be measured against a baseline. This acoustic stimulation can be done in three ways. First, the acoustic stimulant could be applied to the contralateral ear (left, in this case). Second, the acoustic stimulant could be applied to the ipsilateral ear (right, in this case). And finally, the stimulation could be bilateral. If we follow the

afferent and efferent pathways shown in Figure 9–1, it is evident that, in the ipsilateral case, the acoustic stimulant is carried by the afferent nervous system across the midline of the brainstem from the right to the left side. The efferent neurons stimulated on the left side then cross back over to affect their influence on the right cochlea. Thus, in the ipsilateral case, the change seen in OAEs is the result of a *double crossing*, first by the afferents and then by the efferents. In contrast, in the contralateral case, the acoustic stimulant crosses over from the left side via the afferents, but the efferent influence is exerted by the *uncrossed* efferent fibers or the UOCB. Needless to say, both the UOCB and COCB are operational in the binaural case. Liberman, Puria, and Guinan (1996) provided definitive evidence of this circuit configuration by demonstrating that the contralateral effect of the efferent system could be extinguished by sectioning the OCB on the ipsilateral side, thereby eliminating both the UOCB and the COCB. However, sectioning just the COCB did not impact the contralateral effect.

Most laboratory animals have approximately twice the number of crossed fibers as compared to uncrossed MOC fibers. For example, 80% of all MOC neurons in chinchillas and 60% of all MOC neurons in rats are crossed (Robles & Delano, 2008). The ratio between crossed and uncrossed fibers in humans is not yet precisely quantified. The distribution of MOC innervation along the cochlear length also varies by species. In cats, crossed MOC innervation is biased toward the basal turns of the cochlea whereas uncrossed MOC innervation is biased toward the center of the cochlea (Guinan, 1996). In contrast, in mice, the innervation density peaks toward the center of the cochlea (Maison, Adams, & Liberman, 2003). The gradient of MOC innervation across OHC rows also varies by species. The innervation is denser for the first row than for the third row in cats (Liberman, Dodds, & Pierce, 1990), but more uniformly distributed in mice (Maison et al., 2003). General similarities in MOC anatomy and physiology make it convenient to transfer knowledge across species and to apply information obtained in laboratory animals to clinical tools for humans. However, the interspecies differences described above must be considered in making such generalizations.

Cellular Mechanisms

The MOC efferent system exerts its control on the cochlea mainly through the neurotransmitter acetylcholine (ACh) (Eybalin, 1993; Schrott-Fischer et al., 2007). ACh binds to a nicotinic receptor on the basolateral surface of OHCs formed by $\alpha 9$ and $\alpha 10$ subunits (Elgoyhen, Johnson, Boulter, Vetter, & Heinemann, 1994; Elgoyhen et al., 2001; Plazas, Katz, Gomez-Casati, Bouzat, & Elgoyhen, 2005). The inflow of Ca^{2+} current through the ACh-gated cation channel leads to the opening of small-conductance calcium-activated potassium channels leading to hyperpolarization (inhibi-

tion) of OHCs (Blanchet, Erostegui, Sugasawa, & Dulon., 1996; Fuchs & Murrow, 1992a; 1992b; Housley & Ashmore, 1991; Oliver et al., 2000; Yuhas & Fuchs, 1999). The hyperpolarization and conductance changes reduce the electromotility of OHCs on a fast time scale between 10 and 100 ms. On a slower time scale of 10 to 100 sec, the axial stiffness of the OHCs is reduced, thereby increasing their compliance and, hence, electromotility (Cooper & Guinan, 2003; Dalos, He, Lin, Sziklai, Mehta, & Evans,1997). Altering electromotility of OHCs affects their feedback to basilar membrane mechanics and overall cochlear gain. It is still not clear how the in vitro observation of increased OHC eletromotility, due to ACh, results in the in vivo observation that is inhibitory in nature (Dallos et al., 1997). The importance of ACh receptors in affecting the efferent modulation of cochlear function recently was confirmed recently in a genetically engineered mouse, where the $\alpha 9$ units were hypersensitive to ACh. The magnitude and duration of inhibition of various measures of cochlear function by the MOC efferent system was significantly increased in these mice (Taranda et al., 2009).

History

As early as the 1950s, scientists described the influence of efferent stimulation on auditory nerve and cochlear function. A reduction in compound action potential (Galambos, 1956) and endocochlear potential, along with an increase in cochlear microphonic (Fex, 1959), was observed on electrical stimulation of the MOCB. Over the next few decades, changes were documented in basilar membrane motion, IHC potentials, and auditory nerve firing due to MOCB activation (see Guinan, 2006b; Robles & Delano, 2008). Perhaps most pertinent to OAEs, MOCB activation clearly reduces the magnitude of basilar membrane motion and produces a phase change (Cooper & Guinan, 2006). The change in basilar membrane phase due to MOCB activation will become significant later when we discuss the complexities in quantifying the MOC-induced change in OAEs. The very first publications showing alterations in OAE characteristics on efferent stimulation appeared in the early 1980s (Mountain, 1980; Siegel & Kim, 1982).

Terminology

A word about terminology is appropriate as we begin the discussion about the influence of the MOCB on OAEs. First, should the reduction in OAE amplitude due to MOCB activation be categorized as "suppression" or "inhibition?" In much of the auditory literature, the term suppression is used in reference to mechanical interference between two signals. Inhibition, on the other hand, often is associated with neural phenomena. This line of argument would lead to exclusively using the term inhibition when referring to a reduction in OAE amplitude due to MOCB activation. However, in this context, both suppression and inhibition are used interchangeably in the literature. In this chapter, we use the term "inhibition" to describe the reduction in OAE levels caused by MOCB activation.

As with the middle ear muscle reflex, several acoustic signals are involved in the measurement of the influence of the MOCB on OAEs. Acoustic signals are necessary to record OAEs, except in the case of spontaneous OAEs. These signals are invariably referred to as the stimuli. However, acoustic signals also are necessary for the activation of the MOCB in humans. This signal is also freely referred to as a stimulus in the literature. Other terms used to describe the MOCB activating acoustic stimulus include "acoustic stimulus," "suppressor stimulus," "elicitor stimulus," or simply the "elicitor." When the elicitor is presented to the contralateral ear, the paradigm is often referred to as "contralateral acoustic stimulation" or "CAS." In the following discussion, the stimulus used to evoke OAEs will be referred to as the OAE stimulus or simply the stimulus. The sound used to elicit activity of the efferent auditory system will be referred to as the elicitor (e.g., contralateral or ipsilateral elicitor).

IMPORTANCE OF STUDYING EFFERENT MODULATION OF OAES

The extensive and rather complex collection of descending auditory pathways points toward an important role of the efferent system in audi-tory function, hearing, and listening. Indeed, this importance (discussed later in the chapter) has been documented in various experiments on laboratory animals. Measurement of inhibition of otoacoustic emissions offers an indirect, yet non-invasive and clinically feasible, means of assessing function of the efferent auditory system. No other audiologic procedure is readily available for this clinical purpose. Prior to the discovery of OAEs, experimentation on the auditory efferent system was limited largely to laboratory animals. Conclusions from the results of these well-designed experiments could not always be generalized to humans, and essentially there was no opportunity to relate efferent auditory system dysfunction to pathologies encountered clinically. Now, however, ongoing clinical research of OAE findings under different elicitor conditions is yielding information on patterns of efferent system abnormalities associated with a wide variety of disorders and pathologies. Clinical literature on OAE inhibition is reviewed later in this chapter.

Function of the Auditory Efferent System

A fundamental, and interesting, question is: What purpose does the efferent auditory system serve in humans? Carefully controlled experiments in laboratory animals have allowed us to formulate some definitive ideas about the role of the auditory efferents in everyday hearing. However, the extent to which this knowledge is applicable to humans and their environment is uncertain, and is continually examined and debated. Understandably, the level of control over experimental conditions (e.g., noise exposure) and anatomic manipulations (e.g., sectioning of the OCB) possible in experiments involving laboratory animals is not practical for experiments involving human subjects. Therefore, specific knowledge about human subjects has been obtained primarily from normal populations with small pockets of knowledge from special patient populations, such as those with learning disabilities or auditory processing disorders. On occasion, patients undergoing surgical intervention relevant to this area (e.g., sectioning of the vestibular nerve) have been studied with a hope of gaining further insight into the human auditory efferent system. Next, we briefly discuss the main themes about auditory efferent function in the literature.

Protection from Acoustic Trauma

One commonly cited function of the MOC efferent system is protection of the cochlea from trauma, mostly based on data from studies of animals exposed to potentially damaging high intensity sounds. Greater inhibition of OAEs, and the underlying reduction in outer hair cell activity, is associated with less cochlear damage following noise exposure (e.g., Rajan, 2000; Di Girolamo, Napolitano, Alessandrini, & Bruno, 2007). Maison and Liberman (2000) provided perhaps the most striking results supporting the role of the MOC efferent system in protection against noise damage. These authors measured the adaptation of DPOAE level after stimulus onset as a function of time. That is, no extra stimulation of the MOC efferent system was used other than the stimulus tones for the DPOAE. The DPOAE level measured in the ear canal usually showed gradual adaptation over several hundred milliseconds, with the final level often 20 dB lower than the initial level. This adaptation was measured for a variety of stimulus conditions. Specifically, Maison and Liberman (2000) held the level of f_1 constant and varied the level of f_2 in 1-dB steps. For each of these stimulus combinations, the amount of DPOAE adaptation was plotted as a function of varying f_2 level. The authors observed that, as f_2 changed, the amount of adaptation of the DPOAE grew and then suddenly jumped from a negative to a positive value on changing f_2 by as little as 1 dB. The difference in adaptation between these two data points was quantified as the MOC reflex (MOCR). Guinea pigs were divided into three groups with weak, intermediate, and strong MOCR and then exposed to hazardous levels of noise. Interestingly, the animals with the strongest MOCR showed the least change in compound action potential thresholds and those with the weak MOCR showed the most. These data provide very strong evidence in support of a role of the MOC efferent system in protection against noise damage. Although these results created tre-

mendous excitement, attempts to measure similar DPOAE adaptation in humans have not been as successful. Although adaptation is evident, the magnitude of the effect is significantly smaller with the total effect not exceeding a few dB (Kim, Dorn, Neely, & Gorga, 2001).

Support for the role of MOC efferent protection of the cochlea from acoustic trauma is not universal. In a thought-provoking review paper, Christopher Kirk and Smith (2003) conclude that "the paucity of high-intensity noise and the near ubiquity of low-level noise in natural environments supports the hypothesis that the MOC system evolved as a mechanism for unmasking biologically significant acoustic stimuli by reducing the response of the cochlea to simultaneous low-level noise. This suggested role enjoys widespread experimental support" (p. 445). However, critics of this line of thinking abound. The major criticism is that animals with and without auditory efferent systems were not compared in this review. The contention here is that such a comparison across species would have clarified the role of the MOC efferent system.

Hearing in Noise

Another proposed function of the MOC efferent pathways is in improving the capacity for detecting and discriminating sounds, particular brief duration (transient) sounds, in the presence of background noise. Again, evidence in support of the role of the MOC efferent system in sound perception in noise comes mostly from animal studies. Two general experimental approaches are taken to evaluate the impact of the efferent system on hearing. With one approach, detection and discrimination of sound frequency or intensity, or speech sounds, is measured before and then after efferent fibers are severed. Deterioration in auditory performance following surgical ablation of efferent fibers is attributed to the loss of the beneficial effects of the MOC pathways.

The other approach involves measurement of auditory performance before, during, and after stimulation of the efferent system with either an acoustic or electrical signal. Of course, the expectation with this experimental method is enhanced auditory discrimination during activation of the

MOC efferent pathways. Stronger MOC-mediated reductions in outer hair cell activity generally are associated with greater improvement in the detection of signals in the presence of background noise (e.g., Micheyl, Perox, & Collet, 1997). A phenomenon referred to as "unmasking" or "antimasking" is invoked to explain the improvement of sound detection and discrimination in noise. That is, noise acts as a masker in covering up important acoustic cues needed for detection and accurate discrimination of sounds. The MOC efferent pathways reduce the masking effects of noise. With the MOC efferent mediated "release from masking," there is a corresponding improvement in perception of the sounds. The benefits of unmasking apply to performance of relatively simple auditory tasks, such as, discrimination of frequency or intensity and also more complex tasks, such as speech perception.

Results of experiments where MOC suppression strength is compared with speech perception in noise directly in humans are mixed. Speech perception in noise was improved in the presence of contralateral noise in 10 normal-hearing children and the degree of this improvement was correlated with the magnitude of inhibition of TEOAE by contralateral noise (Kumar & Vanaja, 2004). However, in a similar study conducted on adult human subjects, no correlation was found between speech intelligibility in noise and contralateral inhibition of DPOAE (Wagner, Frey, Heppelmann, Plontke, & Zenner, 2007). Thus, the role of the MOC efferent system remains poorly defined and open to debate and further exploration.

Role in Attention

Several research groups have studied the role of the MOC efferent system in attention. The general results seem to indicate that the MOC efferent system is activated during tasks demanding visual attention. Two recent studies in this domain have presented dramatic results worth special mention. First, in human subjects, the contralateral inhibition of TEOAE was shown to be significantly enhanced when subjects were asked to attend to the ipsilateral OAE evoking stimuli, or the contralateral stimuli used to stimulate the MOC efferent system (Harkrider & Bowers, 2009). In another study, behaving chinchilla were implanted with round-window electrodes and compound action potential and cochlear microphonics were measured while the animals performed a visual discrimination task (Delano, Elgueda, Hamame, & Robles, 2007). Accuracy of performance on the task was highly correlated with the reduction of the compound action potential and increase in cochlear microphonics—a signature of MOC efferent system activation.

Role in Auditory Training

A very exciting link between MOC-modulation of OAEs and auditory training was reported recently (de Boer & Thornton, 2008). Adult subjects were trained on a signal-in-noise discrimination task over a 5-day period. Some subjects showed improvement in discrimination after 5 days of training, but others did not. The subjects who showed improvement had significantly lower MOC-induced inhibition of OAEs prior to training. Even more interestingly, the magnitude of MOC inhibition increased in these subjects as they got better at the discrimination task (de Boer & Thornton, 2008). These results not only indicate a role of the MOC efferents in signal detection or discrimination in noise but also demonstrate plasticity in the MOC efferent system.

Modes and Protocols for OAE Suppression

Introduction

Three important questions concerning clinical measurement of OAE inhibition by the MOC efferents need to be answered.

1. What is the optimal OAE type for clinical purposes?
2. Should the (efferent) elicitor be presented ipsilaterally, contralaterally, or bilaterally?
3. What signal type (e.g., tone, noise, etc.) is the best elicitor?

What Is the Optimal OAE Type for Clinical Purposes? Most clinical reports of efferent inhibition of OAEs have used either TEOAE or DPOAE.

However, considerable scientific effort has also been devoted to the study of SFOAE modulation by the MOC efferents. A few studies also have explored the effects of MOC activation on SOAEs. The stimuli used to evoke OAEs undoubtedly add to the activation of the MOC efferents during the measurement process. Moreover, the OAE stimuli along with the efferent elicitor can activate the middle ear muscle reflex (MEMR). This has led to the suggestion that it may be beneficial to use SFOAEs in clinical measures as only one stimulus tone of low or moderate levels is necessary for evoking the OAE (Guinan, 2006). This argument could then be extended to suggest that SOAEs would be the ideal OAE type for clinical measurements of efferent inhibition as they can be recorded without the use of stimulus tones. However, not even all normal-hearing individuals have SOAEs. Moreover neither SFOAEs nor SOAEs are in routine clinical use at this time. Therefore, most if not all of the sparse clinical literature on OAE inhibition by the MOC efferents is built around DPOAEs and TEOAEs.

Should the Efferent Elicitor Be Presented Ipsilaterally, Contralaterally, or Bilaterally? The goal of a clinical tool for measuring efferent inhibition of OAEs is to evoke the MOC efferents and the MOC efferents only. However, stimulating the auditory system for OAE measurement and efferent activation can have some unintended side effects. Any acoustic stimulus that evokes the efferent response can also potentially evoke the MEMR. Additionally, when presented ipsilaterally, the elicitor could cause mechanical interference on the basilar membrane with the OAE stimulus. Thus, for ipsilateral and bilateral presentations of the elicitor, the measurement potentially could reflect the influence of the MOC efferents, the MEMR, and mechanical suppression (akin to masking). Mechanical suppression works on a very fast time scale, faster than either the MEMR or the MOC efferents. However, the time scales of the effects due to MEMR and MOC efferents are similar (Goodman & Keefe, 2006), making them difficult to distinguish. Although using a contralateral elicitor eliminates the influence of mechanical suppression, the problem of potentially activating the MEMR remains. Using contralateral elicitors also makes it easy for the clinician to present this signal from any sound source (such as an audiometer) while measuring OAEs through a dedicated device.

The advantages of using contralateral elicitation of the efferent effect are counterbalanced by the limited size of the effect. It stands to reason that the largest MOC effect is observed when all or most of the MOC efferent neurons are activated. This is achieved by using a bilateral elicitor where both the crossed and uncrossed MOC fibers are activated. Although the bilateral effect can be as large as 10 dB, the contralateral effect often is under 5 dB (Berlin, Hood, Hurley, Wen, & Kemp, 1995).

What Signal Type (Tone, Noise, etc.) Is the Best Elicitor? Both tonal elicitors and noise band elicitors of various bandwidths have been used to examine the effects of the MOC efferents on OAEs. SFOAEs as well as TEOAEs are most effectively inhibited by broadband noise signals. The size of the effect increases with increasing bandwidth of the elicitor noise (Lilaonitkul & Guinan, 2009a). The increase in the size of the inhibition effect with increasing elicitor bandwidth suggests that the efferent effect is integrative along the cochlear length. That is, as the elicitor bandwidth is increased, the amount of overall energy in the elicitor is held constant. Thus, the amount of energy at any particular frequency (or cochlear region) is reduced with increasing elicitor bandwidth. However, the increase in the cochlear region stimulated by the wider band elicitor compensates for the loss of energy at any particular frequency.

Tonal elicitors also can be used to alter OAEs. When tonal elicitors are used, the greatest effect is seen for elicitors 0.5 to 1 octave below the OAE-evoking stimulus (or probe). This pattern is more complicated when ipsilateral, contralateral, and bilateral elicitors are considered (Lilaonitkul & Guinan, 2009b). This pattern of maximal effect slightly below the probe frequency supports a role of the efferent system in aiding signal detection in noise by providing a release from masking.

Measurements in the Clinic

Although integrated apparatus and procedures for measuring OAE inhibition with clinical OAE devices may be just around the corner, none are available at this time. However, the clinician can measure the contralateral efferent inhibition of OAEs by improvising with standard clinical equipment. Measurement of contralateral suppression, using the simplest technique, requires only an OAE device and a separate device for delivering an elicitor signal to the other ear. The elicitor usually is broadband noise produced and delivered manually by either an audiometer or an aural immittance device. The patient is placed near each of the devices. OAEs are first recorded as usual in a quiet condition. The recordings are replicated, and the findings are saved and/or printed. Then, the OAE recording is performed again (and replicated) while a continuous broadband noise of about 60-dB SPL is presented under manual control to the opposite ear. Analysis involves measurement of OAE amplitudes under each condition (quiet vs. contralateral noise) and calculation of a possible reduction (inhibition) in OAE amplitudes (in dB) under the noise condition. Replication of OAE findings under each condition is advisable. To verify results, and the presence or absence of inhibition, the entire process should be repeated two to three times. Alternating the quiet versus noise measurement multiple times is referred to as "interleaving test conditions" (e.g., Hood, 2007). This process of averaging the measurement over several trials helps reduce the biasing effects of a single or a few measurements that may be outliers. This is particularly important as the variability in these results between trials is not fully documented at this time.

Measurements in the Future. We fully expect that some diagnostic OAE devices will include hardware and software solutions to enable the measurement of efferent inhibition in the near future. Imagine in a clinical device an extra transducer (i.e., a second OAE probe assembly) that can be used to deliver the contralateral elicitor and also record OAEs. With specially designed software, these instruments will most likely also allow the measurement of ipsilateral and bilateral inhibition.

Measurement of ipsilateral or binaural inhibition requires some explanation. It would be reasonable to question how OAEs can be recorded with an ipsilateral suppressor without simply masking/suppressing the OAE stimulus as well as the OAE response during the recording. Presentation of a constant noise signal at 50- or 60-dB SPL, while simultaneously attempting to record OAEs from that same ear, will certainly not be fruitful. In the research laboratory, this problem is overcome by bursts of the elicitor noise. The OAE is measurable in the gaps between these noise bursts as the time course of efferent inhibition of OAEs is well studied. Efferent effects are observable in cochlear mechanics as well as OAEs within a few milliseconds (~5 to 10 ms) after the onset of the elicitor. These changes are known as the fast effects. This fast effect has a decay time between 10 and 100 ms. Thus, the OAE level does not return to baseline immediately after the elicitor is turned off. This delay in the offset of the efferent effect allows its measurement in the gaps between the noise bursts. Using a slightly different strategy, clicks can be presented after the offset of the elicitor noise. The TEOAEs recorded with these clicks are compared to those recorded without the preceding noise burst. When using this paradigm, the efferent effect is not evident if the clicks are presented 100 ms or more after the offset of the noise (Berlin et al., 1995).

The efferent system affects cochlear mechanics (and presumably OAEs) over other time scales as well. A slower effect, referred to by that name, is seen over 10 to 100 sec (Sridhar, Liberman, Brown, & Sewell, 1995). Even slower changes, over a matter of 2 to 3 minutes, also have been documented (Larsen & Liberman, 2009). A recently reported clinical investigation demonstrated TEOAE inhibition over even longer durations—15 minutes (van Zyl, Swanepoel, & Hall, 2009). TEOAEs recorded using the linear paradigm with clicks at 60-dB peSPL continued to be inhibited by a contralateral white noise presented at 45-dB SL. A TEOAE amplitude decrease of approximately 2 dB was observed (Figure 9–2). TEOAE levels returned to baseline within a minute of elicitor offset and showed an increase (bounce) after 2 to 3 minutes. This discussion makes it clear that the time course of the MOC effects is varied, complex, and not yet fully documented in humans. However, for clinical application of OAE inhibition, decisions must be made about the optimal duration of the elicitor signal. That issue is tackled next.

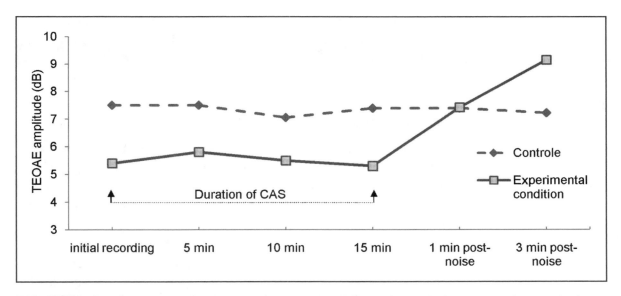

FIGURE 9–2. Inhibition effect of prolonged contralateral sound stimulation on transient evoked otoacoustic emissions. Courtesy of A Van Zyl.

Measurement Variables in OAE Suppression

Several measurement variables in OAE inhibition were just noted, including the duration of the elicitor signal, and the interelicitor interval in case of ipsilateral and bilateral measurements. A variety of other measurement variables influence the degree of OAE inhibition, as summarized in Table 9–1. A few general statements can be confidently made about stimulus and elicitor parameters. Invariably, there is an inverse relation between OAE stimulus intensity and OAE inhibition. That is, in any given inhibition condition, the most robust reduction in OAE amplitude (greatest efferent effect and magnitude of inhibition) occurs at lower stimulus intensity levels and for lower OAE amplitudes. This is most likely related to the increased gain provided by the cochlear amplifier to lower level inputs. As noted by Guinan (2006), "MOC effects are larger as OAE level decreases (presumably because amplifier gain is larger at low sound levels) . . ." (p. 12).

The amount of inhibition increases directly as the level of the elicitor increases. However, very high elicitor levels should be avoided. The use of such high levels increases the likelihood of acoustic crossover, even in contralateral measurements.

If the elicitor crosses over to the ear in which the OAE measurement is being made, the elicitor will contaminate the recording. The likelihood of contamination by the stapedial acoustic reflex also increases with increasing elicitor level. Contraction of the stapedius muscle reduces the amount of OAE stimulus allowed into the cochlea. The amount of OAE signal allowed out of the cochlea is also reduced. Thus, the stapedius reflex will invalidate OAE inhibition measurements, artificially increasing the apparent efferent inhibition effect.

The effect of (OAE) stimulus frequency on efferent inhibition is not well studied. Given that OAEs have a finite bandwidth over which they are routinely recorded, this variance may not be dramatic. OAEs are difficult to record at frequencies much below 1 kHz and are rarely recorded above 8 kHz in the clinic. Within this frequency range, however, the magnitude of inhibition varies as a function of the stimulus frequency or frequency region for TEOAEs and DPOAEs (e.g., Gorga et al., 2008). In preliminary experiments in our laboratories, the maximal inhibition is seen for DPOAEs between 1.5 and 4 kHz. TEOAEs between 1 and 2 kHz also appear to be most sensitive to efferent inhibition.

Although all our discussion up to this point has been related to the DPOAE at $2f_1$-f_2, recent evidence suggests that DPOAEs of other orders

Table 9–1. Measurement (Test Protocol) Factors Influencing Inhibition of Otoacoustic Emissions

Factor	Influence
Type of OAE	
TEOAE	
Stimulus Type	Click stimuli are as commonly employed for recording TEOAEs
Polarity	For measurement of inhibition, click stimuli of the same polarity (a so-called "linear stimulus") are preferred to alternating polarity click stimuli.
Intensity	Inhibition effects are greatest when TEOAEs are evoked by click stimuli between 55 and 60 dB peSPL, rather than the typical higher level clicks around 80 dB peSPL.
DPOAE	
Stimulus Type	Typical measurements have been made at isolated frequencies or at a few frequencies per octave. More recent reports using high density or fine structure recordings reveal more complex effects of efferent activation. Inhibition, enhancement, as well as frequency shifts have been observed (see Figure 9–3).
Intensity	Inhibition is inversely related to intensity level of the OAE stimulus. Greater inhibition occurs for modest stimulus intensity levels (<60 to 65 dB SPL).
Poststimulus Time	
Elicitor ("Suppressor")	
Mode of stimulation	Binaural presentation of inhibition signal (elicitor) invariably produces the greatest amount of OAE (up to 10 dB) inhibition. Ipsilateral inhibition is next in effectiveness. Presentation of an elicitor signal to the ear contralateral to the ear in which OAEs are recorded is the least effective mode with typical inhibition of only 1 to 4 dB.
Type	Noise is more effective as an elicitor than pure tones. Broadband noise is generally more effective as an elicitor than narrowband noise, particularly with TEOAE inhibition paradigms. Low frequency noise band elicitors are least effective. For DPOAEs, the effectiveness of narrow versus broadband noise varies as a function of DPOAE stimulus frequency. Among pure-tone signals, contralateral inhibition is greater for amplitude and/or frequency-modulated tones. Click signals have also been applied in the study of ipsilateral inhibition with a "double click" inhibition paradigm.
Frequency properties, especially for TEOAEs	Inhibition is greater for broadband versus narrowband or tonal elicitors.
Duration	Increased duration of the suppressor signal up to 400 ms produces progressively greater amount of inhibition. Continuous presentation of an elicitor, as in the simple technique often utilized for clinical measurement of contralateral inhibition of OAEs clinically, is effective.
Elicitor versus elicitor	Most inhibition is observed within 5 to 10 ms of elicitor onset and usually 5 to 20 ms after onset.
OAE time interval	With ipsilateral or binaural inhibition, an interval of at least 10 ms is recommended between the offset of the elicitor and presentation of the OAE stimulus in the paradigm similar to a "forward masking" paradigm.

Table 9–1. *continued*

Factor	Influence
Intensity	The amount of inhibition invariably increases with increased elicitor intensity up to about 60- to 75-dB SPL. Also, inhibition is observed for a wider range of OAE frequencies as elicitor intensity increases. Higher elicitor intensity levels (>75 dB) are not advisable as they are associated with greater likelihood of unintended and unwanted contamination of findings by either the acoustic reflex or, for contralateral inhibition paradigms, crossover via bone conduction of the elicitor to the ear used to evoke OAEs.

may be more sensitive to efferent effects. Witte-kindt, Gaese, and Kössl (2009) investigated contralateral inhibition of the quadratic f_2-f_1 distortion product. The magnitude of contralateral inhibition of the f_2-f_1 DP was remarkably large (up to 10 dB), in comparison to the insignificant inhibition of the $2f_1$-f_2 DP, under comparable elicitors.

Suppressor Variables

Any acoustic signal can be used as an elicitor, including broadband noise, narrowband noise, clicks, and pure tones. The effectiveness of elicitors generally increases with bandwidth, as discussed above. Wider elicitor frequency bands correspond with activation of more efferent fibers and, consequently, greater inhibition.

Elicitor *duration* also is a critical variable. For durations less than approximately 400 ms, elicitor effectiveness diminishes proportionally as duration decreases, with the efferent effect disappearing for elicitors shorter than about 100 ms. Longer duration elicitors impose a sustained inhibition of OAEs. Giraud, Coller, & Chery-Croze (1997) showed consistent inhibition of TEOAEs for up to 4 minutes and Moulin and Carrier (1998) documented inhibition of DPOAEs for 20 minutes. The effects of longer elicitors appear to last for a longer period of time after elicitor offset.

Analysis Strategies

Analysis strategies for quantifying the effect of the MOC efferents on OAEs are still being actively researched. Thus, there are no universally accepted clinical analysis schemes at this time. The basic measure employed by most clinicians and researchers is that of inhibition. In human subjects, OAE inhibition appears maximal in the time frame of 10 to 20 ms after the presentation of the elicitor signal. The time frame of analysis is more important in the case of TEOAEs than DPOAEs. In the case of DPOAEs, it is becoming apparent that the frequency of measurement is a critical variable. This emerging but important issue is discussed later under the heading of "Emerging Complexities."

Although the majority of scientific work on inhibition has used changes in OAE level as the metric for efferent effects, other measures are beginning to emerge. Changes to SFOAE phase and TEOAE delay with the activation of MOC efferents was recently reported (Francis & Guinan, 2010). The phase of DPOAE components also appears to be affected by the MOC efferents (Deeter et al., 2009). Finally, the importance of replicating results when making measurements of the efferent effects on OAEs cannot be emphasized enough. Several measurements without the elicitor should be made and the variability between these measures should serve as the baseline. To be considered legitimate, any effect of the MOC efferents must be larger than this inherent fluctuation in OAE level. Optimally, efferent effects on OAEs should be statistically confirmed.

Subject Factors Influencing Suppression of OAEs

Subject factors to be considered in the clinical measurement and analysis of OAE inhibition are summarized in Table 9–2. Acoustic reflex activity certainly can exert a major influence on

Table 9–2. Published Papers Describing the Influence of Various Subject Factors on OAE Inhibition

Factor	Authors (Year)	Findings
Subject task	de Boer & Thornton (2007)	Active attention to tone pips embedded within a click train was associated with reduced inhibition in comparison to inactive conditions. The authors suggest the possibility of a "top-down" control of MOCB activity associated with selective attention.
Attention	Froelich et al. (1993) Michie et al (1996)	Active attention to tone pips embedded within a click train was associated with reduced inhibition in comparison to inactive conditions. The authors suggest the possibility of a "top-down" control of MOCB activity associated with selective attention.
	Zhong et al.(2001)	No evidence of an effect of selective attention on contralateral inhibition of DPOAEs.
Age: developing	Ryan & Piron (1994) Morlet et al. (1999, 2004) Hildesheimer et al. (1999) Goforth et al. (1997) Abdala et al. (1999) Chabert et al. (2006) Gkoritsa et al. (2006, 2007) Yilmaz et al. (2007)	Likelihood of inhibition increases with developmental age, from pre-term, to Infancy, through term birth and older children. Amount of inhibition also increases with for both TEOAEs and DPOAEs.
	Durante & Carvallo (2002) Durante & Carvallo (2008)	Statistically significant decrease in TEOAE inhibition for a group of neonates at risk for hearing loss versus a control group. Authors conclude that medial olivocochlear bundle is reduced, perhaps by neurologic insult.
Age: advancing	Castor et al. (1994) Mazelova et al. (2003) Kim et al. (2002) Parthasarathy (2001)	Decrease in the amount of inhibition with increasing age in older adults. Decrease in inhibition if greater for binaural versus ipsilateral or contralateral elicitor signals. Age-related decrease in inhibition reported for TEOAEs and DPOAEs
	Mukari & Mamat (2008)	Lower contralateral inhibition of DPOAEs in older versus young adults, associated with poorer speech perception in noise.
Gender		No peer-reviewed articles confirming conclusively a gender effect but data suggest the possibility of greater inhibition for females than males.
Ear	Atcherson et al. (2008)	Apparent greater DPOAE inhibition for the right versus left ear, with an interaction between test ear and f_2 frequency
Sensorineural		Reduced OAE levels in cases of sensorineural hearing loss make it difficult to measure inhibition. When inhibition is observed, any differences in magnitude of inhibition from a control population cannot be separated from the general effects of the sensorineural hearing loss.

measurement of OAE inhibition. With elicitors levels sufficient to activate the MEMR, there is the possibility of a reduction in the effective stimulus level and also a reduction in the amplitude of OAEs as they propagate outward through the middle ear system. Because MEMR is consensual, that is, the reflex occurs bilaterally with adequate stimulation of one ear, MEMR activity would be expected to affect ipsilateral, contralateral, and binaural measurement of OAE inhibition. Concerns about the influence of the MEMR on OAE inhibition were regularly expressed in early reviews of the topic (e.g., Guinan, Backus, Lilaonitkul, & Aharonson,, 2003). Now, however, investigators invariably limit elicitor levels (e.g., <70-dB SPL) to minimize the likelihood of eliciting the MEMR. Interestingly, in a recent study Sun (2008) found that the difference in OAE inhibition measurements with and without MEMR activity was less than anticipat-ed. However, the results were complicated in that changes in DPOAE phase were sometimes observed with or without any changes in DPOAE level.

Age consistently affects the inhibition of OAEs. The magnitude of inhibition increases with development from preterm infants through term birth and older infants. At the other end of the age spectrum, it often becomes difficult to isolate the effects of age on efferent inhibition from those on general hearing. That is, when reduced OAE inhibition is observed in older adults, it is difficult to rule out that the result is not an outcome of reduced OAE levels in this population even without efferent activation. However, when general hearing is controlled for, there appears to be a decreased in OAE inhibition with increasing age (Kim, Frisina, & Frisina, 2002). In general, age-related findings are equivalent for TEOAE and DPOAEs. The influence of gender on OAE inhibition has not been evaluated systematically, although there is some suggestion of greater TEOAE inhibition for females versus males (Brashears, Morlet, Berlin, & Hood, 2003). It's important to note that findings reported in most published papers are limited to TEOAEs and to a contralateral elicitor paradigm only.

CLINICAL APPLICATIONS AND FINDINGS

Introduction

The clinical potential of measuring inhibition of OAEs in humans was expressed in the very first publications in this area (e.g., Berlin et al., 1993). This enthusiasm has only increased with time.

Di Girolamo et al. (2007) recently reiterated that inhibition of OAEs "seems to be the only objective and noninvasive method for evaluation of the functional integrity of the medial efferent system" (p. 419). Despite the clinical attraction and appeal of the technique, and clear appreciation of the potential diagnostic value, literature on clinical application of OAE inhibition remains rather limited. Perhaps the single most important constraint for clinical research on OAE inhibition is the lack of clinical instrumentation for adequate presentation of contralateral and ipsilateral elicitor signals and the lack of clinically feasible algorithms for statistical analysis of abnormal versus normal amounts of inhibition.

Children

Published studies of OAE inhibition in different pediatric patient populations are summarized in Table 9–3. For the most part, findings for any given disorder are available from only one or two studies. Some of the papers, however, report promising results with potential clinical application. For example, Attias, Raveh, Ben-Naftali, Zarchi, & Gothlef, (2008) described greater inhibition of TEOAEs for a group of patients with Williams syndrome versus a carefully matched control group. The subjects with Williams syndrome also were less likely to have acoustic reflexes. As the authors point out, "Hyperexcitability of the MOC efferent system coupled with absence of acoustic reflexes may contribute to the hyperacusis in WS . . . " (Attias et al., 2008, p. 193). Patients with auditory neuropathy/dys-synchrony, irrespective of age, showed significantly reduced TEOAE inhibition for ipsilateral, contralateral, and bilateral elicitors (Hood, Berlin, Bordelon, & Rose, 2003).

Adults

Although there are not as many papers describing inhibition of OAEs in adults as in children, the literature includes a wide variety of disorders. The following review focuses on diseases and disorders studied most often. The rather modest literature on OAE suppression in other disorders is summarized in Table 9–3.

Table 9–3. Published Papers Describing OAE Inhibition in Miscellaneous Disorders and Diseases

Disease/Disorder	Study Authors (Date)	Findings
Children		
Diabetes	Ugur et al. (2009)	Reduced magnitude of TEOAE inhibition for children with type I diabetes versus a control group, suggesting the possibility of "early central manifestation of diabetic neuropathy" (p. 555).
Fibromyalgia	Gunduz et al. (2008)	A group of 24 females with fibromyalgia showed no evidence of TEOAE contralateral inhibition in contrast to a control group that did have significant inhibition.
Chemotherapy	Riga et al. (2007)	A reduction in contralateral broadband noise inhibition of DPOAEs in children treated for acute lymphoblastic leukemia with a neurotoxic chemotherapy protocol was interpreted as evidence of toxic effect on the MOCB. Inhibition returned over a 3-year period after chemotherapy.
Williams syndrome	Attias et al. (2008)	Greater TEOAE suppression for persons with Williams syndrome (age 6 to 26 years) than age-matched normally developing subjects.
Learning disabilities	Angeli et al. (2008)	Smaller amount of inhibition for right ear of children with poor versus good academic performance.
Auditory processing disorders (APD)	Burguetti & Carvallo (2008)	No difference in contralateral inhibition n of TEOAEs between children with the diagnosis of APD versus a control group.
	Yalçinkaya et al. (2009)	Significantly lower amounts of TEOAE inhibition for children with "auditory listening problems" including auditory processing problems in background noise than for a control group.
Adults		
Learning disabilities	Garinis et al. (2008)	Greater contralateral TEOAE inhibition for the left ear in a group of Adult subjects with LD, whereas a normal control group showed more inhibition for the right ear. The LD group showed enhancement in the right ear.
Multiple sclerosis	Coelho et al. (2007)	Diminished MOC function (less than normal TEOAE inhibition) was found in patients with MS, including 90% of those with brainstem lesions by MRI and 55% of those without brainstem lesions. The authors suggest contralateral inhibition as a sensitive measure of subtle brainstem dysfunction in MS. There was no control (non-MS) group.
	Lisowska et al. (2008)	Reduced TEOAE inhibition n patients with multiple sclerosis versus a control group.
Cerebellopontine angle tumor	Lisowska et al. (2008)	Reduced TEOAE inhibition patients with unilateral acoustic tumors angle tumor versus a control group.
Migraine	Bolay et al. (2008)	No inhibition of TEOAEs in patients with migraine with and without "aura" versus significant decrease in TEOAE amplitude in control subjects. DPOAE inhibition was found for only one test frequency.

The application of OAEs in the assessment, and even management, of tinnitus was reviewed in Chapter 8. Efferent auditory system dysfunction is sometimes noted in speculation about tinnitus mechanisms. Riga, Papadas, Werner, and Dalchow, (2007) investigated this possibility in a randomized controlled study involving contralateral noise suppression of DPOAEs. Contralateral stimulation with white noise produced significantly less reduction of DPOAE amplitude for subjects with tinnitus than the normal subject group. The authors conclude that: "Patients with normal hearing acuity who have acute tinnitus seem to have a less effective functioning of the cochlear efferent system . . . " (p. 185). Similarly, Favero, Sanchez, Bento, and Nascimento (2006) found "a correlation between diminished effectiveness of the medial olivocochlear bundle and the presence of tinnitus" (p. 223). Other investigators also report evidence implying a role of efferent auditory system dysfunction in tinnitus (e.g., Attias, Zwecker-Lazar, Nageris, Keren, & Groswasser, 2005; Hesse, Andres, Schaaf, & Laubert, 2005; Hesse et al., 2008; Rita & de Azevedo, 2005).

EMERGING COMPLEXITIES

Even the earliest reports of OAE modulation by the MOC efferents showed complexities that were not fully understood. Siegel and Kim (1982), reported reduction, enhancement or no change in OAE levels on efferent activation. These authors argued that the enhancement seen occasionally could be an outcome of selective inhibition of one of the components contributing to the ear canal DPOAE (recall the discussion about multiple DPOAE components in Chapter 3). The full extent of this complexity is only now becoming evident. In the case of DPOAEs in humans, the vast majority of reports have demonstrated a small reduction in DPOAE level (1–3 dB) by the MOC efferents (Abdala, Mishra, & Williams, 2009; Bassim, Miller, Buss, & Smith., 2003). However, much larger enhancements in DPOAE level have sometimes been observed (Maison & Liberman,

2000; Muller, Janssen, Heppelmann, & Wagner, 2005), most often at frequencies where the baseline DPOAE level (without efferent activation) shows a valley or a minimum (Zhang, Boettcher, & Sun, 2007).

The bipolar effect of efferent stimulation on DPOAE level can be observed in the time domain as well. When the MOC efferents are stimulated while monitoring DPOAE level at a specific frequency, a reduction in DPOAE level is observed on activation of the MOC efferents when the monitored frequency is a known level maximum. In contrast, an enhancement is observed when the monitored frequency is a known level minimum (Deeter et al., 2009; Sun, 2008). Because the inhibitory effect of the MOC efferents on DPOAE level is often miniscule, the large enhancement has been an enticing opportunity to exploit clinically. Muller et al. (2005) recommended using the difference between the largest enhancement and the largest reduction as the metric for the efferent effect. This was followed by a recommendation to make the measurement at a DPOAE level minimum to maximize the chances of observing an enhancement (Muller et al., 2005; Wagner, Heppelmann, Muller, Janssen, & Zenner, 2007). However, the variability in the effect of the MOC efferents on DPOAE level is also significantly greater at minima as opposed to maxima. This has tempered the enthusiasm of using measures at level minima (Abdala et al., 2009; Purcell, Butler, Saunders, & Allen, 2008; Zhang et al., 2007).

To make matters even more complicated, DPOAE fine structure appears to shift toward higher frequencies upon efferent activation (Abdala et al., 2009; Deeter et al., 2009; Purcell et al., 2008). As shown in Figure 9–3, a frequency shift in various OAE types can be observed on efferent activation. The effect is perhaps best visualized in case of the SOAE. However, both SFOAE and DPOAE fine structure show clear shifts toward higher frequencies. This shift can give the illusion of an enhancement. Recent reports have acknowledged the greater variability at DPOAE level minima and are recommending clinical measurements at or around level maxima (Abdala et al., 2009; Deeter et al., 2009). Deeter et al. (2009) have even suggested the possibility of using the frequency shift as a metric of the MOC effect.

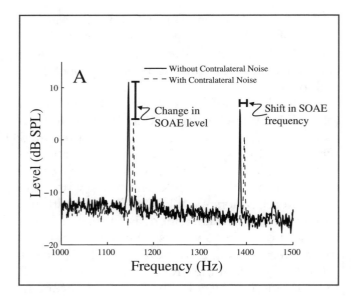

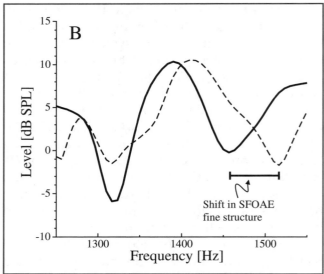

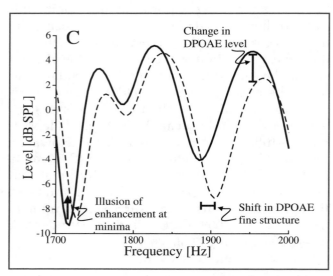

FIGURE 9–3. Examples of frequency shift of SOAE (**A**), SFOAE (**B**), and DPOAE (**C**) fine structure on efferent activation. The baseline measure is always presented using the solid line. The dashed line represents the measurement made with a contralateral broadband noise. These measurements were made from three different normal-hearing, young adult ears. The SOAE recording was made without external stimulation (except the efferent elicitor) over a period of 35 seconds. The SFOAE measurement was made using a probe at 40-dB SPL and a reference tone at 60-dB SPL. The DPOAE measurement was made with stimulus levels of 55 (L_1) and 40 (L_2) dB SPL. The contralateral elicitor was always a broadband noise between 100 and 10,000 Hz presented at 60-dB SPL.

CONCLUSIONS

The clinical evaluation of the MOC efferent system via OAEs holds much promise. The reader is perhaps left with a sense of disappointment that more mature clinical technologies and techniques are not available for this purpose at this time. However, we sense a tremendous urgency in the scientific and clinical communities in this regard. Several laboratories and groups are devoting significant effort in inventing a robust, clean, and efficient measure of efferent modulation of OAEs. A breakthrough is likely to be around the corner.

III 10 III

New Directions in Research and Clinical Application

BEYOND CURRENT PRACTICE

Clinical application of OAEs has reached a stable plateau three decades after their initial discovery. This plateau was reached after the initial burst of research activity that helped define the basic characteristics of OAEs, stimulus parameters permitting consistent measurement, and the physiologic vulnerabilities affecting OAEs. Although this basic information was essential for OAEs to become a mainstream clinical tool, it by no means marks the end of innovation in administration and interpretation of tests involving OAEs. In fact, current OAE research points to a much more incisive and important clinical future for OAEs. It is the future promise of OAEs that we attempt to capture in this chapter. The reader is forewarned, though, that the body of work involving OAEs is large and constantly expanding. What we present here is essentially the tip of the proverbial iceberg. The ultimate success in clinical implementation of any one of the efforts outlined below is unknown. However, each holds significant promise either in isolation or in combination with other methods. We begin by describing innovative methods of recording and analysis of OAEs, and end with a discussion about convergence of instrumentation and techniques.

Variations in TEOAE and SFOAE Measurement

The most commonly employed method for recording click or transient evoked OAEs is referred to as the nonlinear method. The method is so named because the portion of the response that grows nonlinearly with a change in the stimulus level is retained while all linear components of the response are cancelled due to averaging. The linear components are eliminated by using a stimulus train that consists of three clicks of equal magnitude and the same polarity followed by a click that is three times the magnitude of the first three clicks but of the opposite polarity. The polarity is reversed for all of the next train of four clicks, while their magnitude relationship remains unchanged. Once the responses to the four clicks in one stimulus train are added, all linear components are cancelled because the sum of the responses to the first three clicks is equal to that to the fourth click, but of opposite polarity. It is important to note that the clicks are presented 20 ms apart, resulting in a response window of that length. However, the first 2.5 ms of the response window is neglected and the data in this time window is not included in the analysis of the response to eliminate any contamination from stimulus artifact.

Alternative methods for recording TEOAEs have been suggested to capture the linear portion of the response that is lost in the signal averaging when the traditional nonlinear method is employed. In the linear method, the responses to clicks, all of the same polarity and of a given magnitude, are averaged. As responses to clicks of opposing polarities are not averaged, none of the response is cancelled and the "full" response is retained. However, one consequence of not cancelling the linear portions of the response is the retention of a much larger lingering stimulus artifact, often requiring elimination of the first 5 or 6 ms of the response waveform before analysis. Using the linear method appears to reduce the noise floor and to increase the signal-to-noise ratio without significantly altering test performance (Hatzopoulos, Petrucelli, Pelosi, & Martini, 1999).

Another approach for eliminating TEOAE contamination from stimulus artifact is the "double-evoked" method (Keefe, 1998). This method takes advantage of two speakers that are often present in systems used to measure DPOAEs, as well as TEOAEs. Three clicks or chirps are presented in sequence with the two speakers. The first click is presented through the first speaker (p_1), the second through the second speaker (p_2), and the third click is presented simultaneously through both speakers (p_{12}). Because all clicks are presented at the exact same level, the measurement system, and especially the speakers, is not "pushed," thereby preventing system distortion. Importantly, the third click is exactly twice the amplitude of the first two as the click is played through both speakers simultaneously. From the set of recordings to the three clicks, the TEOAE is computed simply as $p_{12} - (p_1+p_2)$. This method has now been used in several research articles (Keefe & Ling, 1998), including one where the characteristics of TEOAEs at frequencies up to 16 kHz were reported (Goodman, Fitzpatrick, Ellison, Jesteadt, & Keefe, 2009). Other relatively recent variations in the measurements of OAEs using click or tone-burst stimuli have included the use of amplitude-modulated tone bursts (Goodman, Withnell, De Boer, Lilly, & Nuttall, 2004) and swept-tones (Bennett & Ozdamar, 2010a, 2010b). The motivation for implementation of these novel methods is often to either record OAEs at frequencies higher than 8 kHz or to extract accurate measures of OAE latency.

The basic premise of scaling, described above as "double evoked," has been used for recording SFOAEs (Kemp, Ryan, & Bray, 1990) as well. Responses at each frequency of interest are recorded using a stimulus at the level of interest. The initial stimulus typically is referred to as the probe. Another recording is then made at the exact same frequency but using a stimulus level that is 6 dB higher (twice in amplitude). The second, higher, stimulus is called the reference. It is assumed that the level of the emission does not change between the probe and reference conditions. Under such circumstances, the emission can be extracted by dividing the reference waveform by two (scaling down) and subtracting it from the probe waveform. This procedure eliminates the stimulus as the scaled down reference is exactly equal to the probe, but exposes the emission, which remains unaffected by the scaling. Whether the 6-dB difference between the probe and the reference is sufficient for capturing the entire emission is still being explored (Siegel, 2005).

Measurements at High Frequencies

Just as TEOAEs have been measured at frequencies as high as 16 kHz (Goodman et al., 2009), so have DPOAEs (Dreisbach & Siegel, 2001, 2005). DPOAEs recorded at these high frequencies appear to be just as reliable as those measured at frequencies in the conventional range (Dreisbach, Long, & Lees, 2006) giving us hope that high frequency clinical measurements may become available with appropriate equipment and calibration techniques.

Maximum Length Sequence (MLS) Technique in OAE Measurement

One of the primary goals during OAE recordings is to achieve the best signal-to-noise ratio possible. The solution is straightforward—averaging the responses to a greater number of stimulus units results in a better signal-to-noise ratio. The improvement in the results follows a very well defined pattern, where every doubling of the averaged responses results in an improvement of the signal-to-noise ratio by 3 dB. It quickly becomes evident that improvements in signal-to-noise ratio come at the expense of test duration. Test time is a precious commodity for many, if not all, patient populations. One method for dramatically improving the signal-to-noise ratio without sacrificing test duration is to use maximum length sequence (MLS) stimulation. MLS stimulation consists of trains of clicks and silences with properties that allow a much shorter interclick interval. Although the stimuli and responses are highly overlapping in the raw waveform, the response to a single click can be extracted using de-convolution algorithms. Successful TEOAE recordings using MLS stimuli were first reported almost two decades ago (Thornton, 1993). The MLS technique for recording TEOAEs has been used successfully in neonates as well (de Boer, Brennan, Lineton, Stevens, & Thornton, 2007).

Simultaneous Stimulus Presentation in DPOAE Recordings

Just as MLS stimuli can be used to reduce TEOAE test duration, reduction of test duration for DPOAE measurement using novel stimulation paradigms has also been attempted. One such attempt is the use of simultaneous rather than sequential tone pairs in DPOAE measurement. With the typical DPOAE test protocol, a pair of tones (say with an f_2 of 1 kHz along with the appropriate f_1) is presented in recording the DPOAEs of interest. Then, the next pair of tones (say with an f_2 of 1.5 kHz) is presented, and the sequence continues. Why not simultaneously present tone pairs to record DPOAEs at more than one frequency at the same time, say using the tone pairs where $f_2 = $ 1, 3, and 8 kHz? It seems reasonable to expect that test duration would be reduced to a third of the original. Using this technique, Kim, Sun, Jung, & Leonard (1997) reported essentially no differences between DPOAE recordings in which one pair of stimulus tones was presented in isolation versus when three pairs were presented simultaneously.

This protocol was subsequently incorporated into a commercial OAE instrument. However, tests using the commercial instrument revealed an unwanted increase in the noise floor for measurements in the simultaneous versus the sequential mode (Schairer, Clukey, & Gould, 2000).

Measurement of Alternative DP Frequencies

As noted often in this book, DPs at the $2f_1$-f_2 frequency are customarily recorded with clinical instrumentation and test protocols. Distortion products, however, also can be found at other, alternative, frequencies. When stimulated with a pair of tones, the healthy cochlea generates a host of distortion products. These distortion products each have a specific arithmetic relationship with the frequencies of the stimulus tones. One family of distortion products has frequencies just lower than those of the stimulus tones. These distortion products can be identified as $2f_1$-f_2, $3f_1$-$2f_2$, $4f_1$-$3f_2$, and so on. In contrast to these distortion products, which are lower in frequency compared to the stimulus tones, another family of distortion products has frequencies just higher than those of the stimulus tones. The higher DPs may be found at frequencies such as $2f_2$-f_1, $3f_2$-$2f_1$, $4f_2$-$3f_1$, and so on. There are yet other DPs, such as the one at f_2-f_1, that have been studied extensively in psychophysics.

Although the complex family of distortion products is generated in the cochlea and many members of this family escape to the ear canal as DPOAE, clinically, we have concentrated almost exclusively on the DPOAE at $2f_1$-f_2. There is good reason for this choice as this DP typically is the most robust and easily recorded. However, the DPOAE at $2f_1$-f_2 also is lower in frequency than the stimulus tones. This poses a practical problem as the possibility of contamination by noise increases with decreasing frequency. Thus, the challenge of recording DPOAE at low frequencies is exacerbated when the DP is lower in frequency than the stimulus tones and embedded within noise.

The relationship is reversed for the DPOAE at $2f_2$-f_1. That is, the DP defined by the expression "two times the higher test frequency (f_2) minus the lower frequency (f_1)" is itself higher in frequency than the two stimulus tones. This has led researchers to suggest that the use of the DPOAE at $2f_2$-f_1 might offer clinical advantages, particularly in noisy patients or test conditions. The optimal stimulus conditions for the generation of the DPOAE at $2f_2$-f_1 are significantly different than those required to generate a robust DPOAE at $2f_1$-f_2. For example, although the optimal stimulus frequency ratio (f_2/f_1) for recording a robust $2f_1$-f_2 is approximately 1.22, the equivalent ratio for recording a robust $2f_2$-f_1 is much smaller. Test performance for $2f_2$-f_1 is significantly better when the optimal stimulus conditions for $2f_2$-f_1 are used as opposed to when the DPOAE at $2f_2$-f_1 is recorded using stimulus parameters optimized for $2f_1$-f_2 (Fitzgerald & Prieve, 2005). However, this test performance still is not as good as that of the regularly used $2f_1$-f_2. More interestingly, test performance has *not* been shown to improve when both $2f_1$-f_2 and $2f_2$-f_1 are used in conjunction to determine presence or absence of hearing loss (Fitzgerald & Prieve, 2005; Gorga, Nelson, Davis, Dorn, & Neely, 2000).

OAE Analysis: Moving Beyond Reliance on Amplitude Alone

One very promising trend that warrants further investigation is the clinical power derived from recording a variety of OAE characteristics in combination. Although we pay attention to signal-to-noise ratio in DPOAE recordings, that metric is used primarily to make a judgment on the quality of the recording. For example, in determining whether the ear is normal or abnormal, analysis of the DPOAE is driven overwhelmingly by DPOAE level (amplitude) alone. In the case of TEOAEs, a few factors such as reproducibility, stability, and so forth also are taken into consideration when making clinical decisions. There may be promise in utilizing more than one measure in combination in clinical DPOAE measurement. DPOAE level and signal-to-noise ratio used in a multivariate analysis appears to improve the predictability of audiometric status (Dorn, Piskorski, Gorga, Neely, & Keefe, 1999). This observation is

exciting because it suggests that improvements in test performance may follow as we become more sophisticated in extracting various characteristics of the DPOAE. Testing the utility of multiple DPOAE characteristics in multivariate models likely will lead to new and more accurate clinical applications.

INTEGRATED INSTRUMENTATION FOR COMBINED RECORDINGS

Since the first generation of clinical OAE instrumentation was introduced in the 1980s, devices for recording transient and distortion product responses have become substantially more sophisticated and powerful. There are close to 20 different OAE devices on the international market. Also, more recently, we have witnessed an expansion of devices approved for use with patients that offer the capability for OAE measurement in combination with other techniques. The degree to which multiple technologies are integrated varies along several dimensions among devices and manufacturers. One dimension is the specific combination of technologies. Currently, a variety of devices exist for combined measurement of TEOAEs and DPOAEs. A handful of manufacturers also offer clinical instrumentation that combines the option for recording OAEs and the auditory brainstem response (ABR). Some of these devices are designed primarily for hearing screening with each technology, whereas others also include features and flexibility for diagnostic application of the two technologies. There also is now instrumentation that combines tympanometry with OAE. Combination devices can also be thought of in terms of the degree that multiple technologies are integrated. For example, is the same probe assembly used for presenting stimulation for ABR and OAE or can the two techniques be recorded simultaneously?

In this section, we summarize trends and advances in the development of instrumentation that permit measurement of OAE with one or more other techniques. A comprehensive review of data based literature on multiple technologies, for example, papers describing clinical trials of

OAE and ABR systems, is beyond the scope of this discussion. We also purposefully refrain from systematically comparing, in any way, devices marketed by different manufacturers. However, examples of devices that combine technologies are offered. The following review will quickly become outdated. With increased clinical demand for small devices appropriate for objective auditory measurement in infants and young children, we anticipate that in coming years additional manufacturers will market instrumentation designed for the three major objective techniques: OAE, ABR, and acoustic immittance measurement. The reader is advised to monitor advances in instrumentation via the Internet and commercial displays at professional meetings and conferences. Before new devices are distributed widely for screening or diagnostic clinical applications, it also is important for the devices and associated software and algorithms to undergo formal clinical trials, with results published in peer-reviewed literature.

COMBINED TEOAE AND DPOAE MEASUREMENT

Not too many years ago, measurement of both TEOAE and DPOAE with a single piece of equipment was physically impossible. Rather, two entirely separate and often large desktop computer-based systems would be purchased and then positioned relatively close to each other in a clinic or, if mobility was necessary, on a very sturdy cart. Today, according to information posted on a Web site devoted to otoacoustic emissions (the Otoacoustic Emissions Portal Zone: http://www.otoemissions.org), approximately 20 devices marketed by 10 different manufacturers permit measurement of both TEOAEs and DPOAEs. Obviously, some manufacturers have more than one device (e.g., designed for screening versus diagnostic purposes or portable versus desktop devices). As an aside, the Otoacoustic Emissions Portal Zone is a rather comprehensive and dynamic source of information on the topic that includes online articles, lectures, and white papers.

With most of the commercially available devices, both types of OAE can be recorded with

the same probe assembly utilizing different software. The user simply selects the appropriate software and proceeds with OAE measurement. As detailed in the review of mechanisms of OAE in Chapter 2, and the review of taxonomy for OAE in Chapter 3, TEOAEs and DPOAEs may target different cochlear mechanisms. We cannot assume, based on what is known about their underlying anatomy and physiology, that findings will be comparable for TEOAE and DPOAE measurement. In fact, it would be advisable to develop clinical protocols with just the opposite assumption. The possibility of distinctly different mechanisms underlying TEOAEs versus DPOAEs argues strongly for combining the two OAE approaches in selected patients to better describe cochlear function and to enhance diagnostic power. Regular application of both OAE technologies with diverse patient populations is very likely to yield interesting clinical data leading to new insights into cochlear function and dysfunction.

OAEs and Middle Ear Function

Earlier in this book, we began our discussion of the application of OAEs in children (Chapter 7) with a reminder about the ongoing importance of the "cross-check principle" in audiology today. Our argument for adherence to a clinical protocol for assessment of auditory function in children that requires recording, analyzing, and interpreting multiple independent audiologic procedures—the cross-check principle—was immediately followed by a rather lengthy review of middle ear diseases and disorders. Considerable space was devoted to the topic of middle ear disorders because they play such a critical role in the analysis and interpretation of OAE findings. Put simply, OAE findings cannot be confidently analyzed and interpreted or applied in a meaningful way clinically, unless middle ear status is known. Perhaps the best evidence in support of this statement is the 2007 Joint Committee on Infant Hearing (JCIH) strong recommendation, really a requirement, for inclusion of both techniques in the pediatric audiology test battery (JCIH, 2007).

Unfortunately, until recently measurement of OAE and middle ear function required two separate devices, often marketed by two different manufacturers. At the very least, recording both OAEs and middle ear function from a single patient meant setting up two devices (e.g., entering demographic patient information), attempting to properly couple to the ear two different probe assemblies (obviously not at the same time), and perhaps transporting the patient to another room in a clinical facility. The logistical problems associated with using two different devices for these two essential measurements no doubt discouraged, or even precluded, their combined application in selected patients. Indeed, for some clinicians (or their respective administrative personnel), the added cost of purchasing two separate pieces of equipment (one for recording OAEs and the other for analyzing middle ear function) was prohibitive.

The recent emergence of clinical instrumentation for combined measurement of OAEs and middle ear function is welcome news for clinicians responsible for auditory assessment of infants and young children. For the most part, integration of technology for OAE measurement and middle ear function into a single handheld device effectively eliminates the above-noted problems. Now, a single device, with one probe assembly and the same general software approach, can be used easily and efficiently for combined OAE recording and tympanometry with any patient in any location. Combined OAE and middle ear measurement is appropriate, and clinically indicated, for a variety of patient populations and settings, ranging from newborn infants undergoing hearing screening in a hospital nursery to preschool in a Head Start program, or school-age children who require annual hearing screening.

An example of a recently introduced device for recording tympanometry and two types of OAEs (TEOAEs and DPOAEs) is shown in Figure 10–1. Recognizing the demand for hearing screening and diagnostic assessment of infants, the tympanometry component of the device includes a low-frequency (226 Hz) and a high-frequency (1000 Hz) probe tone. Other handy features of such systems for quick and technically simple screening of infants and difficult-to-test populations include capacity for storing data for many patients (e.g., 350), automatically sequenced protocols, such as tympanometry followed by either DPOAE or TEOAE, and the option to print the findings following data collection for an immediate hardcopy record with test date and time and/or to later save the data to a computer for storage and further management. Often, two or more versions of a combination device are available, depending on whether its use will be limited to screening or will also include more extensive clinical (e.g., diagnostic) applications. With a screening version of the device, OAE measurement typically is possible for a more restricted frequency region and with fewer stimulus frequencies. Clinical (versus screening) versions of a device also sometimes permit creation of customized protocols and analysis strategies

OAE and Auditory Brainstem Response (ABR)

A variety of manufacturers now offer devices that permit combine OAE and ABR technology. Two representative devices are illustrated in Figure 10–2. The same probe assembly is used for delivery of the stimuli used to generate OAEs and to elicit the ABR. For ABR measurement, the only added feature is a set of three electrodes. Automated analysis is an option for both OAE and ABR findings. There are many clinical advantages associated with combining OAE and ABR technology in infant hearing screening, as recently reviewed in detail in the book *Objective Assessment of Hearing* (Hall & Swanepoel, 2010). One major advantage is the option of deciding at bedside which approach is optimal, given the child's chronological age after birth and medical history. For children born less than 24 hours before the screening, then either an ABR screening approach or a two-step OAE plus ABR combination of technology is essential. Long-standing evidence confirms the unacceptably high failure rates for OAE screening within the first day after birth. However, for an infant who is over 24 hours old and not at risk for hearing loss with screening to be conducted in a well baby nursery or some other nonintensive care setting, OAEs are certainly an appropriate and feasible screening technique.

The Joint Committee on Infant Hearing (JCIH, 2007) provides valuable, and widely accepted,

A

FIGURE 10–1. A. Example of a device that permits the combined measurement of otoacoustic emissions (OAE) and tympanometry. **B.** The EroScan Pro™ used in the assessment of middle ear function (with tympanometry) and cochlear function (with DPOAEs) of an infant returning to a public hospital in South Africa for follow up hearing screening. (Courtesy of Maico EroScan Pro™)

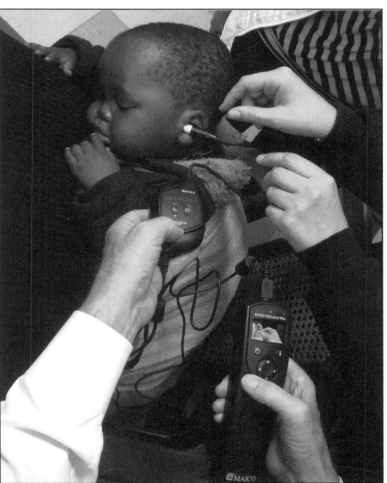

B

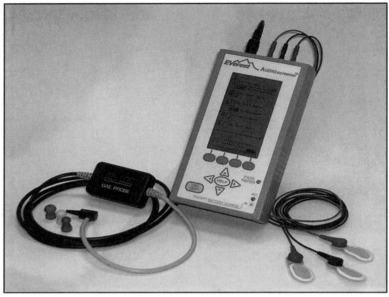

A

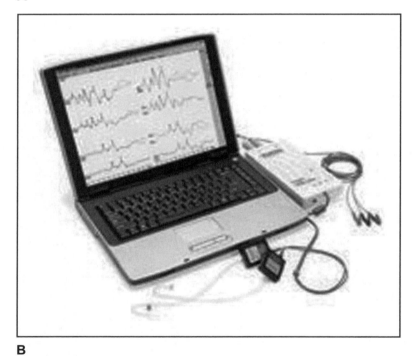

B

FIGURE 10–2. Examples of devices from two manufacturers. **A.** GSI AudioScreener (courtesy of Grason-Stadler). **B.** Intelligent Hearing Systems SmartScreener-Plus2) that permit combined measurement of otoacoustic emissions (OAEs) and the auditory brainstem response (ABR) (courtesy of Intelligent Hearing Systems).

guidance on newborn hearing screening with OAE and ABR. The JCIH endorses both types of technology as appropriate for newborn infant hearing screening. However, the JCIH notes that, "Neural conduction disorders or auditory neuropathy/dys-synchrony without concomitant sensory dys-

function will not be detected by OAE testing" (p. 904). "The JCIH recommends ABR technology as the only appropriate screening technique for use in the NICU." (p. 904). Clearly, OAEs are an accepted technology option for newborn hearing screening. However, OAEs are not an appropriate screening option in the newborn intensive care setting where children with auditory neuropathy or other neurologic auditory dysfunction are most often found. For this reason, ABR is the screening technique of choice in the NICU (or ICN) environment and population or, as summarized in one of the excerpts above and reviewed below, a combination of ABR plus OAE technologies is optimal for hearing screening of children at risk for various types of auditory dysfunction, including neural disorders.

Published evidence from multiple international clinical investigations confirms the value of combining OAE and AABR technologies in some combination for newborn hearing screening (see Hall, 2007, and Hall & Swanepoel, 2010, for review). The techniques can be applied using a "two step" approach to screening or as a combination for each child. With the two-step approach, one technique is used most often and the other technique is additionally applied only in selected cases. For example, with a generally healthy "well baby" population, OAEs are used as the initial screening technique for over 90% of the babies, and AABR is performed only when a child fails OAE screening for one or both ears. The reverse strategy is recommended for infants in an intensive care medical setting and at risk for neurologic dysfunction. That is, ABR screening is relied on for all children, and infants who fail the ABR screening undergo OAE measurement to differentiate possible sensory versus neural auditory dysfunction. Although there typically is a high correlation in the outcome for the two techniques, referral rates typically are lower for automated ABR (AABR) than for OAE techniques, especially within the first 36 hours after birth.

Another approach involving both technologies is more appropriately called "combined OAE/ABR screening," rather than two-step hearing screening. With the combined approach, both OAE and ABR are recorded from each child. Examination of the pattern of findings for both techniques permits differentiation even in the nursery setting of the three most commonly encountered general types of auditory dysfunction involving the: (1) middle ear, (2) cochlea (outer hair cells), or (3) neural pathways (e.g., auditory neuropathy spectrum disorder). According to Hall, Smith, and

Popelka (2004), the advantages of this screening strategy are as follows:

- In ear calibration of signal intensity for OAEs and ABR (selected devices)
- Lower refer (<2%) and false-positive rates (<0.9%) than with either technique alone
- High sensitivity (up to 100%) and specificity (up to 99.7%)
- Fewer diagnostic follow-ups with substantially lower overall cost for early identification of infant hearing loss
- Differentiation at birth of auditory dysfunction as conductive, sensory, or neural (e.g., auditory neuropathy spectrum disorder)
- Faster and more appropriate management based on information on types of auditory dysfunction
- Earlier intervention for hearing impairment
- Higher likelihood of optimal outcome from intervention for hearing loss

OAEs and Auditory Steady State Response (ASSR)

Several manufacturers of devices designed for OAE measurement with ABR also offer software for the auditory steady state response (ASSR). The ASSR is viewed by the JCIH as a supplementary, rather than a required, procedure for diagnostic assessment of infants and young children. However, the inclusion of an option for ASSR measurement can add considerably to the diagnostic flexibility and value of a system. The OAE and ABR instrumentation permits efficient hearing screening, as just noted, and also diagnosis of the type, configuration, and degree of most forms of hearing loss. Careful analysis of OAE and ABR findings, with an appreciation of the principles of electrocochleography (ECochG), generally identifies and sometimes convincingly diagnoses auditory neuropathy spectrum disorder. Frequency-specific estimation of auditory thresholds with an electrophysiologic technique, a requirement for timely and accurate hearing-aid fitting in infants and young children, often is accomplished with ABRs elicited with tone-burst stimuli. In selected cases, OAE and ABR findings will confirm the presence of sensory auditory dysfunction. However, the degree of the hearing loss (greater than about 80 dB HL) precludes estimation of auditory thresholds. For these children, ASSR usually contributes to a reasonably accurate, frequency-specific estimation of auditory threshold or confirmation that there is no electrophysiologic response even to very high tonal stimulus intensity levels (120 to 125 dB HL). The information derived from including the ASSR technique to the test battery contributes to prompt and accurate hearing-aid fitting, and speeds up decisions and management regarding cochlear implantation.

CONCLUDING COMMENTS

None of us can confidently predict future directions of OAE-related research or clinical applications. In fact, in the mid-1970s, as two other objective auditory procedures—acoustic immittance measurement and the auditory brainstem response (ABR)—rapidly assumed important roles in the audiologic test battery, no audiologists anticipated the introduction of yet another objective auditory measure—OAEs. Clinical practice in audiology today is evidenced-based, and founded on information derived from basic and applied hearing research. Therefore, ongoing close review of research findings in the peer-reviewed scientific literature provides a reasonably good predictor of likely directions of OAE-related research or clinical applications.

In years to come, we will certainly witness the inclusion within the clinical test battery OAE instruments with stimulation paradigms and strategies not currently applied in the clinical setting. Recall a discussion earlier in this chapter of a variety of stimulus paradigms for evoking OAE activity, such as maximum length sequence (MLS), atypical stimulus types (e.g., amplitude-modulated tone bursts, noise, chirps), automated input-output functions, and techniques for measurement of DPOAEs elicited by high-frequency stimulation. Which of these nontraditional stimulus techniques will find their way into devices available to clinical audiologists? The general answer to this question is: "Feasible techniques with proven clinical value in the identification and diagnosis of auditory dysfunction." By this criterion, the incorporation of high-frequency OAE capability in clinical devices is almost a certainty. Such devices would immediately offer clinical value for early and sensitive detection of cochlear dysfunction secondary to potentially ototoxic therapeutic drugs. To some extent, your "crystal ball" can be found within the nearest search engine used for a review of scientific literature (e.g., http://www.nlm.nih.gov).

We can follow the same logic for predicting new directions in techniques and paradigms for the analysis of OAEs. Clearly, simple analysis of the amplitude of OAE activity for a single stimulus intensity paradigm, and a modest number of test frequencies, is very valuable in the clinical setting. A vast amount of evidence confirms that OAEs are remarkably sensitive to cochlear dysfunction, when compared to the conventional pure-tone audiogram. As noted earlier in this chapter, analysis of OAEs is clearly moving beyond exclusive reliance on amplitude. A diverse collection of analysis strategies was cited in the earlier discussion, among them measurement of DPs at alternative frequencies (not just the 2f1-f2 frequency), calculation of the latency or phase of OAE activity, logistic regression modeling, adaptive approximations, neural networks, and principle component analysis of TEOAEs. Literature on the detection of subtle cochlear dysfunction in various disorders and disease processes (e.g., noise-induced auditory disorders and diabetes II) suggests that there is a wealth of untapped information available from analysis of OAE fine structure. For example, in selected patient populations, fine structure data reveal deficits not apparent with conventional analysis of OAE amplitude for a limited number of test frequencies per octave.

To be sure, some hot research topics fizzle on the journey from the laboratory to the clinic. We maintain, however, that the surest way to predict future clinical applications of OAE is to "follow the research."

References

Abdala, C. (2003). A longitudinal study of distortion product otoacoustic emission ipsilateral suppression and input/output characteristics in human neonates. *Journal of the Acoustical Society of America, 114,* 3239–3250.

Abdala, C. (2004). Distortion product otoacoustic emission (2f1-f2) suppression in 3-month-old infants: Evidence for postnatal maturation of human cochlear function? *Journal of Acoustical Society of America, 116,* 3572–3580.

Abdala, C. (2005). Effects of aspirin on distortion product otoacoustic emission suppression in human adults: A comparison with neonatal data. *Journal of the Acoustical Society of America, 118,* 1566–1575.

Abdala, C., & Chatterjee, M. (2003). Maturation of cochlear nonlinearity as measured by distortion product otoacoustic emission suppression growth in humans. *Journal of the Acoustical Society of America, 114,* 932–943.

Abdala, C., & Dhar, S. (2010). Distortion product otoacoustic emission phase and component analysis in human newborns. *Journal of the Acoustical Society of America, 127,* 316–325.

Abdala, C., & Fitzgerald, T. S. (2003). Ipsilateral distortion product otoacoustic emission (2f1-f2) suppression in children with sensorineural hearing loss. *Journal of the Acoustical Society of America, 114,* 919–931.

Abdala, C., & Keefe, D. H. (2006). Effects of middle-ear immaturity on distortion product otoacoustic emission suppression tuning in infant ears. *Journal of the Acoustical Society of America, 120,* 3832–3842.

Adunka, O. F., Roush, P. A., Teagle, H. F., Brown, C. J., Zdanski, C. J., Jewells, V., & Buchman, C. A. (2006). Internal auditory canal morphology in children with cochlear nerve deficiency. *Otology & Neurotology, 27,* 793–801.

Akdogan, O., & Ozkan, S. (2006). Otoacoustic emissions in children with otitis media with effusion. *International Journal of Pediatric Otorhinolaryngology, 70,* 1941–1944.

Akman, I., Ozek, E., Kulekci, S., Türkdogan, D., Cebeci, D., & Akdaş, F. (2004). Auditory neuropathy in hyperbilirubinemia: Is there a correlation between serum bilirubin, neuron-specific enolase levels and auditory neuropathy? *International Journal of Audiology, 43,* 516–522.

Aksoy, F., Yildirim, Y. S., Veyseller, B., & Demirhan, H. (2010). Distortion product otoacoustic emissions results in children with middle ear effusion [Article in Turkish]. *Kulak Burun Boğaz Ihtisas Dergisi, 20,* 71–76.

Allen, G. C., Tiu, C., Koike, K., Ritchey, K., Kurs-Lasky, M., & Wax, M. (1998). Transient otoacoustic emissions in children after cisplatin chemotherapy. *Otolaryngology-Head and Neck Surgery, 118,* 584–588.

Amedee, R. G. (1995). The effects of chronic otitis media with effusion on the measurement of transiently evoked otoacoustic emissions. *Laryngoscope, 105,* 589–595.

American Academy of Audiology (AAA). (1997). Identification of hearing loss and middle ear dysfunction in preschool and school aged children [Position statement]. *Audiology Today, 9,* 21–23.

American Academy of Pediatrics Task Force on Newborn and Infant Hearing. (1999). Newborn and infant hearing loss: Detection and intervention. *Pediatrics, 103,* 527–529.

American Speech Language Hearing Association (ASHA). (1993). Audiologic management of individuals receiving cochleotoxic drug therapy. *ASHA 2002 Desk Reference, 2,* 81–92.

American Speech Language Hearing Association (ASHA). (1997). *Guidelines for audiological screening.* Rockville Pike, MD: ASHA.

American Speech-Language-Hearing Association (ASHA). (2002). *Guidelines for audiology service provision in and for schools.* Rockville Pike, MD: ASHA.

American Standards Institute. (1988). Occluded ear stimulator. ANSI S3.25-1989-R1995. New York, NY: American Standards Institute.

Angeli, M. L., Almeida, C. I., & Sens, P. M. (2008). Comparative study between school performance on first grade children and suppression of otoacoustic transient emission. *Brazilian Journal of Otorhinolaryngology, 74,* 112–117.

Anioł-Borkowska, M., Namysłowski, G., Lisowska, G., Kwiek, S., & Hajduk, A. (2005). Evaluation of the cochlear efferent system in patients with cerebello-pontine angle tumor. *Pol Merkur Lekarski, 111,* 283–285.

Arruda, P. O., & Silva, I. M. (2008). Study of otoacoustic emissions during the female hormonal cycle. *Review of Brasilian Otorrinolaringology (Engl. ed.), 74,* 106–111.

ASHA Audiologic Assessment Panel 1996. (1997, 2002). Guidelines for screening infants and children for outer and middle ear disorders, birth through 18 years. In *ASHA 2002 Desk Reference* (Vol. 4, pp. 342–349). Rockville, MD: American Speech-Language-Hearing Association.

Aslan, S., Serarslan, G., Teksoz, E., & Dagli, S. (2010). Audiological and transient evoked otoacoustic emission findings in patients with vitiligo. *Otolaryngology-Head and Neck Surgery, 142,* 409–414.

Atcherson, S. R., Martin, M. J., & Lintvedt, R. (2008). Contralateral noise has possible asymmetric frequency-sensitive effect on the 2F1-F2 otoacoustic emission in humans. *Neuroscience Letters, 438,* 107–110.

Attias, J., Horovitz, G., El-Hatib N., & Nageris, B. (2001). Detection and clinical diagnosis of noise-Induced hearing loss by otoacoustic emissions. *Noise and Health, 3,* 19–31.

Attias, J., Raveh, E., Ben-Naftali, N. F., Zarchi, O., & Gothelf, D. (2008). Hyperactive auditory efferent system and lack of acoustic reflexes in Williams syndrome. *Journal of Basic and Clinical Physiology and Pharmacology, 19,* 193–207.

Attias, J., Sapir, S., Bresloff, I., Reshef-Haran, I., & Ising, H. (2004). Reduction in noise-induced temporary threshold shift in humans following oral magne-

sium intake. *Clinical Otolaryngology and Allied Science, 29,* 635–641.

Attias, J., Zwecker-Lazar, I., Nageris, B., Keren, O., & Groswasser, Z. (2005). Dysfunction of the auditory efferent system in patients with traumatic brain injuries with tinnitus and hyperacusis. *Journal of Basic and Clinical Physiology and Pharmacology, 16,* 117–126.

Avan, P., & Bonfils, P. (2005). Distortion-product otoacoustic emission spectra and high-resolution audiometry in noise-induced hearing loss. *Hearing Research, 209,* 68–75.

Avan, P., Bonfils, P., & Mom, T. (2001). Correlations among distortion product otoacoustic emissions, thresholds and sensory cell impairments. *Noise and Health, 3,* 1–18.

Babac, S., Djerić, D., & Ivanković, Z. (2007). Newborn hearing screening. [Article in Serbian]. *Srp Arh Celok Lek, 135,* 264–268.

Baggio, C. L., Silveira, A. F., Hyppolito, M. A., Salata, F. F., & Rossato, M. (2010). A functional study on gentamicin-related cochleotoxicity in its conventional dose in newborns. *Brazilian Journal of Otorhinolaryngology, 76,* 91–95.

Baguley, D. M. (2002). Mechanisms of tinnitus. *British Medical Bulletin, 63,* 195–212.

Baguley, D. M., Jones, S. E., & Moffat, D. A. (2003). A small vestibular schwannoma arising from the inferior vestibular nerve. *Journal of Laryngology and Otology, 117,* 498–500.

Balatsouras, D. G. (2004). The evaluation of noise-induced hearing loss with distortion product otoacoustic emissions. *Medical Science Monitor, 10,* CR218–CR222.

Balatsouras, D., Kaberos, A., Karapantzos, E., Homsioglou, E., Economou, N. C., & Korres, S. (2004). Correlation of transiently evoked otoacoustic emission measures to auditory thresholds. *Medical Science Monitor, 10,* MT24–MT30.

Balatsouras, D. G., Kaberos, A., Kloutsos, G., Economou, N. C., Sakellariadis, V., Fassolis, A., & Korres, S. G. (2006). Correlation of transiently evoked to distortion-product otoacoustic emission measures in healthy children. *International Journal of Pediatric Otorhinolaryngology, 70,* 89–93.

Balatsouras, D. G., Kaberos, A., Korres, S., Kandiloros, D., Ferekidis, E., & Economou, C. (2003). Detection of pseudohypacusis: A prospective, randomized study of the use of otoacoustic emissions. *Ear and Hearing, 24,* 518–527.

Balatsouras, D. G., Korres, S., Manta, P., Panousopoulou, A., & Vassilopoulos, D. (2007). Cochlear function in facioscapulohumeral muscular dystrophy. *Otology & Neurotology, 28,* 7–10.

Balatsouras, D. G., Rallis, E., Homsioglou, E., Fiska, A., & Korres, S. G. (2007). Ramsay Hunt syndrome in a 3-month-old infant. *Pediatric Dermatology, 24,* 34–37.

Balatsouras, D. G., Tsimpiris, N., Korres, S., Karapantzos, I., Papadimitriou, N., & Danielidis, V. (2005). The effect of impulse noise on distortion product otoacoustic emissions. *International Journal of Audiology, 44,* 540–549.

Baldwin, S. M., Gajewski, B. J., & Widen, J. E. (2010). An evaluation of the cross-check principle using visual reinforcement audiometry, otoacoustic emissions, and tympanometry. *Journal of the American Academy of Audiology, 21,* 187–196.

Balkany, T. J., Berman, S. A., Simmons, M. A., & Jafek, B. (1978). Middle ear effusions in neonates. *Laryngoscope, 88,* 398–405.

Bamiou, D. E., Ceranic, B., Cox, R., Watt, H., Chadwick, P., & Luxon, L. M. (2008). Mobile telephone use effects on peripheral audiovestibular function: A case-control study. *Bioelectromagnetics, 29,* 108–117.

Bar-Haim, Y., Henkin, Y., Ari-Even-Roth, D., Tetin-Schneider, S., Hildesheimer, M., & Muchnik, C. (2004). Reduced auditory efferent activity in childhood selective mutism. *Biological Psychiatry, 55,* 1061–1068.

Barros, S. M., Frota, S., Atherino, C. C., & Osterne, F. (2007). The efficiency of otoacoustic emissions and pure-tone audiometry in the detection of temporary auditory changes after exposure to high sound pressure levels. *Review of Brasilian Otorrinolaringology, 73,* 592–598.

Bartnik, G., Hawley, M., Rogowski, M., Raj-Koziak, D., Fabijanska, A., & Formby, C. (2009). [Distortion product otoacoustic emission levels and input/output growth functions in normal-hearing individuals with tinnitus and/or hyperacusis]. *Otolaryngology Polska, 63,* 171–181.

Bartnik, G., Rogowski, M., Fabijańska, A., Raj-Koziak, D., & Borawska, B. (2004). Analysis of the distortion product otoacoustic emission (DPOAE) and input/output function (I/O) in tinnitus patient with normal hearing [Article in Polish]. *Otolaryngology Polska, 58,* 1127–1132.

Bathelier, C., François, M., & Lucotte, G. (2004). Neonatal detection of the 35delG mutation of the GJB2 gene in families at risk for deafness. *Genetic Counseling, 15,* 61–66.

Bayazit, Y. A., Yilmaz, M., Gunduz, B., Altinyay, S., Kemaloglu, Y. K., Onder, M., & Gurer, M. A. (2007). Distortion product otoacoustic emission findings in Behçet's disease and rheumatoid arthritis. *ORL Journal of Otorhinolaryngology and Related Specialties, 69,* 233–238.

Beattie, R. C., Kenworthy, O. T. & Luna, C. A. (2003). Immediate and short-term reliability of distortion-product otoacoustic emissions. *International Journal of Audiology, 42,* 348–354.

Benito-Orejas, J. I., Ramírez, B., Morais, D., Almaraz, A., & Fernández-Calvo, J. L. (2008). Comparison of two-step transient evoked otoacoustic emissions (TEOAE) and automated auditory brainstem response (AABR) for universal newborn hearing screening programs. *International Journal of Pediatric Otorhinolaryngology, 72,* 1193–1201.

Bennett, C. L., & Ozdamar, O. (2010a). High-frequency transient evoked otoacoustic emissions acquisition with auditory canal compensated clicks using swept-tone analysis, *Journal of the Acoustical Society of America, 127,* 2410–2419.

Bennett, C. L., & Ozdamar, O. (2010b). Swept-tone transient-evoked otoacoustic emissions, *Journal of the Acoustical Society of America, 128,* 1833–1844.

Berg, A. L., Olson, T. J., & Feldstein, N. A. (2005). Cerebellar pilocytic astrocytoma with auditory presentation: Case study. *Journal of Child Neurology, 20,* 914–915.

Berlin, C. I. (1997). *The efferent auditory system: Basic science and clinical applications.* San Diego, CA: Singular Publishing Group.

Berlin, C .I., Hood, L.J., Hurley, A., Wen, H., & Kemp, D. T. (1995). Bilateral noise suppresses click-evoked otoacoustic emissions more than ipsilateral or contralateral noise. *Hearing Research, 87,* 96–103.

Berlin, C. I., Hood, L. J., Morlet, T., Wilensky, D., Li, L., Mattingly, K. R., Taylor-Jeanfreau, J., . . . Frisch, S. A. (2010). Multi-site diagnosis and management of 260 patients with auditory neuropathy/dys-synchrony (auditory neuropathy spectrum disorder). *International Journal of Audiology, 49,* 30–43.

Berlin, C. I., Hood, L. J., Morlet, T., Wilensky, D., St. John, P., Montgomery, E., & Thibodaux, M. (2005). Absent or elevated middle ear muscle reflexes in the presence of normal otoacoustic emissions: A universal finding in 136 cases of auditory neuropathy/dys-synchrony. *Journal of the American Academy of Audiology, 16,* 546–553. Comment in: *Journal of the American Academy of Audiology, 2007, 18,* 187–189; author reply: 189–190.

Berlin, C. I., Hood, L. J., Wen, H., Szabo, P., Ceola, R. P., Rigby, P., & Jackson, D. F. (1993). Contralateral suppression of non-linear click-evoked otoacoustic emissions. *Hearing Research, 71,* 1–11.

Berlin, C. I., Morlet, T., & Hood, L. J. (2003). Auditory neuropathy/dyssynchrony: its diagnosis and management. *Pediatric Clinics of North America, 50,* 331–340.

Berninger, E. (2007). Characteristics of normal newborn transient-evoked otoacoustic emissions: Ear asymmetries and sex effects. *International Journal of Audiology, 46,* 661–669.

Beutner, D., Foerst, A., Lang-Roth, R., von Wedel, H., & Walger, M. (2007). Risk factors for auditory neuropathy/auditory synaptopathy. *ORL Journal of Otorhinolaryngology and Related Specialties, 69,* 239–244.

Bian, L., & Chen, S. (2011). Behaviors of cubic distortion product otoacoustic emissions evoked by amplitude modulated tones. *Journal of the Acoustical Society of America, 129,* 828–839.

Biró, K., Noszek, L., Prekopp, P., Nagyiványi, K., Géczi, L., Gaudi, I., & Bodrogi, I. (2006a). Characteristics and risk factors of cisplatin-induced ototoxicity in testicular cancer patients detected by distortion product otoacoustic emission. *Oncology, 70,* 177–184.

Biró, K., Noszek, L., Prekopp, P., Nagyiványi, K., Géczi, L., Gaudi, I., & Bodrogi, I. (2006b). Detection of late ototoxic side effect of cisplatin by distortion otoacoustic emission (DPOAE) [Article in Hungarian]. *Magy Onkology, 50,* 329–335.

Bockstael, A., Keppler, H., Dhooge, I., D'haenens, W., Maes, L., Philips, B., & Vinck, B. (2008). Effectiveness of hearing protector devices in impulse noise verified with transiently evoked and distortion product otoacoustic emissions. *International Journal of Audiology, 47,* 119–133.

Boege, P., & Janssen, T. (2002). Pure-tone threshold estimation from extrapolated distortion product otoacoustic emission I/O-functions in normal and cochlear hearing loss ears. *Journal of the Acoustical Society of America, 111,* 1810–1818.

Boegli, H., Wunderli, J. M., & Brink, M. (2008). Assessment of military shooting noise. *Journal of the Acoustical Society of America, 123,* 3820.

Bolay, H., Bayazit, Y.A., Gündüz, B., Ugur, A. K., Akçali, D., Altunyay, S., . . . Babacan, A. (2008). Subclinical dysfunction of cochlea and cochlear efferents in migraine: An otoacoustic emission study. *Cephalalgia, 28,* 309–317.

Boleas-Aguirre, M. S., Vazquez, F., & Perez, N. (2007). Progressive cochleo-vestibular labyrinthitis. *Review of Laryngology Otology and Rhinology (Bordeaux), 128,* 63–64.

Bonfils, P. (1989). Spontaneous otoacoustic emissions: clinical interest. *Laryngoscope, 99,* 752–756.

Bonfils, P., Bertrand, Y., & Uziel, A. (1988). Evoked otoacoustic emissions: Normative data and presbycusis. *Audiology, 27,* 27–35.

Bonfils, P., & Uziel, A. (1988). Evoked otoacoustic emissions in patients with acoustic neuromas. *American Journal of Otolaryngology, 9,* 412–417.

Boone, R. T., Bower, C. M., & Martin, P. F. (2005). Failed newborn hearing screens as presentation for otitis media with effusion in the newborn population. *International Journal of Pediatric Otorhinolaryngology, 69,* 393–397.

Borin, A., & Cruz, O. L. (2008). Study of distortion-product otoacoustic emissions during hypothermia in humans [Article in Portuguese]. *Review of Brasilian Otorrinolaringology,* (Engl. ed.), *74,* 401–409.

Bravo, O., Ballana, E., & Estivill, X. (2006). Cochlear alterations in deaf and unaffected subjects carrying the deafness-associated A1555G mutation in the mitochondrial 12S rRNA gene. *Biochemical and Biophysical Research in Communication, 344,* 511–516.

Brock, P. R., Bellman, S. C., Yeomans, E. C.; Pinkerton, C. R., & Pritchard, J. (1991). Cisplatin ototoxicity in children: A practical grading system. *Medicine and Pediatric Oncololgy, 19,* 295–300.

Brown, A., Williams, D., & Gaskill, S. (1993). The effect of aspirin on cochlear mechanical tuning. *Journal of the Acoustical Society of America, 93,* 3298–3307.

Burguetti, F. A., & Carvallo, R. M. (2008). Efferent auditory system: Its effect on auditory processing. *Brazilian Journal of Otorhinolaryngology, 74,* 737–745.

Buchman, C. A., Roush, P. A., Teagle, H. F., Brown, C. J., Zdanski, C. J., & Grose, J. H. (2006). Auditory neuropathy characteristics in children with cochlear nerve deficiency. *Ear and Hearing, 27,* 399–408.

Burch-Sims, G. P.. & Matlock, V. R. (2005). Hearing loss and auditory function in sickle cell disease. *Journal of Communicative Disorders, 38,* 321–329.

Burke, S .R., Rogers, A. R., Neely, S. T., Kopun, J. G. Tan, H., & Gorga, M. P. (2010). Influence of calibration method on distortion-product Otoacoustic emission measurement: I. Test performance. *Ear and Hearing, 31,* 533–545.

Campbell, K. C. M. (2006). *Pharmacology and ototoxicity for audiologists.* New York, NY: Delmar Cengage Learning.

Canale, A., Lacilla, M., Giordano, C., De Sanctis, A., & Albera, R. (2005). The prognostic value of the otoacoustic emission test in low frequency sudden hearing loss. *European Archives of Otorhinolaryngology, 262,* 208–212.

Cane, M. A., O'Donoghue, G. M., & Lutman, M. E. (1992). The feasibility of using evoked otoacoustic emissions to monitor cochlear function during acoustic neuroma surgery. *Scandinavian Audiology, 21,* 173–176.

Castor, X., Veuillet, E., Morgon, A., & Collet, L. (1994). Influence of aging on active cochlear micromechanical properties and on the medial olivocochlear system in humans. *Hearing Research, 77,* 1–8.

Ceranic, J. B., Prasher, D. K., & Luxon, L. M. (1995). Tinnitus and otoacoustic emissions. *Clinical Otolaryngology, 20,* 192–200.

Cetin, T., Yetiser, S., Cekin, E., Durmus, C., Nevruz, O., & Oktenli, C. (2004). Outer hair cell activity of the cochlea in patients with iron deficiency anemia. *Auris Nasus Larynx, 31,* 389–394.

Chabert, R., Guitton, M. J., Amram, D., Uziel, A., Pujol, R., Lallemant, J. G., & Puel, J. L. (2006). Early maturation of evoked otoacoustic emissions and medial olivocochlear reflex in preterm neonates. *Pediatric Research, 59,* 305–308.

Chan, V. S., Wong, E. C., & McPherson, B. (2004). Occupational hearing loss: screening with distortion-product otoacoustic emissions. *International Journal of Audiology, 43,* 323–329.

Chang S. O., Jang, Y. J., & Rhee, C. K. (1998). Effects of middle ear effusion on transient evoked otoacoustic emissions in children. *Auris Nasus Larynx, 25,* 243–247.

Chao, T. K., & Chen, T. H. (2006). Distortion product otoacoustic emissions as a prognostic factor for idiopathic sudden sensorineural hearing loss. *Audiology & Neurootology, 11,* 331–338.

Charlier, K., & Debruyne, F. (2004). The effect of ventilation tubes on otoacoustic emissions. A study of 106 ears in 62 children. *Acta Otorhinolaryngologica Belgium, 58,* 67–71.

Chen, C. N., & Young, Y. H. (2006). Differentiating the cause of acute sensorineural hearing loss between Ménière's disease and sudden deafness. *Acta Otolaryngologica, 126,* 25–31.

Cheng, X., Li, .L, Brashears, S., Morlet, T., Ng, S. S., Berlin, C., Hood, L., & Keats, B. (2005). Connexin 26 variants and auditory neuropathy/dys-synchrony among children in schools for the deaf. *American Journal of Medical Genetics, 139,* 13–18.

Cho, H. H., & Cho, Y. B. (2005). Otoacoustic emissions in an aberrant internal carotid artery: A case report. *European Archives of Otorhinolaryngology, 262,* 213–216.

Chomicki, A., Sakka, L., Avan, P., Khalil, T., Lemaire, J. J., & Chazal, J. (2007). Derivation of cerebrospinal fluid: consequences on inner ear biomechanics in adult patients with chronic hydrocephalus [Article in French]. *Neurochirurgie, 53,* 265–271.

Christopher Kirk, E., & Smith, D. W. (2003)."Protection from acoustic trauma is not a primary function of the medial olivocochlear efferent system, *Journal of the Association for Research in Otolaryngology, 4,* 445–465.

Chu, P. L., Wu, C. C., Hsu, C. J., Wang, Y. T., & Wu, K. D. (2007). Potential ototoxicity of aluminum in hemodialysis patients. *Laryngoscope, 117,* 137–141.

Cianfrone, G., Ralli, G., Fabbricatore, M., Altissimi, G., & Nola, G. (2000). Distortion product otoacoustic emmissions in Ménière's disease. *Scandinavian Audiology, 29,* 111–119.

Cilento, B. W., Norton, S. J., & Gates, G. A. (2003). The effects of aging and hearing loss on distortion product otoacoustic emissions. *Otolaryngology-Head and Neck Surgery, 129,* 382–389.

Clark, W. W., Kim, D. O., Zurek, P. M., & Bohne, B. A. (1984). Spontaneous otoacoustic emissions in chinchilla ear canals: Correlation with histopathology and suppression by external tones. *Hearing Research, 16,* 299–314.

Clarke, E. M., Ahmmed, A., Parker, D., & Adams, C. (2006). Contralateral suppression of otoacoustic emissions in children with specific language impairment. *Ear and Hearing 27,* 153–160.

Coelho, A., Ceranić, B., Prasher, D., Miller, D. H., & Luxon, L. M. (2007). Auditory efferent function is affected in multiple sclerosis. *Ear and Hearing, 28,* 593–604.

Collet, L., Kemp, D. T., Veuillet, E., Duclaux, R., Moulin, A., & Morgon, A. (1990). Effect of contralateral auditory stimuli on active cochlear micro-mechanical properties in human subjects. *Hearing Research, 43,* 251–262.

Conti, G., & Sergi, B. (2003). Auditory and vestibular findings in Fabry disease: A study of hemizygous males and heterozygous females. *Acta Paediatrica Supplement, 92,* 33–37.

Coradini, P. P., Cigana, L., Selistre, S. G., Rosito, L. S., & Brunetto, A.L. (2007). Ototoxicity from cisplatin therapy in childhood cancer. *Journal of Pediatric Hematology and Oncology, 29,* 355–360.

Csanády, M., Tóth, F., Hogye, M., Vass, A., Sepp, R., Csanády, M. Jr., . . . Forster, T. (2007). Hearing disturbances in hypertrophic cardiomyopathy. Is the sensorineural disorder neurogenic or myogenic? *International Journal of Cardiology, 116,* 53–56.

(CTCAE). Common Terminology Criteria for Adverse Events. National Institutes of Health National Cancer Institute (NCI). U.S. Department of Health and Human Services. (2009). [http://ctep.cancer.gov/protocolDevelopment/electronic_applications/ctc.htm]

Cullington H., Kumar, B., & Flood, L. (1998). Feasibility of otoacoustic emissions as a hearing screen following grommet insertion. *British Journal of Audiology, 32,* 57–62.

Dagli, M., Sivas Acar, F., Karabulut, H., Eryilmaz, A., & Erkol Inal, E. (2007). Evaluation of hearing and cochlear function by DPOAE and audiometric tests in patients with ankylosing spondilitis. *Rheumatology International, 27,* 511–516.

Davidson, P. W., Weiss, B., Beck, C., Cory-Slechta, D. A., Orlando, M., Loiselle, D., . . . Myers, G. J. (2006). Development and validation of a test battery to assess subtle neurodevelopmental differences in children. *Neurotoxicology, 27*, 951–969.

Davilis, D., Korres, S. G., Balatsouras, D. G., Gkoritsa, E., Stivaktakis, G., & Ferekidis, E. (2005). The efficacy of transiently evoked otoacoustic emissions in the detection of middle-ear pathology. *Medical Science Monitor, 11*, MT75–MT78.

Dean, J. B., Hayashi, S. S., Albert, C. M., King, A. A., Karzon, R., & Hayashi, R.J. (2008). Hearing loss in pediatric oncology patients receiving carboplatin-containing regimens. *Journal of Pediatric Hematology and Oncology, 30*, 130–134.

de Azevedo, R. F., Chiari, B. M., Okada, D. M., & Onishi, E. T. (2007). Impact of acupuncture on otoacoustic emissions in patients with tinnitus. *Review of Brasilian Otorrinolaringology (Engl. ed.), 73*, 599–607.

de Boer, J., Brennan, S., Lineton, B., Stevens, J., & Thornton, A. R. (2007). Click-evoked otoacoustic emissions (CEOAEs) recorded from neonates under 13 hours old using conventional and maximum length sequence (MLS) stimulation, *Hearing Research, 233*, 86–96.

de Boer, J., & Thornton, A. R. (2007). Effect of subject task on contralateral suppression of click evoked otoacoustic emissions. *Hearing Research, 233*, 117–123.

De Felice, C., De Capua, B., Costantini, D., Martufi, C., Toti, P., Tonni, G., . . . Latini, G. (2008). Recurrent otitis media with effusion in preterm infants with histologic chorioamnionitis—A 3 year follow-up study. *Early Human Development, 84*, 667–671.

de Kleine, E., Mateijsen, D. J., Wit, H. P., & Albers, F. W. (2002). Evoked otoacoustic emissions in patients with Ménière's disease. *Otology & Neurotology, 23*, 510–516.

De Leenheer, E. M., Bosman, A. J., Kunst, H. P., Huygen, P. L., & Cremers, C. W. (2004). Audiological characteristics of some affected members of a Dutch DFNA13/COL11A2 family. *Annals of Otology, Rhinology, and Laryngology, 113*, 922–929.

Delehaye, E., Capobianco, S., Bertetto, I. B., & Meloni, F. (2008). Distortion-product otoacoustic emission: early detection in deferoxamine induced ototoxicity. *Auris Nasus Larynx, 35*, 198–202.

Deltenre, P., Mansbach, A. L., Bozet, C., Christiaens, F., Barthelemy, P., Paulissen, D., & Renglet, T. (1999). Auditory neuropathy with preserved cochlear microphonics and secondary loss of otoacoustic emissions. *Audiology, 38*, 187–195.

de Magalhães, S. L., Fukuda, Y., Liriano, R. I., Chami, F. A., Barros, F., & Diniz, F. L. (2003). Relation of hyper-acusis in sensorineural tinnitus patients with normal audiological assessment. *International Tinnitus Journal, 9*, 79–83.

Desai, A., Reed, D., Cheyne, A., Richards, S., & Prasher, D. (1999). Absence of otoacoustic emissions in subjects with normal audiometric thresholds implies exposure to noise. *Noise and Health, 1*, 58–65.

de Waal, R., Hugo, R., Soer, M., & Krüger, J. J. (2002). Predicting hearing loss from otoacoustic emissions using an artificial neural network. *South African Journal of Communication Disorders, 49*, 28–39.

Dhar, S., & Abdala, C. (2007). A comparative study of distortion-product-otoacoustic-emission fine structure in human newborns and adults with normal hearing. *Journal of the Acoustical Society of America, 122*, 2191–2202.

Dhooge, I. J., De Vel, E., Verhoye, C., Lemmerling, M., & Vinck, B. (2005). Otologic disease in turner syndrome. *Otology & Neurotology, 26*, 145–150.

Dhooge, I., Dhooge, C., Geukens, S., De Clerck, B., De Vel, E., & Vinck, B. M. (2006). Distortion product otoacoustic emissions: An objective technique for the screening of hearing loss in children treated with platin derivatives. *International Journal of Audiology, 45*, 337–343.

Digiovanni, J. J., & Nair, P. (2006). Spontaneous recovery of sudden sensorineural hearing loss: Possible association with autoimmune disorders. *Journal of the American Academy of Audiology, 17*, 498–505.

Di Girolamo, S., Napolitano, B., Alessandrini, M., & Bruno, E. (2007). Experimental and clinical aspects of the efferent auditory system, *Acta Neurochirurgie Supplement 97*, 419–424.

Dille, M., Glattke, T. J., & Earl, B. R. (2007). Comparison of transient evoked otoacoustic emissions and distortion product otoacoustic emissions when screening hearing in preschool children in a community setting. *International Journal of Pediatric Otorhinolaryngology, 71*, 1789–1795.

Dorn, P. A., Piskorski, P., Gorga, M. P., Neely, S. T., & Keefe, D. H. (1999). Predicting audiometric status from distortion product otoacoustic emissions using multivariate analysis. *Ear and Hearing, 20*, 149–163.

Doyle, K. J., McLaren, C. E., Shanks, J. E., Galus, C. M., & Meyskens, F. L. (2001). Effects of difluoromethylornithine chemoprevention on audiometry thresholds and otoacoustic emissions. *Archives of Otolaryngology-Head and Neck Surgery, 127*, 553–558.

Dragicević, D., Vlaski, L., Komazec, Z., & Jović, R. M. (2010). Transient evoked otoacoustic emissions in young children with otitis media with effusion before and after surgery. *Auris Nasus Larynx 37*, 281–285.

Dreisbach, L. E., Kramer, S. J., Cobos, S., & Cowart, K. (2007). Racial and gender effects on pure-tone thresholds and distortion-product otoacoustic emissions (DPOAEs) in normal-hearing young adults. *International Journal of Audiology, 46,* 419–426.

Dreisbach, L. E., Long, K. M., & Lees, S. E. (2006). Repeatability of high-frequency distortion-product otoacoustic emissions in normal-hearing adults. *Ear and Hearing, 27,* 466–479.

Dreisbach, L. E., & Siegel, J. H. (2001). Distortion-product otoacoustic emissions measured at high frequencies in humans, *Journal of the Acoustical Society of America, 110,* 2456–2469.

Dreisbach, L. E., & Siegel, J. H. (2005). Level dependence of distortion-product otoacoustic emissions measured at high frequencies in humans. *Journal of the Acoustical Society of America, 117,* 2980–2988.

Drexl, M., Henke, J., & Kössl, M. (2004). Isoflurane increases amplitude and incidence of evoked and spontaneous otoacoustic emissions. *Hearing Research, 194,* 135–142.

Driscoll, C., Kei, J., Bates, D., & McPherson, B. (2002). Transient evoked otoacoustic emissions in children studying in special schools. *International Journal of Pediatric Otorhinolaryngology, 64,* 51–60.

Driscoll, C., Kei, J., & McPherson, B. (2002). Handedness effects on transient evoked otoacoustic emissions in schoolchildren. *Journal of the American Academy of Audiology, 13,* 403–406.

Driscoll, C., Kei, J., Shyu, J., & Fukai, N. (2004). The effects of body position on distortion-product otoacoustic emission testing. *Journal of the American Academy Audiology, 15,* 566–573.

Duman, K., Ayçiçek, A., Sargin, R., Kenar, F., Yilmaz, M. D., & Dereköy, F. S. (2008). Incidence of auditory neuropathy among the deaf school students. *International Journal of Pediatric Otorhinolaryngology, 72,* 1091–1095.

Dunckley, K. T., & Dreisbach, L. E. (2004). Gender effects on high frequency distortion product otoacoustic emissions in humans. *Ear and Hearing, 25,* 554–564.

Durante, A. S., & Carvallo, R. M. (2002). Contralateral suppression of otoacoustic emissions in neonates. *International Journal of Audiology, 41,* 211–215.

Durante, A. S., & Carvallo, R. M. (2008). Contralateral suppression of linear and nonlinear transient evoked otoacoustic emissions in neonates at risk for hearing loss. *Journal of Communicative Disorders, 41,* 70–83.

Duvdevany, A., & Furst, M. (2006). Immediate and long-term effect of rifle blast noise on transient-evoked otoacoustic emissions. *Journal of Basic Clinical Physiology and Pharmacology, 17,* 173–185.

Duvdevany, A., & Furst, M. (2007). The effect of longitudinal noise exposure on behavioral audiograms and transient-evoked otoacoustic emissions. *International Journal of Audiology, 46,* 119–127.

Eggermont, J. J. (2007). Pathophysiology of tinnitus. (Langguth, B., Hajak, G., Kleinjung, T., Cacace, A. & Moller, A.R., Eds.). *Progress in Brain Research, 166,* 19–32.

Eiserman, W. D., Hartel, D. M., Shisler, L., Buhrmann, J., White, K. R., & Foust, T. (2008).Using otoacoustic emissions to screen for hearing loss in early childhood care settings. *International Journal of Pediatric Otorhinolaryngology, 72,* 475–482.

El-Badry, M. M., & McFadden, S. L. (2009). Evaluation of inner hair cell and nerve fiber loss as sufficient pathologies underlying auditory neuropathy. *Hearing Research, 255,* 84–90.

el-Kady, M. A., Durrant, J. D., Tawfik, S., Abdel-Ghany, S., & Moussa, A. M. (2006). Study of auditory function in patients with chronic obstructive pulmonary diseases. *Hearing Research, 212,* 109–116.

Ellison, J. C., & Keefe, D. H. (2005). Audiometric predictions using stimulus-frequency otoacoustic emissions and middle ear measurements. *Ear and Hearing, 26,* 487–503.

Emanuel, D. C. (2002). The auditory processing battery: Survey of common practices. *Journal of the American Academy of Audiology, 13,* 93–117.

Emara, A. A., & Gabr, T. A. (2010). Auditory steady state response in auditory neuropathy. *Journal of Laryngology and Otology, 124,* 950–956.

Engdahl, B. (2007). Otoacoustic emissions in the general adult population of Nord-Trøndelag, Norway: I. Distributions by age, gender, and ear side. *International Journal of Audiology, 41,* 64–77.

Engdahl, B., & Tambs, K. (2002). Otoacoustic emissions in the general adult population of Nord-Trøndelag, Norway: II. Effects of noise, head injuries, and ear infections. *International Journal of Audiology, 41,* 78–87.

Engdahl, B., Tambs, K., Borchgrevink, H. M., & Hoffman, H. J. (2005). Otoacoustic emissions in the general adult population of Nord-Trøndelag, Norway: III. Relationships with pure-tone hearing thresholds. *International Journal of Audiology, 44,* 15–23.

Engel-Yeger, B., Zaaroura, S., Zlotogora, J., Shalev, S., Hujeirat, Y., Carrasquillo, M., . . . Pratt, H. (2003). Otoacoustic emissions and brainstem evoked potentials in compound carriers of connexin 26 mutations. *Hearing Research, 175,* 140–151.

Epstein, M., Buus, S., & Florentine, M. (2004). The effects of window delay delinearization, and frequency on tone-burst otoacoustic emission input/

output measurements. *Journal of the Acoustical Society of America, 116,* 1160–1167.

Erbek, S. S., Erbek, H. S., Yilmaz, S., Topal, O., Yucel, E., & Ozluoglu, L. N. (2006). Cochleovestibular dysfunction in ankylosing spondylitis. *Audiology & Neurootology, 11,* 294–300.

Erdem, T., Ozturan, O., Miman, M.C., Ozturk, C., & Karatas, E. (2003). Exploration of the early auditory effects of hyperlipoproteinemia and diabetes mellitus using otoacoustic emissions. *European Archives of Otorhinolaryngology, 260,* 62–66.

Esparza, C. M., Jáuregui-Renaud, K., Morelos, C. M., Muhl, G. E., Mendez, M. N., Carillo, N. S., . . . Cardenas, M. (2007). Systemic high blood pressure and inner ear dysfunction: A preliminary study. *Clinical Otolaryngology, 32,* 173–178.

Fausti, S. A., Larson, V. D., Noffsinger, D., Wilson, R. H., Phillips, D. S., & Fowler, C. G. (1994). High-frequency audiometric monitoring strategies for early detection of ototoxicity. *Ear and Hearing, 15,* 232–239.

Fávero, M. L., Sanchez, T. G., Bento, R. F., & Nascimento, A. F. (2006). Contralateral suppression of otoacoustic emission in patients with tinnitus [Article in Portuguese]. *Review Brasilian Otorrinolaringology, 72,* 223–226.

Ferré Rey, J., & Morelló-Castro, G. (2003). Validation of the otoacoustic emissions in presbyacusis [Article in Spanish]. *Acta Otorrinolaringology Espana, 54,* 177–182.

Ferri, G. G., Modugno, G. C., Calbucci, F., Ceroni, A. R., & Pirodda, A. (2009). Hearing loss in vestibular schwannomas: Analysis of cochlear function by means of distortion-product otoacoustic emissions. *Auris Nasus Larynx, 36,* 644–648.

Fetoni, A. R., Garzaro, M., Ralli, M., Landolfo, V., Sensini, M., Pecorari, G., . . . Giordano, C. (2009). The monitoring role of otoacoustic emissions and oxidative stress markers in the protective effect of antioxidant administration in noise-exposed subjects: A pilot study. *Medical Science Monitor, 15,* PR1–8.

Fetterman, B. L. (2001). Distortion-product otoacoustic emissions and cochlear microphonics: Relationships in patients with and without endolymphatic hydrops. *Laryngoscope, 111,* 946–954.

Filipo, R., Attanasio, G., Barbaro, M., Viccaro, M., Musacchio, A., Cappelli, G., & De Seta, E. (2007). Distortion product otoacoustic emissions in otosclerosis: Intraoperative findings. *Advances in Otorhinolaryngology, 65,* 133–136.

Fitzgerald, T., & Prieve, B. (2005). Detection of hearing loss using 2f2-f1 and 2f1-f2 distortion-product otoacoustic emissions. *Journal of Speech, Language and Hearing Research, 48,* 1165–1186.

Foerst, A., Beutner, D., Lang-Roth, R., Huttenbrink, K. B., von Wedel, H., & Walger, M. (2006). Prevalence of auditory neuropathy/synaptopathy in a population of children with profound hearing loss. *International Journal of Pediatric Otorhinolaryngology, 70,* 1415–1422.

Forli, F., Mancuso, M., Santoro, A., Dotti, M. T., Siciliano, G., & Berrettini, S. (2006). Auditory neuropathy in a patient with mitochondrial myopathy and multiple mtDNA deletions. *Journal of Laryngology and Otology, 120,* 888–891.

Fortnum, H. M., Summerfield, A. Q., Marshall, D. H., Davis, A. C., & Bamford, J. M. (2001). Prevalence of permanent childhood hearing impairment in the United Kingdom and implications for universal neonatal hearing screening: Questionnaire based ascertainment study. *British Medical Journal, 323,* 536–554.

Fraenkel, R., Freeman, S., & Sohmer, H. (2003). Use of ABR threshold and OAEs in detection of noise induced hearing loss. *Journal of Basic Clinical Physiolology and Pharmacology, 14,* 95–118.

François, M., Laccourreye, L., Huy, E. T., & Narcy, P. (1997). Hearing impairment in infants after meningitis: detection by transient evoked otoacoustic emissions. *Journal of Pediatrics, 130,* 712–717.

Frederiksen, B. L., Cayé-Thomasen, P., Lund, S. P., Wagner, N., Asal, K., Olsen, N. V., & Thomsen, J. (2007). Does erythropoietin augment noise induced hearing loss? *Hearing Research, 223,* 129–137.

Fridman, V. L. (2003). Registration of various classes of otoacoustic emission in definition of acoustic sensitivity in different forms of neurosensory hypoacusis [Article in Russian]. *Vestnik Otorinolaringology, 6,* 20–23.

Frisina, S. T., Mapes, F., Kim, S., Frisina, D. R., & Frisina, R. D. (2006). Characterization of hearing loss in aged type II diabetics. *Hearing Research, 211,* 103–113.

Fritsch, M. H., Wynne, M. K., & Diefendorf, A. O. (2002). Transient-evoked otoacoustic emissions from ears with tympanostomy tubes. *International Journal of Pediatric Otorhinolaryngology, 66,* 29–36.

Froelich, P., Collet, L., & Morgon, A. (1993). Transiently evoked otoacoustic emission amplitudes change with changes in directed attention. *Physiology and Behavior, 53,* 679–682.

Fukai, N., Shyu, J., Driscoll, C., & Kei, J. (2005). Effects of body position on transient evoked otoacoustic emissions: the clinical perspective. *International Journal of Audiology, 44,* 8–14.

Galambos, R. (1956). Suppression of auditory nerve activity by stimulation of efferent fibers to the cochlea. *Journal of Neurophysiology, 19,* 424–437.

Gallo-Terán, J., Morales-Angulo, C., Sánchez, N., Manrique, M., Rodríguez-Ballesteros, M., Moreno-

Pelayo, M.A., . . . del Castillo, I. (2006). Auditory neuropathy due to the Q829X mutation in the gene encoding otoferlin (OTOF) in an infant screened for newborn hearing impairment [Article in Spanish]. *Acta Otorrinolaringologica Espana, 57*, 333–335.

Garinis, A. C., Glattke, T., & Cone-Wesson, B. K. (2008). TEOAE suppression in adults with learning disabilities. *International Journal of Audiology, 47*, 607–614.

Garner, C. A., Neely, S. T., & Gorga, M. P. (2008). Sources of variability in distortion product otoacoustic emissions. *Journal of the Acoustical Society of America, 124*, 1054–1067.

Gawron, W., Pośpiech, L., Noczyńska, A., & Orendorz-Fraczkowska, K. (2004). Electrophysiological tests of the hearing organ in Hashimoto's disease. *Journal of Pediatric Endocrinology & Metabolism, 17*, 27–32.

Gawron, W., Wikiera, B., Rostkowska-Nadolska, B., Orendorz-Fraczkowska, K., & Noczyńska, A. (2008). Evaluation of hearing organ in patients with Turner syndrome. *International Journal of Pediatric Otorhinolaryngology, 72*, 575–579.

Georgalas, C., Xenellis, J., Davilis, D., Tzangaroulakis, A., & Ferekidis, E. (2008). Screening for hearing loss and middle-ear effusion in school-age children, using transient evoked otoacoustic emissions: A feasibility study. *Journal of Laryngology and Otology, 21*, 1–6.

Germain, D. P., Avan, P., Chassaing, A., & Bonfils, P. (2002). Patients affected with Fabry disease have an increased incidence of progressive hearing loss and sudden deafness: An investigation of twenty-two hemizygous male patients. *BMC Medical Genetics, 3*, 10.

Gibson, W. P., & Sanli, H. (2007). Auditory neuropathy: An update. *Ear and Hearing, 28* (Suppl. 2),102S–106S.

Gierek, T., Markowski, J., Kokot, F., Paluch, J., Wiecek, A., & Klimek, D. (2002). Electrophysiological examinations (ABR and DPOAE) of hearing organ in hemodialysed patients suffering from chronic renal failure [Article in Polish]. *Otolaryngology Polska, 56*, 189–194.

Gierek, T., Smółka, W., Zbrowska-Bielska, D., Klimczak-Gołab, L., & Majzel, K. (2004). The evaluation of distortion product otoacoustic emissions after stapedotomy [Article in Polish]. *Otolaryngology Polska, 58*, 817–820.

Giraud, A., Collet, L., & Chery-Croze, S. (1997). Suppression of otoacoustic emissions is unchanged after several minutes of contralateral acoustic stimulation. *Hearing Research, 109*, 78–82.

Gkoritsa, E., Korres, S., Psarommatis, I., Tsakanikos, M., Apostolopoulos, N., & Ferekidis, E. (2007). Maturation of the auditory system: 1. Transient otoacoustic emissions as an index of inner ear maturation. *International Journal of Audiology, 46*, 271–276.

Gkoritsa, E., Korres, S., Segas, I., Xenelis, I., Apostolopoulos, N., & Ferekidis, E. (2007). Maturation of the auditory system: 2. Transient otoacoustic emission suppression as an index of the medial olivocochlear bundle maturation. *International Journal of Audiology, 46*, 277–286.

Gkoritsa, E., Tsakanikos, M., Korres, S., Dellagrammaticas, H., Apostolopoulos, N., & Ferekidis, E. (2006). Transient otoacoustic emissions in the detection of olivocochlear bundle maturation. *International Journal of Pediatric Otorhinolaryngology, 70*, 671–676.

Glattke, T. J., Pafitis, I. A., Cummiskey, C., & Herer, G. R. (1995). Identification of hearing loss in children using measures of transient otoacoustic emission reproducibility. *American Journal of Audiology, 4*, 71–86.

Goforth, L., Hood, L. J., & Berlin, C. I. (1997). Efferent suppression of transient-evoked otoacoustic emissions in human infants. *ARO Abstracts, 20*, 166.

Gold, T J. (1989). New ideas in science. *Journal of Scientific Exploration, 3*(2), 103–112.

Goodman, S. S., Fitzpatrick, D. F., Ellison, J. C., Jesteadt, W., & Keefe, D. H. (2009). High-frequency click-evoked otoacoustic emissions and behavioral thresholds in humans, *Journal of the Acoustical Society of America, 125*, 1014–1032.

Goodman, S. S., Withnell, R. H., De Boer, E., Lilly, D. J., & Nuttall, A. L. (2004). Cochlear delays measured with amplitude-modulated tone-burst-evoked OAEs. *Hearing Research, 188*, 57–69.

Gopal, K. V., Carney, L., & Bishop, C. E. (2004). Auditory measures in clinically depressed individuals. I. Basic measures and transient otoacoustic emissions. *International Journal of Audiology, 43*, 493–498.

Gopal, K. V., Herrington, R., & Pearce, J. (2009). Analysis of auditory measures in normal hearing young male adult cigarette smokers using multiple variable selection m methods with predictive validation assessments. *International Journal of Otolaryngology.* E-pub ahead of print 745151.

Gorga, M. P., Dierking, D. M., Johnson, T. A., Beauchaine, K. L., Garner, C. A., & Neely, S. T. (2005). A validation and potential clinical application of multivariate analyses of distortion-product otoacoustic emission data. *Ear and Hearing, 26*, 593–607.

Gorga M. P., Neely, S. T, Dierking, D. M., Kopun, J., Jolkowski, K., Groenenboom, K., Tan, H., & Stiegemann, B. (2008). Low-frequency and high-frequency distortion product otoacoustic emission suppression in humans. *Journal of the Acoustical Society of America, 123*, 2172–2190.

Gorga, M. P., Neely, S. T., Dorn, P. A., & Hoover, B. M. (2003). Further efforts to predict pure-tone thresholds from distortion product otoacoustic emission input/output functions. *Journal of the Acoustical Society of America, 113,* 3275–3284.

Gorga, M. P., Nelson, K., Davis, T., Dorn, P. A., & Neely, S. T. (2000). Distortion product otoacoustic emission test performance when both 2f1-f2 and 2f2-f1 are used to predict auditory status, *Journal of the Acoustical Society of America, 107,* 2128–2135.

Gossow-Müller-Hohenstein, E., Hirschfelder, A., Scholz, G., & Mrowinski, D. (2003). Aural fullness and endolymphatic hydrops [Article in German]. *Laryngorhinootologie., 82,* 97–101.

Gouveris, H., & Mann, W. (2004). Increased amplitudes of distortion product otoacoustic emissions in patients with unilateral acoustic neuroma. *ORL Journal of Otorhinolaryngology and Related Specialties, 66,* 302–325.

Gouveris, H., Maurer, J., & Mann, W. (2005). DPOAE-grams in patients with acute tonal tinnitus. *Otolaryngology-Head and Neck Surgery, 132,* 550–553.

Gouveris, H., Victor, A., & Mann, W. (2006). Transient evoked otoacoustic emissions in vestibular neuritis. *Annals of Otology Rhinology Laryngology, 115,* 908–911.

Gouveris, H. T., Victor, A., & Mann, W. J. (2007). Cochlear origin of early hearing loss in vestibular schwannoma. *Laryngoscope, 117,* 680–683.

Granjeiro, R. C., Kehrle, H. M., Bezerra, R. L., Almeida, V. F., Sampaio, A. L., & Oliveira, C. A. (2008). Transient and distortion product evoked oto-acoustic emissions in normal hearing patients with and without tinnitus. *Otolaryngology-Head and Neck Surgery, 138,* 502–506.

Gravel, J. S., Dunn, M., Lee, W. W., & Ellis, M. A. (2006). Peripheral audition of children on the autistic spectrum. *Ear and Hearing, 27,* 299–312.

Grewe, T. S., Danhauer, J. L., Danhauer, K. J., & Thornton, A. R. (1994). Clinical use of otoacoustic emissions in children with autism. *International Journal of Pediatric Otorhinolaryngology, 30,* 123–132.

Griffiths, H., James, D., Davis, R., Hartland, S., & Molony, N. (2007). Hearing threshold assessment post grommet insertion. Is it reliable? *Journal of Laryngology and Otology, 121,* 431–434.

Griz, S., Cabral, M., Azevedo, G., & Ventura, L. (2007). Audiologic results in patients with Moebiüs sequence. *International Journal of Pediatric Otorhinolaryngology, 71,* 1457–1463.

Groh, D., Pelanova, J., Jilek, M., Popelar, J., Kabelka, Z., & Syka, J. (2006). Changes in otoacoustic emissions and high-frequency hearing thresholds in children and adolescents. *Hearing Research, 212,* 90–98.

Gruss, I., Berlin, M., Greenstein, T., Yagil, Y., & Beiser, M. (2007). Etiologies of hearing impairment among infants and toddlers: 1986–1987 versus 2001. *International Journal of Pediatric Otorhinolaryngology, 71,* 1585–1589.

Guimaraes, P., Frisina, S. T., Mapes, F., Tadros, S. F., Frisina, D. R., & Frisina, R. D. (2006). Progestin negatively affects hearing in aged women. *Proceedings of the National Academy of Sciences U S A, 103,* 14246–14249.

Guinan, J. J., Jr. (2006). Olivocochlear efferents: Anatomy, physiology, function, and the measurement of efferent effects in humans. *Ear and Hearing, 27,* 589–607.

Guinan, J. J., Jr., Backus, B. C., Lilaonitkul, W., & Aharonson, V. (2003). Medial olivocochlear efferent reflex in humans: otoacoustic emission (OAE) measurement issues and the advantages of stimulus frequency OAEs. *Journal of the Association for Research in Otolaryngology, 4,* 521–540.

Gunduz, B., Bayazit, Y. A., Celenk, F., Saridoğan, C., Guclu, A. G., Orcan, E., & Meray, J. (2008). Absence of contralateral suppression of transiently evoked otoacoustic emissions in fibromyalgia syndrome. *Journal of Laryngology and Otology, 122,* 1047–1051.

Guo, Y. K., Yang, X. M., Xie, D. H., Tang, Q. L., & Lu, Y. D. (2005). Clinical observation of sensorineural hearing loss in patients suffering from nasopharyngeal carcinoma after radiotherapy [Article in Chinese]. *Zhonghua Er Bi Yan Hou Tou Jing Wai Ke Za Zhi, 40,* 805–809.

Guven, S., Tas, A., Adali, M. K., Yagiz, R., Alagol, A., Uzun, C., . . . Karasalihoglu, A. R. (2006). Influence of anaesthetic agents on transient evoked otoacoustic emissions and stapedius reflex thresholds. *Journal of Laryngology and Otology, 120,* 10–15.

Hajduk, A., Lisowska, G., Namysłowski, G., Szprynger, K., Szczepańska, M., & Widziszowska, A. (2005). Evaluation of cochlear function in children with chronic renal failure [Article in Polish]. *Polski Merkuriusz Lekarski, 19,* 304–306.

Hall, J. W. III. (2000). *Handbook of otoacoustic emissions.* San Diego, CA: Singular Publishing Group.

Hall, J. W. III. (2004). An ounce of prevention is worth a pound of cure. *Tinnitus Today: Journal of the American Tinnitus Association, 29,* 14–16.

Hall, J. W. III. (2006, April). *Hearing screening of kindergarten children: Increasing efficiency and accuracy.* Presentation at the annual Convention of the American Academy of Audiology, Minneapolis, MN.

Hall, J. W. III, & Johnston, K. N. (2007). Electroacoustic and electrophysiologic auditory measures in the assessment of (central) auditory processing disorder. In F. E. Musiek & G. D. Chermak (Eds.), *Handbook of (central) auditory processing disorder. Volume I. Auditory neuroscience and diagnosis.* San Diego, CA: Plural Publishing.

Hall, J. W. III, Smith , S., & Popelka, G. (2004). Newborn hearing screening with combined otoacoustic emissions and auditory brainstem response. *Journal of the American Academy of Audiology, 15,* 414–425.

Hall, J. W. III, & Swanepoel, D. W. (2010). *Objective assessment of hearing.* San Diego, CA: Plural Publishing.

Hallenbeck, H., & Dancer, J. (2003). Distortion-product otoacoustic emissions in ears with normal hearing sensitivity: test-retest variability. *Perceptual Motor Skills, 97,* 990–992.

Halligan, C. S., Bauch, C. D., Brey, R. H., Achenbach, S. J., Bamlet, W. R., McDonald, T. J., & Matteson, E. L. (2006). Hearing loss in rheumatoid arthritis. *Laryngoscope, 116,* 2044–2049.

Halloran, D. R., Hardin, J. M., & Wall, T. C. (2009). Validity of pure-tone hearing screening at well-child visits. *Archives of Pediatric and Adolescent Medicine, 163,* 158–163.

Halloran, D. R., Wall, T. C., Evans, H. H., Hardin, J. M., & Woolley, A. L. (2005). Hearing screening at well-child visits. *Archives of Pediatric and Adolescent Medicine, 159,* 949–955.

Hamdan, A. L., Abouchacra, K. S., Zeki Al Hazzouri, A. G., & Zaytoun, G. (2008). Transient-evoked otoacoustic emissions in a group of professional singers who have normal pure-tone hearing thresholds. *Ear and Hearing, 29,* 360–377.

Hamed, S. A., & El-Attar, A. M. (2010). Cochlear dysfunction in hyperuricemis: Otoacoustic emission analysis. *American Journal of Otolaryngology, 31,* 154–161.

Hamed, S. A., Elattar, A. M., & Hamed, E. A. (2006). Irreversible cochlear damage in myasthenia gravis —otoacoustic emission analysis. *Acta Neurolologica Scandinavia, 113,* 46–54.

Han, J., Li, F., Zhao, C., Zhang, Z., & Ni, D. (2003). Study on distortion product otoacoustic emissions and expanded high frequency audiometry in noise exposure workers [Article in Chinese]. *Lin Chuang Er Bi Yan Hou Ke Za Zhi, 17,* 16–19.

Harding, G. W., Bohne, B. A., Lee, S. C., & Salt, A. N. (2007). Effect of infrasound on cochlear damage from exposure to a 4 kHz octave band of noise. *Hearing and Research, 225,* 128–138.

Harkrider, A. W., & Bowers, C. D. (2009). Evidence for a cortically mediated release from inhibition in the human cochlea. *Journal of the American Academy of Audiology, 20,* 208–215.

Harkrider, A. W., Champlin, C. A., & McFadden, D. (2001). Acute effect of nicotine on non-smokers: I. OAEs and ABRs. *Hearing Research, 160,* 73–88.

Harkrider, A. W., & Tampas, J. W. (2006). Differences in responses from the cochleae and central nervous systems of females with low versus high acceptable noise levels. *Journal of the American Academy of Audiology, 17,* 667–676.

Harris, F. P., & Probst, R. (1992). Transiently evoked otoacoustic emissions in patients with Menière's disease. *Acta Otolaryngologica, 112,* 36–44.

Hatzopoulos, S., Amoroso, C., Aimoni, C., Lo Monaco, A., Govoni, M., & Martini, A. (2002). Hearing loss evaluation of Sjögren's syndrome using distortion product otoacoustic emissions. *Acta Otolaryngologica Supplement, 548,* 20–25.

Hatzopoulos, S., Ciorba, A., Petruccelli, J., Grasso, D., Sliwa, L., Kochanek, K., . . . Martini, A. (2009). Estimation of pure-tone thresholds in adults using extrapolated distortion product otoacoustic emission input/output-functions and auditory steady state responses. *International Journal of Audiology, 48,* 625–631.

Hatzopoulos, S., Petrucelli, J., Morlet, T., & Martini, A. (2003). TEOAE recording protocols revised: data from adult subjects, *International Journal of Audiology, 42,* 339–347.

Hatzopoulos, S., Petruccelli, J., Pelosi, G., & Martini, A. (1999). A TEOAE screening protocol based on linear click stimuli: Performance and scoring criteria, *Acta Otolaryngologica, 119,* 135–139.

Heitmann, J., Waldman, B., Schnitzler, H. U., Plinkert, P. K., & Zenner, H. P. (1998). Suppression of distortion product otoacoustic emissions (DPOAE) near f1-f2 removes DP-gram fine structure—Evidence for secondary generator. *Journal of the Acoustical Society of America, 103,* 1527–1531.

Helleman, H. W., Jansen, E. J., & Dreschler, W. A. (2010). Otoacoustic emissions in a hearing conservation program: General applicability in longitudinal monitoring and the relation to changes in pure-tone thresholds. *International Journal of Audiology, 49,* 410–419.

Herzog, M., Shehata-Dieler, W. E., & Dieler, R. (2001). Transient evoked and distortion product otoacoustic emissions following successful stapes surgery. *European Archives of Otorhinolaryngology, 258,* 61–66.

Hess, C., Rosanowski, F., Eysholdt, U., & Schuster, M. (2006). Hearing impairment in children and adolescents with Down's syndrome [Article in German]. *HNO, 54,* 227–232.

Hesse, G., Andres, R., Schaaf, H., & Laubert, A. (2008). DPOAE and lateral inhibition in chronic tinnitus [Article in German]. *HNO, 56,* 694–700.

Hesse, G., Schaaf, H., & Laubert, A. (2005). Specific findings in distortion product otoacoustic emissions and growth functions with chronic tinnitus. *International Tinnitus Journal, 11,* 6–13.

Hild, U., Hey, C., Baumann, U., Montgomery, J., Euler, H. A., & Neumann, K. (2008). High prevalence of hearing disorders at the Special Olympics indicate need to screen persons with intellectual disability. *Journal of Intellectual Disability Research, 52,* 520–528.

Hildesheimer, M., Hamburger, A., Ari-Even Roth, D., Muchnik, C., & Kuint, J. (1999). The maturation of the auditory efferent system in neonates tested by suppression effect of transient evoked otoacoustic emission (TEOAE). *IERASG Abstracts, XVI,* 1–5.

Hirschfelder, A., Gossow-Müller-Hohenstein, E., Hensel, J., Scholz, G., & Mrowinski D. (2005). Diagnosis of endolymphatic hydrops using low frequency modulated distortion product otoacoustic emissions [Article in German]. *HNO, 53,* 597–599, 612–617.

Ho, V., Daly, K. A., Hunter, L. L., & Davey, C. (2002). Otoacoustic emissions and tympanometry screening among 0–5 year olds. *Laryngoscope, 112,* 513–519.

Hof, J. R., Anteunis, L. J., Chenault, M. N., & van Dijk, P. (2005). Otoacoustic emissions at compensated middle ear pressure in children. *International Journal of Audiology, 44,* 317–320.

Hof, J. R., Dijk, P., Chenault, M. N., & Anteunis, L. J. (2005). A two-step scenario for hearing assessment with otoacoustic emissions at compensated middle ear pressure (in children 1–7 years old). *International Journal of Pediatric Otorhinolaryngology, 69,* 649–655.

Hof-Duin, N. J., & Wit, H. P. (2007). Evaluation of low-frequency biasing as a diagnostic tool in Ménière patients. *Hearing Research, 231,* 84–89.

Hoffmann, J., Ihrig, A., Hoth, S., & Triebig, G. (2006). Field study to explore possible effects of styrene on auditory function in exposed workers. *Indian Health, 44,* 283–286.

Hofstetter, P., Ding,, D., & Salvi, R. (1997). Magnitude and pattern of inner and outer hair cell loss in chinchilla as a function of carboplatin dose. *Audiology, 36,* 301–311.

Holenweg, A., & Kompis, M. (2010). Non-organic hearing loss: New and confirmed findings. *European Archives of Otorhinolaryngology, 267,* 1213–1219.

Hood, L. J., Berlin, C. I., Bordelon, J., & Rose, K. (2003). Patients with auditory neuropathy/dys-synchrony lack efferent suppression of transient evoked otoacoustic emissions. *Journal of the American Academy of Audiology, 14,* 302–313.

Hoth, S. (2005). On a possible prognostic value of otoacoustic emissions: a study on patients with sudden hearing loss. *European Archives of Otorhinolaryngology, 262,* 217–224.

Hoth, S., & Bönnhoff, S. (1993). Clinical use of transitory otoacoustic evoked emissions in therapeutic follow-up [Article in German]. *HNO, 41,* 135–145.

Hoth, S., Gudmundsdottir, K., & Plinkert, P. (2010). Age dependence of otoacoustic emissions: The loss of amplitude is primarily caused by age-related hearing loss and not by aging alone. *European Archives or Otorhinolaryngology, 267,* 679–690.

Hunter, L. L., Davey, C. S., Kohtz, A., & Daly, K. A. (2007). Hearing screening and middle ear measures in American infants and toddlers. *International Journal of Pediatric Otorhinolaryngology, 71,* 1429–1438.

Hutchinson, K. M., Alessio, H., & Baiduc, R. R. (2010). Association between cardiovascular health and hearing function: Pure-tone and distortion product otoacoustic emission measures. *American Journal of Audiology, 19,* 26–35.

Hwang, J. H., Tan, C. T., Chiang, C. W., & Liu, T. C. (2003). Acute effects of alcohol on auditory thresholds and distortion product otoacoustic emissions in humans. *Acta Otolaryngologica, 123,* 936–940.

Ibargüen, A. M., Montoya, F. S., del Rey, A. S., & Fernandez, J. M. (2008). Evaluation of the frequency selectivity of contralateral acoustic stimulation on the active mechanisms of the organ of Corti by analyzing the changes in the amplitude of transitory evoked otoacoustic emissions and distortion products. *Journal of Otolaryngology-Head and Neck Surgery, 37,* 457–462.

Ikiz, A. O., Unsal, E., Kirkim, G., Erdag, T. K., & Guneri, E. A. (2007). Hearing loss and middle ear involvement in patients with juvenile idiopathic arthritis. *International Journal of Pediatric Otorhinolaryngology, 71,* 1079–1085.

Inoue, Y., Ogawa, K., & Kanzaki, J. (2003). Hearing improvement after tumor removal in a vestibular schwannoma patient with severe hearing loss. *European Archives of Otorhinolaryngology, 260,* 487–489.

International Electrotechnical Commission. (1981). Occluded ear-stimulator for the measurement of earphones couple to the ear by inserts. IEC 60711. Geneva, Switzerland: Author.

International Electrotechnical Commission. (2009). Electroacoustics—Audiometric equipment—Part 6: Instruments for the measurement of otoacoustic emissions. IEC 60645-6. Geneva, Switzerland: Author.

Ishida, I. M., Sugiura, M., Teranishi, M., Katayama, N., & Nakashima, T. (2008). Otoacoustic emissions, ear fullness and tinnitus in the recovery course of sudden deafness. *Auris Nasus Larynx, 35,* 41–46.

Ishikawa, K., Tamagawa, Y., Takahashi, K., Kimura, H., Kusakari, J., Hara, A., & Ichimura, K. (2002). Non-syndromic hearing loss caused by a mitochondrial T7511C mutation. *Laryngoscope, 112,* 1494–1499.

Ismail, H., & Thornton, A. R. (2003). The interaction between ear and sex differences and stimulus rate. *Hearing Research, 179,* 97–103.

Ito, K., Suzuki, S., Murofushi, T., Ishimoto, S., Iwasaki, S., & Karino, S. (2005). Neuro-otologic findings in unilateral isolated narrow internal auditory meatus. *Otology & Neurotology, 26,* 767–772.

Jabłonka, A., Pośpiech, L., & Orendorz-Fraczkowska, K. (2003). Evaluation of glycerol test in Meniere's disease with pure tone audiometry and distortion product otoacoustic emission [Article in Polish]. *Otolaryngology Polska, 57,* 731–737.

Jacob, L. C., Aguiar, F. P., Tomiasi, A. A., Tschoeke, S. N., & Bitencourt, R. F. (2006). Auditory monitoring in ototoxicity. *Review of Brasilian Otorrinolaringology, 72,* 836–844.

Jansen, E. J., Helleman, H. W., Dreschler, W. A., & de Laat, J. A. (2009). Noise induced hearing loss and other hearing complaints among musicians of symphony orchestras. *International Archives of Occupational & Environmental Health, 82,* 153–164.

Janssen, T., Boege, P., von Mikusch-Buchberg, J., & Raczek, J. (2005). Investigation of potential effects of cellular phones on human auditory function by means of distortion product otoacoustic emissions. *Journal of the Acoustical Society of America, 117,* 1241–1247.

Janssen, T., Gehr, D. D., Klein, A., & Müller, J. (2005). Distortion product otoacoustic emissions for hearing threshold estimation and differentiation between middle-ear and cochlear disorders in neonates. *Journal of the Acoustical Society of America, 117,* 2969–2979.

Janssen, T., Klein, A., & Gehr, D. D. (2003). Automated hearing threshold estimation in newborns using extrapolated DPOAE input/output functions [Article in German]. *HNO, 51,* 971–980.

Janssen, T., Niedermeyer, H. P., & Arnold, W. (2006). Diagnostics of the cochlear amplifier by means of distortion product otoacoustic emissions. *ORL Journal of Otorhinolaryngology and Related Specialties, 68,* 334–339.

Jedrzejczak, W. W., Blinowska, K. J., & Konopka, W. (2005). Time-frequency analysis of transiently evoked otoacoustic emissions of subjects exposed to noise. *Hearing Research, 205,* 249–255.

Jedrzejczak, W. W., Blinowska, K. J., & Konopka, W. (2006). Resonant modes in transiently evoked otoacoustic emissions and asymmetries between left and right ear. *Journal of the Acoustical Society of America, 119,* 2226–2231.

Jerger, J. F., & Hayes, D. (1976). The cross-check principle in pediatric audiometry. *Archives of Otolaryngology, 102,* 614–620.

Jerger, J., & Musiek, F. E. (2000). Report of the consensus conference on the diagnosis of auditory processing disorders in school-aged children. Consensus Development Conference. *Journal of the American Academy of Audiology, 11,* 467–474.

Jerger, J., & Musiek, F. E. (2002). On the diagnosis of auditory processing disorder: A reply to "Clinical and research concerns regarding the 2000 APD consensus report and recommendations. *Audiology Today, 14,* 19–21.

Jiang, Z. D., Zhang, Z., & Wilkinson, A. R. (2005). Distortion product otoacoustic emissions in term infants after hypoxia-ischaemia. *European Journal of Pediatrics, 164,* 84–87.

Job, A., Cian, C., Esquivié, D., Leifflen, D., Trousselard, M., Charles, C., & Nottet, J. B. (2004). Moderate variations of mood/emotional states related to alterations in cochlear otoacoustic emissions and tinnitus onset in young normal hearing subjects exposed to gun impulse noise. *Hearing Research, 193,* 31–38.

Job, A., & Nottet, J. B. (2002). DPOAEs in young normal-hearing subjects with histories of otitis media: Evidence of sub-clinical impairments. *Hearing Research, 167,* 28–32.

Job, A., Raynal, M., & Kossowski, M. (2007). Susceptibility to tinnitus revealed at 2 kHz range by bilateral lower DPOAEs in normal hearing subjects with noise exposure. *Audiology & Neurootology, 12,* 137–414.

Jock, B. M., Hamernik, R. P., Aldrich, L. G., Ahroon, W. A., Patriello, K. L., & Johnson, A. R. (1996). Evoked-potential thresholds and cubic distortion product otoacoustic emissions in the chinchilla following carboplatin treatment and noise exposure. *Hearing Research, 96,* 179–190.

Johansson, M. S., & Arlinger, S. D. (2003). Otoacoustic emissions and tympanometry in a general adult population in Sweden. *International Journal of Audiology, 42,* 448–464.

Johnsen, N. J., Bagi, P., & Elberling, C. (1983). Evoked acoustic emissions from the human ear: III. Findings in neonates. *Scandinavian Audiology, 12,* 17–24.

Johnson, A. C., Morata, T. C., Lindblad, A. C., Nylén, P. R., Svensson, E. B., Krieg, E., . . . Prasher, D. (2006). Audiological findings in workers exposed to styrene alone or in concert with noise. *Noise and Health, 8,* 45–57.

Johnson, T. A. (2010). Cochlear sources and otoacoustic emissions. *Journal of the American Academy of Audiology, 21,* 176–186.

Johnson, T. A., Neely, S. T., Garner, C. A., & Gorga, M. P. (2006). Influence of primary-level and primary-frequency ratios on human distortion product otoacoustic emissions. *Journal of the Acoustical Society of America, 119,* 418–428.

Johnson, T. A., Neely, S. T., Kopun, J. G., Dierking, D. M., Tan, H., & Gorga, M. P. (2010). Clinical test performance of distortion-product otoacoustic emissions using new stimulus conditions. *Ear and Hearing, 31,* 74–83.

Joint Committee of Infant Hearing. (2007). Year 2007 position statement: Principles and guidelines for early hearing detection and intervention programs. *Pediatrics,,120*(4), 898–921.

Jutras, B., Russell, L. J., Hurteau, A. M., & Chapdelaine, M. (2003). Auditory neuropathy in siblings with Waardenburg's syndrome. *International Journal of Pediatric Otorhinolaryngology, 67,* 1133–1142.

Kalcioglu, M. T., Bayindir, T., Erdem, T., & Ozturan, O. (2005). Objective evaluation of the effects of intravenous lidocaine on tinnitus. *Hearing Research, 199,* 81–88.

Kalluri, R., & Shera, C. A. (2007). Comparing stimulus-frequency otoacoustic emissions measured by compression, suppression, and spectral smoothing. *Journal of the Acoustical Society of America, 122,* 3562–3575.

Karabulut, H., Dagli, M., Ates, A., & Karaaslan, Y. (2010). Results for audiology and distortion product and transient evoked otoacoustic emissions in patients with systemic lupus erythematosus. *Journal of Laryngology and Otology, 124,* 137–140.

Karatas, E., Göksu, S., Durucu, C., Isik, Y., & Kanlikama, M. (2006). Evaluation of hearing loss after spinal anesthesia with otoacoustic emissions. *European Archives of Otorhinolaryngology, 263,* 705–710

Karatas, E., Miman, M. C., Ozturan, O., Erdem, T., & Kalcioglu, M. T. (2007). Contralateral normal ear after mastoid surgery: Evaluation by otoacoustic emissions (mastoid drilling and hearing loss). *ORL Journal of Otorhinolaryngology and Related Specialties, 69,* 18–24.

Karlsson, K. K., Berninger, E., & Alvan, G. (1991). The effect of quinine on psychoacoustic tuning curves, stapedius reflexes and evoked otoacoustic emissions in healthy volunteers. *Scandinavian Audiology, 20,* 83–90.

Katbamna, B., Homnick, D. N., & Marks, J. H. (1998). Contralateral suppression of distortion product otoacoustic emissions in children with cystic fibrosis: Effects of tobramycin. *Journal of the American Academy of Audiology, 9,* 172–178.

Katbamma, B., Homnick, D. N., & Marks, J. H. (1999). Effects of chronic tobramycin treatment on distortion product otoacoustic emissions. *Ear and Hearing, 20,* 393–402.

Katz, J., Johnson, C. D., Brandner, S., Delagrange, T., Ferre, J., King, J., . . . Tillery, K. (2002). Clinical and research concerns regarding the 2000 APD consensus report and recommendations. *Audiology Today, 14,* 14–17.

Ke, X., Yu, H., Liu, Y., Gu, Z., Lu, Y., & Li, L. (2000). Examinations of distortion product otoacoustic emission in hereditary progressive non-syndromic hearing loss [Article in Chinese]. *Zhonghua Er Bi Yan Hou Ke Za Zhi, 35,* 102–124.

Keefe, D. H. (1997). Otorefectance of the cochlea and middle ear. *Journal of the Acoustical Society of America, 102,* 2849–2859.

Keefe, D. H. (1998). Double-evoked otoacoustic emissions. I. Measurement theory and nonlinear coherence. *Journal of the Acoustical Society of America, 103,* 3489–3498.

Keefe, D.H. (2002). Spectral shapes of forward and reverse transfer functions between ear canal and cochlea estimated using DPOAE input/output functions. *Journal of the Acoustical Society of America, 111,* 249–260.

Keefe, D. H. (2007). Influence of middle-ear function and pathology on otoacoustic emissions. In M. S. Robinette & T. J. Glattke (Eds.), *Otoacoustic emissions: Clinical applications* (3rd ed., pp. 163–196). New York, NY: Thieme Medical Publishers.

Keefe, D. H., Gorga, M. P., Jesteadt, W., & Smith, L. M. (2008). Ear asymmetries in middle-ear, cochlear, and brainstem responses in human infants. *Journal of the Acoustical Society of America, 123,* 1504–1512.

Keefe, D. H., Gorga, M. P., Neely, S. T., Zhao, F., & Vohr, B. R. (2003). Ear-canal acoustic admittance and reflectance measurements in human neonates. II. Predictions of middle-ear in dysfunction and sensorineural hearing loss. *Journal of the Acoustical Society of America, 113,* 407–422.

Keefe, D. H., & Ling, R. (1998). Double-evoked otoacoustic emissions. II. Intermittent noise rejection, calibration and ear-canal measurements, *Journal of the Acoustical Society of America, 103,* 3499–3508.

Keefe, D. H., Zhao, F., Neely, S. T., Gorga, M. P., & Vohr, B. R. (2003). Ear-canal acoustic admittance and reflectance effects in human neonates. I. Predictions of otoacoustic emissions and auditory brainstem responses. *Journal of the Acoustical Society of America, 113,* 389–406.

Kei, J., Brazel, B., Crebbin, K., Richards, A., & Willeston, N. (2007). High frequency distortion product otoacoustic emissions in children with and without middle ear dysfunction. *International Journal of Pediatric Otorhinolaryngology, 71,* 125–133.

Kei, J., Robertson, K., Driscoll, C., Smyth, V., McPherson, B., Latham, S., & Loscher, J. J. (2002). Seasonal effects on transient evoked otoacoustic emission screening outcomes in infants versus 6-year-old children. *Journal of the American Academy of Audiology, 13,* 392–399.

Kei, J., Sockalingam, R., Holloway, C., Agyik, A., Brinin, C., & Baine, D. (2003). Transient evoked otoacoustic emissions in adults: a comparison between two test protocols. *Journal of the American Academy of Audiology, 14,* 536.

Kemp, D. T. (1978). Stimulated acoustic emissions from within the human auditory system. *Journal of the Acoustical Society of America, 64,* 1386–1391.

Kemp, D. T. (1979). The evoked cochlear mechanical response and the auditory microstructure—evidence for a new element in cochlear mechanics. *Scandinavian Audiology Supplement, 9,* 35–47

Kemp, D. T. (1980). Towards a model for the origin of cochlear echoes. *Hearing Research, 2,* 533–548

Kemp, D. T., & Brown, A. M. (1983). An integrated view of cochlear mechanical nonlinearities observable from the ear canal. In E. De Boer & M. A. Viergever (Eds.), *Mechanics of hearing* (pp. 75–82). The Hague, The Netherlands:: Martinus Nijhoff.

Kemp, D. T., Ryan, S., & Bray, P. (1990). A guide to the effective use of otoacoustic emissions. *Ear and Hearing, 11,* 93–105.

Keppler, H., Dhooge, I., Corthals, P., Maes, L., D'haenens, W., Bockstael, A., . . . Vinck, B. (2010). The effects of aging on evoked otoacoustic emissions and efferent suppression of transient evoked otoacoustic emissions. *Clinical Neurophysiology, 121,* 359–365.

Keppler, H., Dhooge, I., Maes, L., D'haenens, W., Bockstael, A., Philips, B., . . . Vinck, B. (2010). Short-term auditory effects of listening to an MP3 player. *Archives of Otolaryngology Head and Neck Surgery, 136,* 538–548.

Khalfa, S., Bruneau, N., Rogé B., Georgieff, N., Veuillet, E., Adrien, J. L., . . .Collet, L. (2001). Peripheral auditory asymmetry in infantile autism. *European Journal of Neuroscience, 13,* 628–638.

Khechinaschvili, S., Metreveli, D., Svanidze, N., Knothe, J., & Kevanishvili, Z. (2007). The hearing system under thyroid hypofunction. *Georgian Medical News.,144,* 30–33.

Kim, A. H., Edwards, B. M., Telian, S. A., Kileny, P. R., & Arts, H. A. (2006). Transient evoked otoacoustic emissions pattern as a prognostic indicator for hearing preservation in acoustic neuroma surgery. *Otology & Neurotology, 27,* 372–379.

Kim, D. O., Sun, X. M., Jung, M. D., & Leonard, G. (1997). A new method of measuring distortion product otoacoustic emissions using multiple tone pairs: Study of human adults. *Ear and Hearing, 18,* 277–285.

Kim, J. S., Nam, E. C., & Park, S. I. (2005). Electrocochleography is more sensitive than distortion-product otoacoustic emission test for detecting noise-induced temporary threshold shift. *Otolaryngology-Head and Neck Surgery, 133,* 619–624.

Kim, S., Frisina, D. R., & Frisina, R. D. (2002). Effects of age on contralateral suppression of distortion product otoacoustic emissions in human listeners with normal hearing. *Audiology & Neurootology, 7,* 348–357.

Kirkim, G., Erdağ, T. K., & Serbetçioğlu, M. B. (2008). [Audiological evaluation of a child with Klippel Feil syndrome]. *Kulak Burun Boğaz Ihtisas Dergisi, 18,* 171–174.

Kirkim, G., Serbetçioğlu, M. B., & Ceryan, K. (2005). Auditory neuropathy in children: diagnostic criteria and audiological test results. *Kulak Burun Boğaz Ihtisas Dergisi, 15,* 1–8.

Knight, K. R., Kraemer, D. F., Winter, C., & Neuwelt, E. A. (2007). Early changes in auditory function as a result of platinum chemotherapy: Use of extended high-frequency audiometry and evoked distortion product otoacoustic emissions. *Journal of Clinical Oncology, 25,* 1190–1195.

Knight R. D. (2004). Diplacusis, hearing threshold and otoacoustic emissions in an episode of sudden, unilateral cochlear hearing loss. *International Journal of Audiology, 43,* 45–53.

Knight, R. D., & Kemp, D. T. (2000). Indications of different distortion product otoacoustic emission mechanisms from a detailed f1, f2 area study. *Journal of the Acoustical Society of America, 107,* 457–473.

Knight, R. D., & Kemp D. T. (2001). Wave and place fixed DPOAE maps of the human ear. *Journal of the Acoustical Society of America, 109,* 1513–1525.

Koike, K. J., & Wetmore, S. (1999) Interactive effects of middle ear pathology and the associated hearing loss on transient-evoked otoacoustic emission measures. *Otolaryngology-Head and Neck Surgery, 121,* 238–244.

Koivunen, P., Uhari, M., Laitakari, K., Ahlo, O. P., & Luotonen, J. (2000). Otoacoustic emissions and tympanometry in children with otitis media. *Ear and Hearing, 21,* 212–217.

Kok, M., van Zanten, G., Brocaar, M., & Jongejan, H. (1994). Click-evoked oto-acoustic emissions in very low birthweight infants: A cross-sectional data analysis. *Audiology, 33,* 152–164.

Komazec, Z., Lemajić, S., & Vlaski, L. (2004). Audiologic diagnostics of vestibular schwannoma [Article in Serbian]. *Medicinski Pregled, 57,* 81–85.

Konopka, W., & Olszewski, J. (2005). Otoacoustic emissions and recruitment [Article in Polish]. *Otolaryngology Polska, 59,* 731–736.

Konopka, W., Olszewski, J., Pietkiewicz, P., & Mielczarek, M. (2005). Impulse noise influence on hearing [Article in Polish]. *Polski Merkuriusz Lekarski, 19,* 296–297.

Konopka, W., Olszewski, J., Pietkiewicz, P., & Mielczarek, M. (2006). Distortion product otoacoustic emissions before and after one year exposure to impulse noise [Article in Polish]. *Otolaryngology Polska, 60,* 243–247.

Konopka, W., Olszewski, J., & Straszyński, P. (2006). Evaluation of the risk on hearing loss at soldiers [Article in Polish]. *Otolaryngology Polska, 60,* 249–253.

Konopka, W., Pawlaczyk-Luszczyńska, M., Straszyński, P., & Sliwińska-Kowalska, M. (2004). Assessment of acoustic environment and its effect on hearing in jet engine technical personnel [Article in Polish]. *Medical Progress, 55,* 329–335.

Konopka, W., Pawlaczyk-Luszczynska, M., Sliwinska-Kowalska, M., Grzanka, A., & Zalewski, P. (2005). Effects of impulse noise on transiently evoked otoacoustic emission in soldiers. *International Journal of Audiology, 44,* 3–7.

Konopka, W., Zalewski, P., & Pietkiewicz, P. (2001). Evaluation of transient and distortion product otoacoustic emissions before and after shooting practice. *Noise and Health, 3,* 29–37.

Konradsson, K. S., Svensson, O., Carlborg, B., & Grenner, J. (1999). Tympanic pressure gradients and otoacoustic emissions. *Ear and Hearing, 20,* 403–409.

Kornblum, C., Broicher, R., Walther, E., Herberhold, S., Klockgether, T., Herberhold, C., & Schröder, R. (2005). Sensorineural hearing loss in patients with chronic progressive external ophthalmoplegia or Kearns-Sayre syndrome. *Journal of Neurology, 252,* 1101–1107.

Korres, G. S., Balatsouras, D. G., Tzagaroulakis, A., Kandiloros, D., Ferekidou, E., & Korres, S. (2009). Distortion product otoacoustic emissions in an industrial setting. *Noise and Health, 11,* 103–110.

Korres, S., Balatsouras, D., Manta, P., Economou, C., Yiotakis, I., & Adamopoulos, G. (2002). Cochlear dysfunction in patients with mitochondrial myopathy. *ORL Journal of Otorhinolaryngology and Related Specialties, 64,* 315–320.

Korres, S. G., Balatsouras, D. G., Nikolopoulos, T., Korres, G. S., Economou, N. C., & Ferekidis, E. (2006). The effect of the number of averaged responses on the measurement of transiently evoked otoacoustic emissions in newborns. *International Journal of Pediatric Otorhinolaryngology, 70,* 429–433.

Kothe, C., Fleischer, S., Blank, M., & Hess, M. (2004). Pronounced bilateral mesocochlear hearing impairment from 40–45 dB in spite of regular TEOAE, DPOAE and inconspicuous click-BERA [Article in German]. *HNO, 52,* 557–560.

Kothe, C., Fleischer, S., Breitfuss, A., & Hess, M. (2006). Unilateral auditory neuropathy. A rare differential diagnosis of unilateral deafness [Article in German]. *HNO, 54,* 215–220.

Kramer, S., Dreisbach, L., Lockwood, J., Baldwin, K., Kopke, R., Scranton, S., & O'Leary, M. (2006). Efficacy of the antioxidant N-acetylcysteine (NAC) in protecting ears exposed to loud music. *Journal of the American Academy of Audiology, 17,* 265–278.

Krisztina, B. (2009). [Detection of ototoxic effect of cisplatin with otoacoustic emission in testicular cancer patients]. *Magyar Onkológia, 53,* 279–283.

Krueger, W. W., & Ferguson, L. (2002). A comparison of screening methods in school-aged children. *Otolaryngology-Head and Neck Surgery, 127,* 516–519.

Kumar, A., Mathew, K., Alexander, S. A., & Kiran, C. (2009). Output sound pressure levels of personal music systems and their effect on hearing. *Noise and Health, 11,* 132–140.

Kumar, A. U., Hegde, M., & Mayaleela. (2010). Perceptual learning of non-native speech contrast and functioning of the olivocochlear bundle. *International Journal of Audiology, 49,* 488–496.

Kumar, U. A., & Jayaram, M. M. (2006). Prevalence and audiological characteristics in individuals with auditory neuropathy/auditory dys-synchrony. *International Journal of Audiology, 45,* 360–366.

Kumar Sinha, A., Montgomery, J. K., Herer, G. R., & McPherson, D. L. (2008). Hearing screening outcomes for persons with intellectual disability: A preliminary report of findings from the 2005 Special Olympics World Winter Games. *International Journal of Audiology, 47,* 399–403.

Kummer, P., Schuster, E. M., Rosanowski, F., Eysholdt, U., & Lohscheller, J. (2006). The influence of conductive hearing loss on DPOAE-threshold. The effect of an individually optimized stimulation [Article in German]. *HNO, 54,* 457–464, 466–467.

Kuroda, T. (2007). Clinical investigation on spontaneous otoacoustic emission (SOAE) in 447 ears. *Auris Nasus Larynx, 34,* 29–38.

Kuroda, T., Chida, E., Kashiwamura, M., Matsumura, M., & Fukuda, S. (2008). Changes to spontaneous otoacoustic emissions (SOAEs) due to cisplatin administration. *International Journal of Audiology, 47,* 695–701.

Lalaki, P., Markou, K., Tsalighopoulos, M. G., & Daniilidis, I. (2001). Transiently evoked otoacoustic emissions as a prognostic indicator in idiopathic sudden hearing loss. *Scandinavian Audiology Supplement, 30,* 141–145.

Lapsley Miller, J. A., Marshall, L., & Heller, L. M. (2004). A longitudinal study of changes in evoked otoacoustic emissions and pure-tone thresholds as measured in a hearing conservation program. *International Journal of Audiology, 43,* 307–322.

Lapsley Miller, J. A., Marshall, L., Heller, L. M., & Hughes, L. M. (2006). Low-level otoacoustic emissions may predict susceptibility to noise-induced hearing loss. *Journal of the Acoustical Society of America, 120,* 280–296.

Lee, C., Robinson, P., & Chelladurai, J. (2002). Reversible sensorineural hearing loss. *International Journal of Pediatric Otorhinolaryngology, 66,* 297–301.

Lehmann, D., Weeks, S., Jacoby, P., Elsbury, D., Finucane, J., Stokes, A., . . . Komrp, K. O. (2008). Absent otoacoustic emissions predict otitis media in young Aboriginal children: A birth cohort study in Aboriginal and non-Aboriginal children in an arid zone of Western Australia. *BMC Pediatrics, 28,* 32

Lesperance, M. M., Hall, J. W., III, Bess, F. H., Jain, P., Ploplis, B., San Agustin, T. B., . . . Wilcox, E.R. (1995). A gene for autosomal dominant nonsyndromic hereditary hearing impairment maps to 4p16.3. *Human Molecular Genetics, 4,* 1967–1972.

Levina, Iu.V. (2004). Evoked otoacoustic emission in conduction of dehydration tests in patients with Meniere's disease [Article in Russian]. *Vestn Otorinolaringology, 4,* 11–14.

Li, F., Liang, R., Chen, C., & Song, W. (2006). Application of distortion product otoacoustic emissions in detection of pseudohypacusis and exaggerated hearing loss Article in Chinese]. *Lin Chuang Er Bi Yan Hou Ke Za Zhi, 20,* 303–305.

Li, F., Wang, H., Chen, J., & Liang, R. (2005). Auditory neuropathy in children(analysis of 14 cases) [Article in Chinese]. *Lin Chuang Er Bi Yan Hou Ke Za Zhi, 19,* 19–21.

Li, Q., & Zhong, N. (1999). Diagnostic significance of transiently evoked otoacoustic emissions in Meniere's disease [Article in Chinese]. *Lin Chuang Er Bi Yan Hou Ke Za Zhi., 13,* 435–437.

Lichtenstein, V., & Stapells, D. R. (1996). Frequency-specific identification of hearing loss using transient-evoked otoacoustic emissions to clicks and tones. *Hearing Research, 98,* 125–136.

Lin, J., Liu, R., & Chen, Q. (2008). Relationship between otoacoustic emissions and blood-lead levels in school children [Article in Chinese]. *Lin Chung Er Bi Yan Hou Tou Jing Wai Ke Za Zhi, 22,* 446–448.

Lindhardt, B. O. (2008). Hearing disorders and rock music. [Article in Danish]. *Ugeskr Laeger, 15,* 4233–4235.

Lisowska, G., Namysłowski, G., Hajduk, A., Polok, A., Tomaszewska, R., & Misiołek, M. (2005). Hearing evaluation in children during chemotherapy [Article in Polish]. *Polski Merkuriusz Lekarski, 19,* 340–342.

Lisowska, G., Namysłowski, G., Hajduk, A., Polok, A., Tomaszewska, R., & Misiołek, M. (2006). Otoacoustic emissions measurements in children during the chemotherapy because of the acute lymphoblastic leukemia [Article in Polish]. *Otolaryngology of Poland, 60,* 415–420.

Lisowska, G., Namysłowski, G., Misiołek, M., Scierski W., Orecka, B., Czecior, E., & Dziendziel, A. (2008). Efferent suppression test--sensitivity and specificity [Article in Polish]. *Otolaryngology Polska, 62,* 747–754.

Lisowska, G., Namyslowski, G., Morawski, K., & Strojek, K. (2001). Early identification of hearing impairments in patients with type 1 diabetes mellitus. *Otology & Neurotology, 22,* 316–320.

Lisowska, G., Namysłowski, G., Morawski. K., & Strojek, K. (2002). Otoacoustic emissions and auditory brain stem responses in insulin dependent diabetic patients [Article in Polish]. *Otolaryngology Polska, 56,* 217–225.

Lisowska, G., Namysłowski, G., Orecka, B., Misiołek, M., Scierski, W., & Czecior, E. (2007). The influence of aging on otoacoustic emissions in normally hearing subjects [Article in Polish]. *Otolaryngology Polska, 61,* 796–800.

Liu, A., Cui, Y., Ge, X., & Wang, C. (2003). Distortion product otoacoustic emission test of cochlear deafness [Article in Chinese]. *Lin Chuang Er Bi Yan Hou Ke Za Zhi, 17,* 588–590.

Loehlin, J. C., & McFadden, D. (2003). Otoacoustic emissions, auditory evoked potentials, and traits related to sex and sexual orientation. *Archives of Sex Behavior, 32,* 115–127.

Long, G. R., & Tubis, A. (1988). Modification of spontaneous and evoked otoacoustic emissions and associated psychoacoustic microstructure by aspirin consumption. *Journal of the Acoustical Society of America, 84,* 1343–1353.

Lonsbury-Martin, B. L., Martin, G. K., McCoy, M. J., & Whitehead, M. (1994). Otoacoustic emissions testing in young children: middle ear influences. *American Journal of Otology, 15*(Suppl.1), 13–20.

Lonsbury-Martin, B. L., Martin, G. K., Probst, R., & Coats, A. C. (1988). Spontaneous otoacoustic emissions in a nonhuman primate. II. Cochlear anatomy. *Hearing Research, 33,* 69–93.

López-Díaz-de-León, E., Silva-Rojas, A., Ysunza, A., Amavisca, R., & Rivera, R. (2003). Auditory neuropathy in Friedreich ataxia. A report of two cases. *International Journal of Pediatric Otorhinolaryngology, 67,* 641–648.

Lotfi, Y., & Mehrkian, S. (2007). The prevalence of auditory neuropathy in students with hearing impairment in Tehran Iran. *Archives of Iranian Medicine, 10,* 233–235.

Loundon, N., Marcolla, A., Roux, I., Rouillon, I., Denoyelle, F., Feldmann, D., . . . Garabedian, E. N. (2005).

Auditory neuropathy or endocochlear hearing loss? *Otology & Neurotology, 26,* 748–754.

Lucertini, M., Moleti, A., & Sisto, R. (2002). On the detection of early cochlear damage by otoacoustic emission analysis. *Journal of Acoustical Society of America, 111,* 972–978.

Lucertini, M., Viaggi, F., Pasquazzi, F., & Cianfrone, G. (2001). Medico-legal use of otoacoustic emissions in the audiological selection process for employment. *Scandinavian Audiology, 52,* 146–147.

Luo, R., Mai, J., Chen, Q., Yang, S., & Zhong, J. (2001). Evaluation of sensorineural hearing loss in childhood [Article in Chinese]. *Zhonghua Er Bi Yan Hou Ke Za Zhi, 36,* 346–351.

Lyons, A., Kei, J., & Driscoll, C. (2004). Distortion product otoacoustic emissions in children at school entry: A comparison with pure-tone screening and tympanometry results. *Journal of the American Academy of Audiology, 15,* 702–715.

Madasu, R., Ruckenstein, M. J., Leake, F., Steere, E., & Robbins, K. T. (1997). Ototoxic effects of supradose cisplatin with sodium thiosulfate neutralization. *Archives of Otolaryngology-Head and Neck Surgery, 123,* 978–981.

Madden, C., Rutter, M., Hilbert, L., Greinwald, J. H., Jr., & Choo, D. I. (2002). Clinical and audiological features in auditory neuropathy. *Archives of Otolaryngology-Head and Neck Surgery, 128,* 1026–1030.

Magliulo, G., Cianfrone, G., Gagliardi, M., Cuiuli, G., & D'Amico, R. (2004). Vestibular evoked myogenic potentials and distortion-product otoacoustic emissions combined with glycerol testing in endolymphatic hydrops: Their value in early diagnosis. *Annals of Otology, Rhinology and Laryngology, 113,* 1000–1005.

Maia, J. R., & Russo, I. C. (2008). Study of the hearing of rock and roll musicians [Article in Portuguese]. *Pro Fono, 20,* 49–54.

Mao, X., Zheng, C., Zheng, R., Lin, X., & Shen, Z. (2005). Tinnitus with normal hearing and evoked otoacoustic emissions [Article in Chinese]. *Lin Chuang Er Bi Yan Hou Ke Za Zhi, 19,* 14–16.

Marler, J. A., Sitcovsky, J. L., Mervis, C. B., Kistler, D. J., & Wightman, F. L. (2010). Auditory function and hearing loss in children and adults with Williams syndrome: Cochlear impairment in individuals with otherwise normal hearing. *American Journal of Medical Genetics C Seminars in Medical Genetics, 154C,* 248–265.

Marlin, S., Feldmann, D., Nguyen, Y., Rouillon, I., Loundon, N., Jonard, L., . . . Denoyelle, F. (2010). Temperature-sensitive auditory neuropathy associated with an otoferlin mutation: Deafening fever!

Biochemical Biophysical Research Communications, 394, 737–742.

Marques, F. P., & da Costa, E. A. (2006). Exposure to occupational noise: otoacoustic emissions test alterations. *Review of Brasilian Otorrinolaringology (Engl. ed.), 72,* 362–366.

Marshall, L., & Heller, L. M. (1998). Transient-evoked otoacoustic emissions as a measure of noise-induced threshold shift. *Journal of Speech, Language and Hearing Research, 41,* 1319–1334.

Marshall, L., Lapsley Miller, J. A., & Heller, L. M. (2001). Distortion-product otoacoustic emissions as a screening tool for noise-induced hearing loss. *Noise and Health, 3,* 43–60.

Martin, G. K., Ohms, L. A., Franklin, D. J., Harris, F. P., & Lonsbury-Martin, B. L. (1990). Distortion product emissions in humans: III. Influence of sensorineural hearing loss. *Annals of Otology, Rhinology, and Laryngology Supplement, 147,* 30–42.

Martin, G. K., Stagner, B. B., & Lonsbury-Martin, B. L. (2010). Evidence for basal distortion-product otoacoustic emission components. *Journal of the Acoustical Society of America, 127,* 2955–2972.

Martins, L. M., Camargos, P. A., Becker, H. M., Becker, C. G., & Guimarães, R. E. (2010). Hearing loss in cystic fibrosis. *International Journal of Pediatric Otorhinolaryngology, 74,* 469–473.

Massinger, C., Gawehn, J., & Keilmann, A. (2003). Acoustic schwannoma with progressive hearing loss in children. A case report [Article in German]. *Laryngorhinootologie, 82,* 92–96.

Matsunaga, T., Kumanomido, H., Shiroma, M., Goto, Y., & Usami, S. (2005). Audiological features and mitochondrial DNA sequence in a large family carrying mitochondrial A1555G mutation without use of aminoglycoside. *Annals of Otology, Rhinology and Laryngology, 114,* 153–160.

Mauermann, M., Uppenkamp, S., van Hengel, P. W., & Kollmeier, B. (1999). Evidence for the distortion product frequency place as a source of distortion product otoacoustic emission (DPOAE) fine structure in humans. I. Fine structure and higher-order DPOAE as a function of the frequency ratio f2/f1. *Journal of the Acoustical Society of America, 106,* 3473–3483.

Mazelova, J., Popelar, J., & Syka, J. (2003). Auditory function in presbycusis: Peripheral and central changes. *Experimental Gerontology, 38,* 87–94.

McFadden, D. (1993). A masculinizing effect on the auditory systems of human females having male co-twins. *Proceedings of the National Academy of Sciences U S A, 90,* 11900–11904.

McFadden, D. (2002). Masculinization effects in the auditory system. *Archives of Sex Behavior, 31,* 99–111.

McFadden, D., & Mishra, R. (1993). On the relation between hearing sensitivity and otoacoustic emissions. *Hearing Research, 71,* 208–213.

McPherson, B., Li, S. F., Shi, B. X., Tang, J. L., & Wong, B. Y. (2006). Neonatal hearing screening: Evaluation of tone-burst and click-evoked otoacoustic emission test criteria. *Ear and Hearing, 27,* 256–262.

Méndez-Ramírez Mdel, R., & Altamirano-González, A. (2006). Transient evoked otoacoustic emissions (TEOAEs) in patients with acute middle ear pathology [Article in Spanish]. *Cir Cir, 74,* 9–14.

Mertz, K., Bencsik, B., Büki, B., & Avan, P. (2004). Noninvasive testing of intracranial pressure changes due to body position in infants [Article in Hungarian]. *Orvosi Hetilap, 145,* 1427–1430.

Mezzalira, R., Maudonnet, O. A., Pereira, R. G., & Ninno, J. E. (2004). The contribution of otoneurological evaluation to tinnitus diagnosis. *International Tinnitus Journal, 10,* 65–72.

Micheyl, C., & Collet, L. (1994). Interrelations between psychoacoustical tuning curves and spontaneous and evoked otoacoustic emissions. *Scandinavian Audiology, 23,* 171–178.

Michie, P. T., LePage, E. L., Solowij, N., Haller, M., & Terry, L. (1996). Evoked otoacoustic emissions and auditory selective attention. *Hearing Research, 98,* 54–67.

Mills, D. M., Feeney, M. P., & Gates, G. A. (2007). Evaluation of cochlear hearing disorders: Normative distortion product otoacoustic emission measurements. *Ear and Hearing, 28,* 778–792.

Mo, L., Yan, F., Liu, H., Han, D., & Zhang, L. (2010). Audiological results in a group of children with auditory neuropathy spectrum disorder. *ORL Journal of Otorhinolaryngology and Related Specialties., 72,* 75–79.

Mo, W., Zhang, Y. Y., Lei, Y. Q., Sun, W., Shao, P. F., Sun, Y. F., . . . Zhao, Z. Y. (2005). Hearing screening in infants with congenital cytomegalovirus infection [Article in Chinese]. *Zhejiang Da Xue Xue Bao Yi Xue Ban, 34,* 358–360.

Mobley, S. R., Odabasi, O., Ahsan, S., Martin, G., Stagner, B., & Telischi, F. F. (2002). Distortion-product otoacoustic emissions in nonacoustic tumors of the cerebellopontine angle. *Otolaryngology-Head and Neck Surgery, 126,* 115–120.

Moleti, A., Sisto, R., & Lucertini, M. (2002). Linear and nonlinear transient evoked otoacoustic emissions in humans exposed to noise. *Hearing Research, 174,* 290–295.

Monnery, P. M., Srouji, E. I., & Bartlett, J. (2004). Is cochlear outer hair cell function affected by mobile telephone radiation? *Clinical Otolaryngology and Allied Sciences, 29,* 747–749.

Montoya, F. S., Ibargüen, A. M., del Rey, A. S., & Fernández, J. M. (2007). Evaluation of cochlear function in patients with tinnitus using spontaneous and transitory evoked otoacoustic emissions. *Journal of Otolaryngology, 36,* 296–302.

Montoya, F. S., Ibargüen, A. M., Vences, A. R., del Rey, A. S., & Fernandez, J. M. (2008). Evaluation of cochlear function in normal-hearing young adults exposed to MP3 player noise by analyzing transient evoked otoacoustic emissions and distortion products. *Journal of Otolaryngology-Head and Neck Surgery, 37,* 718–724.

Moore, B. C., Glasberg, B., & Schlueter, A. (2010). Detection of dead regions in the cochlea: relevance for combined electric and acoustic stimulation. *Advances in Otorhinolaryngology, 67,* 43–50.

Mora, R., Crippa, B., Mora, F., & Dellepiane, M. (2006). A study of the effects of cellular telephone microwave radiation on the auditory system in healthy men. *Ear, Nose and Throat Journal., 85,* 160–163.

Morawski, K., Namyslowski, G., Lisowska, G., Bazowski, P., Kwiek, S., & Telischi, F. F. (2004). Intraoperative monitoring of cochlear function using distortion product otoacoustic emissions (DPOAEs) in patients with cerebellopontine angle tumors. *Otology & Neurotology, 25,* 818–825.

Morimoto, N., Tanaka, T., Taiji, H., Horikawa, R., Naiki, Y., Morimoto, Y., & Kawashiro, N. (2006). Hearing loss in Turner syndrome. *Journal of Pediatrics, 149,* 697–701.

Morita, S., Masanobu, S., & Iizuka, K. (2010). Nonorganic hearing loss in childhood. *International Journal of Pediatric Otorhinolaryngology, 74,* 441–446.

Morlet, T., Durrant, J. D., Lapillonne, A., Putet, G., Collet, L., & Duclaux, R. (2003). Development of auditory asymmetry in transient evoked otoacoustic emissions in pre-term infants. *Journal of the American Academy of Audiology, 14,* 339–346.

Morlet, T., Goforth, L., Hood, L. J., Ferber, C., Duclaux, R., & Berlin, C. I. (1999). Development of human cochlear active mechanism asymmetry: Involvement of the medial olivocochlear system? *Hearing Research, 134,* 153–162.

Morlet, T., Hamburger, A., Kind, J., Ari-Even Roth, D., Gartner, M., Muchnik, C., . . . Hildesheimer, M. (2004). Assessment of medial olivocochlear system function in pre-term and full-term newborns using a rapid test of transient otoacoustic emissions. *Clinical Otolaryngology and Allied Sciences, 29,* 183–190.

Moulin, A. (2000). Influence of primary frequencies ratio on distortion product otoacoustic emissions amplitude. II. Interrelations between multicomponent DPOAEs, tone-burst-evoked OAEs, and spontaneous OAEs. *Journal of the Acoustical Society of America, 107,* 1471–1486.

Moulin, A., & Carrier, S. (1998). Time course of the medial olivocochlear efferent effect on otoacoustic emissions in humans. *NeuroReport, 9,* 3741–3744.

Mozota Núñez, J. R., Mozota Núñez, M. L., Mozota Núñez, M. J., Vila Mendiburu, I., & Mozota Ortiz, J. R. (2003). Otoacoustic emissions in sudden deafness [Article in Spanish]. *Anales Otorrinolaringologicos Ibero Americanos, 30,* 289–300.

Muchnik, C., Ari-Even Roth, D., Othman-Jebara, R., Putter-Katz, H., Shabtai, E. L., & Hildesheimer, M. (2004). Reduced medial olivocochlear bundle system function in children with auditory processing disorders. *Audiology & Neurootology, 9,* 107–114.

Mukari, S. Z., & Mamat, W. H. (2008). Medial olivocochlear functioning and speech perception in noise in older adults. *Audiology & Neurootology, 13,* 328–334.

Mukhamedova, G. R. (2003). Cochlear function study by registration of otoacoustic emission of different classes in persons exposed to intensive industrial noise [Article in Russian]. *Vestn Otorinolaringology, 6,* 24–28.

Mulheran, M., & Degg, C. (1997). Comparison of distortion product OAE generation between a patient group requiring frequency gentimicin therapy and control subjects. *British Journal of Audiology, 31,* 5–9.

Mulheran, M., Wiselka, M., & Johnston, M. N. (2004). Evidence of subtle auditory deficit in a group of patients recovered from bacterial meningitis. *Otology & Neurotology, 25,* 302–307.

Muluk, N. B., & Birol, A. (2007). Effects of Behçet's disease on hearing thresholds and transient evoked otoacoustic emissions. *Journal of Otolaryngology, 36,* 220–226.

Muluk, N. B., Yilmaz, E., & Dinçer, C. (2006). Effects of extracorporeal shock wave lithotripsy treatment on transient evoked otoacoustic emissions in patients with urinary lithiasis. *Journal of Otolaryngology, 35,* 320–326.

Murdin, L., Patel, S., Walmsley, J., & Yeoh, L.H. (2008). Hearing difficulties are common in patients with rheumatoid arthritis. *Clinical Rheumatology, 27,* 637–640.

Murnane, O. D., & Kelly, J. K. (2003). The effects of high-frequency hearing loss on low-frequency components of the click-evoked otoacoustic emission. *Journal of the American Academy of Audiology, 14,* 525–533.

Murray, N., LePage, E., & Butler, T. (2004). Hearing health of New South Wales prison inmates. *Australian New Zealand Journal of Public Health, 28,* 537–541.

Muszyński, P., Składzień, J., Reroń, E., Strek, P., Popielski, L., Dutsch-Wicherel, M., . . . Bartnik, J. (2007). Transient evoked otoacustic emission in children with juvenile idiopathic arthritis [Article in Polish]. *Otolaryngology Polska, 61,* 972–978.

Naeve, S. L., Margolis, R. H., Levine, S. C., & Fournier, E. M. (1992). Effect of ear-canal air pressure on evoked otoacoustic emissions. *Journal of the Acoustical Society of America, 91,* 2091–2095.

Nagy, A. L., Tóth, F., Vajtai, R., Gingl, Z., Jóri, J., & Kiss, J. G. (2002). Effects of noise on the intensity of distortion product otoacoustic emissions. *International Tinnitus Journal, 8,* 94–96.

Nakashima, T., Ueda, H., Misawa, H., Suzuki, T., Tominaga, M., Ito, A., . . . Meikle, M. B. (2002). Transmeatal low-power laser irradiation for tinnitus. *Otology & Neurotology, 23,* 296–300.

Namysłowski, G., Lisowska, G., Misiołek, M., Scierski, W., Orecka, B., & Czecior, E. (2007). The influence of gender on otoacoustic emissions in normally hearing subjects [Article in Polish]. *Otolaryngology Polska, 61,* 792–795.

Namysłowski, G., Morawski, K., Trybalska, G., & Urbaniec, P. (2004). The latencies of the 2f1-f2 DPOAE measured using phase gradient method in young adults and in workers chronically exposed to noise. *Otolaryngology Polska, 58,* 131–138.

Namysłowski, G., Morawski, K., Urban, I., Lisowska, G., & Skalski, J. (2003). Influence of hypothermia and extracorporeal circulation on transiently evoked otoacoustic emission (TEOAE) in children operated on for various heart defects (I) [Article in Polish]. *Otolaryngology Polska, 57,* 263–269.

Namysłowski, G., Trybalska, G., Scierski, W., Mrówka-Kata, K., Bilińska-Pietraszek, E., & Kawecki, D. (2003). Hearing evaluation in patients suffering from hypercholesterolemia [Article in Polish]. *Otolaryngology Polska, 57,* 725–730.

Nataletti, P., Sisto, R., Pieroni, A., Sanjust, F., & Annesi, D. (2007). Pilot study of professional exposure and hearing functionality of orchestra musicians of a national lyric theatre [Article in Italian]. *Giornale Italiano di Medicina del Lavoro ed Ergonomia, 29,* 496–498.

Neely, S. T., & Gorga, M. P. (1998). Comparison between intensity and pressure as measures of sound level in the ear canal. *Journal of the Acoustical Society of America, 104,* 2925–2934.

Neely, S. T., Johnson, T. A., & Gorga, M. P. (2005). Distortion-product otoacoustic emission measured with continuously varying stimulus level. *Journal of the Acoustical Society of America, 117,* 1248–1259.

Neely, S. T., Johnson, T. A., Kopun, J., Dierking, D. M., & Gorga, M. P. (2009). Distortion-product otoacoustic emission input/output characteristics in normal-hearing and hearing-impaired human ears. *Journal of the Acoustical Society of America, 126,* 728–738.

Negley, C., Katbamna, B., Crumpton, T., & Lawson, G. D. (2007). Effects of cigarette smoking on distortion product otoacoustic emissions. *Journal of the American Academy of Audiology, 18,* 665–674.

Nelson, J. J., Giraud, A., Walsh, R., & Mortelliti, A.J. (2010). Impact on hearing of routine ear suctioning at the tympanic membrane. *American Journal of Otolaryngology,* Epub 2010 Apr 13.

Neumann, K., Dettmer, G., Euler, H. A., Giebel, A., Gross, M., Herer, G., . . . Montgomery, J. (2006). Auditory status of persons with intellectual disability at the German Special Olympic Games. *International Journal of Audiology, 45,* 83–90.

Ngo, R. Y., Tan, H. K., Balakrishnan, A., Lim, S. B., & Lazaroo, D. T. (2006). Auditory neuropathy/auditory dys-synchrony detected by universal newborn hearing screening. *International Journal of Pediatric Otorhinolaryngology, 70,* 1299–1306.

Ni, D. F., Li, F. R., Zhang, Z. Y., Xu, C. X., & Zhao, C. X. (2000). Analyse of the site of lesion in auditory neuropathy [Article in Chinese]. *Lin Chuang Er Bi Yan Hou Ke Za Zhi, 14,* 293–295.

Nicolas-Puel, C., Akbaraly, T., Lloyd, R., Berr, C., Uziel, A., Rebillard, G., & Puel, J. L. (2006). Characteristics of tinnitus in a population of 555 patients: Specificities of tinnitus induced by noise trauma. *International Tinnitus Journal, 12,* 64–70.

Niedzielska, G., & Katska, E. (2002). TEOAE after treatment of otitis media with effusion. *Annales Universitatis Mariae Curie Sklodowska [Med]., 57,* 58–61.

Niedzielska, G., Katska, E., & Kusa, W. (2001). Hearing loss in chronic alcoholics. *Annales Universitatis Mariae Curie Sklodowska [Med], 56,* 99–101.

Nivoloni Kde, A., da Silva-Costa, S. M., Pomílio, M. C., Pereira, T., Lopes Kde, C., de Moraes, V. C., . . . Sartorato, E. L. (2010). Newborn hearing screening and genetic testing in 8974 Brazilian neonates. *International Journal of Pediatric Otorhinolaryngology, 74,* 926–929.

Norris, V. W., Arnos, K. S., Hanks, W. D., Xia, X., Nance, W. E., & Pandya, A. (2006). Does universal newborn hearing screening identify all children with GJB2 (Connexin 26) deafness? Penetrance of GJB2 deafness. *Ear and Hearing, 27,* 732–741.

Norton, S. J., Gorga, M. P., Widen, J. E., Folsom, R. C., Sininger, Y., Cone-Wesson, B., . . . Fletcher, K. l. (2000). Identification of neonatal hearing impairment: Evaluation of transient evoked otoacoustic emissions, distortion product otoacoustic emissions, and auditory brainstem response test performance. *Ear and Hearing, 21,* 508–528.

Nottet, J. B., Moulin, A., Brossard, N., Suc, B., & Job, A. (2006). Otoacoustic emissions and persistent tinnitus after acute acoustic trauma. *Laryngoscope, 116,* 970–975.

Nozza, R. J., Sabo, D. L., & Mandel, E. M. (1997). A role for otoacoustic emissions in screening for hearing impairment and middle ear disorders in school-age children. *Ear and Hearing, 18,* 227–239.

Nuttall, A. L., Grosh, K., Zheng, J., de Boer, E., Sou, Y., & Ren, T. (2004). Spontaneous basilar membrane oscillation and otoacoustic emission at 15 kHz in guinea pig. *Journal of the Association for Research in Otolaryngology: JARO, 5,* 337–348.

O'Brien, I., Wilson, W., & Bradley, A. (2008). Nature of orchestral noise. *Journal of the Acoustical Society of America, 124,* 926–939.

Odabasi, A. O., Telischi, F. F., Gomez-Marin, O., Stagner, B., & Martin, G. (2002). Effect of acoustic tumor extension into the internal auditory canal on distortion-product otoacoustic emissions. *Annals of Otology, Rhinology, and Laryngology, 111,* 912–915.

Oeken, J. (1998). Distortion product otoacoustic emissions in acute acoustic trauma. *Noise and Health, 1,* 56–66.

Oeken, J., Stumpf, R., & Bootz, F. (2002). DPOAEs and vestibular function in different types of autosomal-dominant non-syndromal hearing impairment. *Auris Nasus Larynx, 29,* 29–34.

Ohlms, L. A., Lonsbury-Martin, B. L., & Martin, G. K. (1991). Acoustic-distortion products: separation of sensory from neural dysfunction in sensorineural hearing loss in human beings and rabbits. *Otolaryngology-Head and Neck Surgery, 104,* 159–174.

Oishi, N., Kanzaki, S., Kataoka, C., Tazoe, M., Takei, Y., Nagai, K., . . . Ogawa, K. (2009). Acute-onset unilateral psychogenic hearing loss in adults: Report of six cases and diagnostic pitfalls. *ORL, 71,* 279–283.

Olszewski, J., Miłoński, J., & Olszewski, S. (2006). Analysis of hearing threshold shift measured by otoacoustic emissions in soldiers after shooting with use hearing protectors [Article in Polish]. *Medical Progress, 57,* 109–114.

Olszewski, J., Miłoński, J., Olszewski, S., & Majak, J. (2007). Hearing threshold shift measured by otoacoustic emissions after shooting noise exposure in soldiers using hearing protectors. *Otolaryngology-Head and Neck Surgery, 136,* 78–81.

Olszewski, J., Miłoński, J., Sułkowski, W. J., Majak, J., & Olszewski, S. (2005). Temporary hearing threshold shift measured by otoacoustic emissions in subjects exposed to short-term impulse noise. *International Journal of Occupational Medical and Environmental Health, 18,* 375–379.

Olusanya, B. O. (2010). Is undernutrition a risk factor for sensorineural hearing loss in early infancy? *British Journal of Nutrition 103,* 1296–1301.

Olusanya, B. O., & Somefun, A. O. (2009). Place of birth and characteristics of infants with congenital and early-onset hearing loss in a developing country. *International Journal of Pediatric Otorhinolaryngology, 73,* 1263–1269.

Onishi, E. T., Fukuda, Y., & Suzuki, F. A. (2004). Distortion product otoacoustic emissions in tinnitus patients. *International Tinnitus Journal, 10,* 13–16.

Osterhammel, P. A., Nielsen, L. H., & Rasmussen, A. N. (1993). Distortion product otoacoustic emissions. *Scandinavian Audiology, 22,* 111–116.

Ottaviani, F., Dozio, N., Neglia, C.B., Riccio, S., & Scavini, M. (2002). Absence of otoacoustic emissions in insulin-dependent diabetic patients: Is there evidence for diabetic cochleopathy? *Journal of Diabetes Complications, 16,* 338–343.

Ottaviani, F., Neglia, C. B., Scotti, A., & Capaccio, P. (2002). Arachnoid cyst of the cranial posterior fossa causing sensorineural hearing loss and tinnitus: A case report. *European Archives of Otorhinolaryngology, 259,* 306–308.

Ottaviani, F., Neglia, C.B., Ventrella, L., Giugni, E., & Motti, E. (2002). Hearing loss and changes in transient evoked otoacoustic emissions after gamma knife radiosurgery for acoustic neurinomas. *Archives of Otolaryngology-Head and Neck Surgery, 128,* 1308–1312.

Owens, J. J., McCoy, M. J., Lonsbury-Martin, B. L., & Martin, G. K. (1993). Otoacoustic emissions in children with normal ears, middle ear dysfunction, and ventilation tubes. *American Journal of Otology, 14,* 34–40.

Oysu, C., Aslan, I., Ulubil, A., & Baserer, N. (2002). Incidence of cochlear involvement in hyperbilirubinemic deafness. *Annals of Otology, Rhinology and Laryngology, 111,* 1021–1025.

Ozimek, E., Wicher, A., Szyfter, W., & Szymiec, E. (2006). Distortion product otoacoustic emission (DPOAE) in tinnitus patients. *Journal of the Acoustical Society of America, 119,* 527–538.

Ozturan, O., Erdem, T., Miman, M. C., Kalcioglu, M. T., & Oncel, S. (2002). Effects of the electromagnetic field of mobile telephones on hearing. *Acta Otolaryngologica, 122,* 289–293.

Paglialonga, A., Fiocchi, S., Del Bo, L., Ravazzani, P., & Tognola, G. (2010). Quantitative analysis of cochlear active mechanisms in tinnitus subjects with normal hearing sensitivity: Time-frequency analysis of transient evoked otoacoustic emissions and contralateral suppression. *Auris Nasus Larynx.* Epub ahead of print 2010 June 16.

Paglialonga, A., Tognola, G., Parazzini, M., Lutman, M. E., Bell, S. L., Thuroczy, G., & Ravazzani, P. (2007). Effects of mobile phone exposure on time frequency fine structure of transiently evoked otoacoustic emissions. *Journal of the Acoustical Society of America, 122,* 2174–2182.

Pan, H. G., Cui, Y. H., Gao, Q. X., Wang, C. F., & Ge, X. (2000). The study of the distortion product emission in the people with hypertriglyceride [Article in Chinese]. *Lin Chuang Er Bi Yan Hou Ke Za Zhi, 14,* 299–300.

Panda, N. K., Jain, R., Bakshi, J., & Munjal, S. (2010). Audiologic disturbances in long-term mobile phone users. *Journal of Otolaryngology-Head and Neck Surgery, 39,* 5–11.

Parazzini, M., Bell, S., Thuroczy, G., Molnar, F., Tognola, G., Lutman, M. E., & Ravazzani, P. (2005). Influence on the mechanisms of generation of distortion product otoacoustic emissions of mobile phone exposure. *Hearing Research, 208,* 68–78.

Parazzini, M., Brazzale, A. R., Paglialonga, A., Tognola, G., Collet, L., Moulin, A., . . . Ravazzani, P. (2007). Effects of GSM cellular phones on human hearing: the European project "GUARD." *Radiation Research, 168,* 608–613.

Parazzini, M., Hall, A. J., Lutman, M. E., & Kapadia, S. (2005). Effect of aspirin on phase gradient of 2F1-F2 distortion product otoacoustic emissions. *Hearing Research, 205,* 44–52.

Parazzini, M., Lutman, M. E., Moulin, A., Barnel, C., Sliwinska-Kowalska, M., Zmyslony, M., . . . Ravazzani, P. (2010). Absence of short-term effects of UMTS exposure on the human auditory system. *Radiation Research, 173,* 91–97.

Parazzini, M., Ravazzani, P., Medaglini, S., Weber, G., Fornara, C., Tognola, G., . . . Grandori, F. (2002). Click-evoked otoacoustic emissions recorded from untreated congenital hypothyroid newborns. *Hearing Research, 166,* 136–142.

Park, S. N., Park, K. H., Park, S. Y., Jeon, E. J., Chang, K. H., & Yeo, S. W. (2007). Clinical and biochemical factors that affect DPOAE expressions in children with middle ear effusion. *Otolaryngology-Head and Neck Surgery, 136,* 23–26.

Parthasarathy, T. K. (2001). Aging and contralateral suppression effects on transient evoked otoacoustic emissions. *Journal of the American Academy of Audiology, 12,* 80–85.

Paschoal, C. P., & Azevedo, M. F. (2004). Transient otoacoustic emissions in infants of diabetic mothers [Article in Portuguese]. *Pro Fono, 16,* 197–202.

Pawlaczyk-Luszczyńska, M., Dudarewicz, A., Bak, M., Fiszer, M., Kotyło, P., & Sliwińska-Kowalska, M. (2004). Temporary changes in hearing after exposure to shooting noise. *International Journal of Occupational Medicine and Environmental Health, 17,* 285–293.

Pawlak-Osińska, K., Kaźmierczak, H., Kaźmierczak, W., & Szpoper, M. (2004). Ozone therapy and pressure-pulse therapy in Ménière's disease. *International Tinnitus Journal, 10,* 54–57.

Peng, J. H., Tao, Z. Z., & Huang, Z. W. (2007). Long-term sound conditioning increases distortion product otoacoustic emission amplitudes and decreases olivocochlear efferent reflex strength. *NeuroReport, 18,* 1167–1170.

Penner, M. J. (1995). Frequency variation of spontaneous otoacoustic emissions during a naturally occurring menstrual cycle, amenorrhea, and oral contraception: A brief report. *Ear and Hearing, 16,* 428–432.

Penner, M. J. (1996). The emergence and disappearance of one subject's spontaneous otoacoustic emissions. *Ear and Hearing, 17,* 116–119.

Perez, N., Boleas, M. S., & Martin, E. (2005). Distortion product otoacoustic emissions after intratympanic gentamicin therapy for unilateral Ménière's disease. *Audiology & Neurootology, 10,* 69–78.

Petrescu, N. (2008). Loud music listening. *McGill Journal of Medicine, 11,* 169–176.

Plantinga, R. F., Cremers, C. W., Huygen, P. L., Kunst, H. P., & Bosman, A. J. (2007). Audiological evaluation of affected members from a Dutch DFNA8/12 (TECTA) family. *Journal of the Association for Research in Otolaryngology, 8,* 1–7.

Plinkert, P. K., Bootz, F., & Vossieck, T. (1994). Influence of static middle ear pressure on transiently evoked otoacoustic emissions and distortion products. *European Archives of Otolaryngology, 251,* 95–99.

Plomp, R. (1976). *Aspects of tone sensation: A psychophysical study.* London, UK: Academic Press

Podwall, A., Podwall, D., Gordon, T. G., Lamendola, P., & Gold, A. P. (2002). Unilateral auditory neuropathy: Case study. *Journal of Child Neurology, 17,* 306–309.

Polak, M., Eshraghi, A.A., Nehme, O., Ahsan, S., Guzman, J., Delgado, R. E., . . . Van De Water, T. R. (2004). Evaluation of hearing and auditory nerve function by combining ABR, DPOAE and eABR tests into a single recording session. *Journal of Neuroscience Methods, 134,* 141–149.

Pośpiech, L., Sztuka-Pietkiewicz, A., Jabłonka, A., & Orendorz-Fraczkowska, K. (2003). DPOAE otoemission in patients with tinnitus and normal hearing [Article in Polish]. *Otolaryngology Polska, 57,* 905–910.

Pouyatos, B., Gearhart, C., Nelson-Miller, A., Fulton, S., & Fechter, L. (2007). Oxidative stress pathways in the potentiation of noise-induced hearing loss by acrylonitrile. *Hearing Research, 224,* 61–74.

Prasher, D., Al-Hajjaj, H., Aylott, S., & Aksentijevic, A. (2005). Effect of exposure to a mixture of solvents and noise on hearing and balance in aircraft maintenance workers. *Noise and Health, 7,* 31–39.

Prieve, B. A., Calandruccio, L., Fitzgerald, T., Mazevski, A., & Georgantas, L. M. (2008). Changes in transient-evoked otoacoustic emission levels with negative tympanometric peak pressure in infants and toddlers. *Ear and Hearing, 29,* 533–542.

Prieve, B. A., & Falter, S. R. (1995). COAEs and SSOAEs in adults with increased age. *Ear and Hearing, 16,* 521–528.

Prieve, B. A., Gorga, M. P., Schmidt, A., Neely, S., Peters, J., Schultes, L., & Jesteadt, W. (1993). Analysis of transient-evoked otoacoustic emissions in normal hearing and hearing-impaired ears. *Journal of the Acoustical Society of America, 93,* 3308–3319.

Probst, R. (2007). Audiological evaluation of patients with otosclerosis. *Advances in Otorhinolaryngology, 65,* 119–126.

Probst, R., Lonsbury-Martin, B., & Martin, G. (1991). A review of otoacoustic emissions. *Journal of the Acoustical Society of America, 89,* 2027–2067.

Psarommatis, I., Kontorinis, G., Kontrogiannis, A., Douniadakis, D., & Tsakanikos, M. (2009). Pseudohypacusis: The most frequent etiology of sudden hearing loss in children. *European Archives of Otorhinolarygology, 266,* 1857–1867.

Psarommatis, I., Riga, M., Douros, K., Koltsidopoulos, P., Douniadakis, D., Kapetanakis, I., & Apostolopoulos, N. (2006) Transient infantile auditory neuropathy and its clinical implications. *International Journal of Pediatric Otorhinolaryngology, 70,* 1629–1637.

Psarommatis, I., Valsamakis, T., Raptaki, M., Kontrogiani, A., & Douniadakis, D. (2007). Audiologic evaluation of infants and preschoolers: A practical approach. *American Journal of Otolaryngology, 28,* 392–396.

Psillas, G., Psifidis, A., Antoniadou-Hitoglou, M., & Kouloulas, A. (2006). Hearing assessment in preschool children with speech delay. *Auris Nasus Larynx, 33,* 259–263.

Punnett, A., Bliss, B., Dupuis, L. L., Abdolell, M., Doyle, J., & Sung, L. (2004). Ototoxicity following pediatric hematopoietic stem cell transplantation: a prospective cohort study. *Pediatric Blood Cancer, 42,* 598–603.

Puria, S. (2003). Measurements of middle ear forward and reverse acoustics: Implications for otoacoustic emissions. *Journal of Acoustical Society of America, 113,* 2773–2789.

Quaranta, A., Scaringi, A., Sallustio, V., & Quaranta, N. (2002). Cochlear function in ears with immunomediated inner ear disorder. *Acta Otolaryngology Supplement., 548,* 15–19.

Quaranta, A., Zini, C., Gandolfi, A., Piazza, F., De Thomasis, G., Frisina, A., . . . Uccelli, M. (2001). Acoustic neuroma: clinical-functional finding, results and surgical complication [Article in Italian]. *Acta Otorhinolaryngolica Italica, 21,* 1–20.

Rabinowitz, P. M., Pierce Wise, J., Sr., Hur Mobo, B., Antonucci, P. G., Powell, C., & Slade, M. (2002). Antioxidant status and hearing function in noise-exposed workers. *Hearing Research, 173,* 164–171.

Radomskij, P., Schmidt, M. A., Heron, C. W., & Prasher, D. (2002). Effect of MRI noise on cochlear function. *Lancet, 359,* 1485.

Rajan, R. (2000). Centrifugal pathways protect hearing sensitivity at the cochlea in noisy environments that exacerbate the damage induced by loud sound. *Journal of Neuroscience, 20,* 6684–6693.

Ramsebner, R., Volker, R., Lucas, T., Hamader, G., Weipoltshammer, K., Baumgartner, W. D., . . . Frei, K. (2007). High incidence of GJB2 mutations during screening of newborns for hearing loss in Austria. *Ear and Hearing, 28,* 298–301.

Rapin, I., & Gravel, J. (2003). Auditory neuropathy": Physiologic and pathologic evidence calls for more diagnostic specificity. *International Journal of Pediatric Otorhinolaryngology, 67,* 707–728.

Rapin, I., & Gravel, J.S. (2006). Auditory neuropathy: A biologically inappropriate label unless acoustic nerve involvement is documented. *Journal of the American Academy of Audiology, 17,* 147–150.

Reavis, K. M., Phillips, D. S., Fausti, S. A., Gordon, J. S., Helt. W. J., Wilmington, D., . . . Konrad-Martin, D. (2008). Factors affecting sensitivity of distortion-product otoacoustic emissions to ototoxic hearing loss. *Ear and Hearing, 29*(6), 875–893.

Reisser, C. F., Kimberling, W. J., & Otterstedde, C. R. (2002). Hearing loss in Usher syndrome type II is nonprogressive. *Annals of Otology, Rhinology and Laryngology, 111,* 1108–1111.

Reron, A., Reron, E., Strek, P., & Trojnar-Podlesny, M. (2002). The effect of hormonal factors on the hearing organ in women, after surgical castration. Preliminary report. *Neuro Endocrinology Letters, 23,* 455–458.

Ress, B. D., Sridhar, K. S., Balkany, T. J., Waxman, G. M., Stagner, B. B., & Lonsbury-Martin, B. L.. (1999). Effects of cis-platinum chemotherapy on otoacoustic emissions: The development of an objective screening protocol. *Otolaryngology-Head and Neck Surgery, 121,* 693–701.

Reuter, K., & Hammershøi, D. (2007). Distortion product otoacoustic emission of symphony orchestra musicians before and after rehearsal. *Journal of the Acoustical Society of America, 121,* 327–336.

Richardson, H. C., Elliott, C., & Hill, J. (1996). The feasibility of recording transiently evoked otoacoustic emissions immediately following grommet insertion. *Clinical Otolaryngology and Allied Health Sciences, 21,* 445–448.

Riga, M., Korres, S., Psarommatis, I., Varvutsi, M., Giotakis, I., Papadis, T., . . . Apostolopoulos, N. (2007). Neurotoxicity of BMF-95 on the medial olivocochlear bundle assessed by means of contralateral suppression of $2f_1$–$2f_2$ distortion product otoacoustivemissions *Otology & Neurotology, 28,* 208–212.

Riga, M., Korres, S., Varvutsi, M., Kosmidis, H., Douniadakis, D., Psarommatis, I., . . . Ferekidis, E. (2007). Long-term effects of chemotherapy for acute lymphoblastic leukemia on the medial olivocochlear bundle: Effects of different cumulative doses of gentamicin. *International Journal of Pediatric Otorhinolaryngology, 71,* 1767–1773.

Riga, M., Papadas, T., Werner, J. A., & Dalchow, C. V. (2007). A clinical study of the efferent auditory system in patients with normal hearing who have acute tinnitus. *Otology & Neurotology, 28,* 185–190.

Riga, M., Psarommatis, I., Korres, S., Lyra, Ch., Papadeas, E., Varvutsi, M., Ferekidis, E., & Apostolopoulos, N. (2006). The effect of treatment with vincristine on transient evoked and distortion product otoacoustic emissions. *International Journal of Pediatric Otorhinolaryngology, 70,* 1003–1008.

Rita, M., & de Azevedo, M. F. (2005). Otoacoustic emissions and medial olivocochlear system: patients with tinnitus and no hearing loss [Article in Portuguese]. *Pro Fono, 17,* 283–292.

Robinette, M.S. (1992). Clinical observations with transient evoked otoacoustic emissions with adults. *Seminars in Hearing, 13,* 23–36.

Robinette, M. S., & Bauch, C. B. (1991, April). *Pre- and postoperative EOAE and ABR results on selected patients with eighth nerve tumors.* Poster presented at the Third Annual Convention of the American Academy of Audiology, Denver, CO.

Robinette, M. S., & Glattke, T. J. (Eds.). (2007). *Otoacoustic emissions: Clinical applications.* New York, NY: Thieme.

Rocha, E. B., Frasson de Azevedo, M., & Ximenes Filho, J. A. (2007). Study of the hearing in children born from pregnant women exposed to occupational noise: Assessment by distortion product otoacoustic emissions. *Review of Brasilian Otorrinolaringology (Engl. ed.), 73,* 359–369.

Rodríguez Domínguez, F. J., Cubillana Herrero, J. D., Cañizares Gallardo, N., & Pérez Aguilera, R. (2007).

Prevalence of auditory neuropathy: Prospective study in a tertiary-care center [Article in Spanish]. *Acta Otorrinolaringologica Espana, 58,* 239–245.

Rogers, A. R., Burke, S. R., Kopun, J. G., Tan, H., Neely, S. T., & Gorga, M. P. (2010). Influence of calibration method on distortion-product Otoacoustic emission measurements: II. Threshold prediction. *Ear and Hearing, 31,* 546–554.

Rosanowski, F., Eysholdt, U., & Hoppe, U. (2006). Influence of leisure-time noise on outer hair cell activity in medical students. *International Archives of Occupational Environmental Health, 80,* 25–31.

Rózańska-Kudelska, M., Chodynicki, S., Kinalska, I., & Kowalska, I. (2002). Hearing loss in patients with diabetes mellitus type II [Article in Polish]. *Otolaryngology Polska, 56,* 607–610.

Rubens, D. D., Vohr, B. R., Tucker, R., O'Neil, C. A., & Chung, W. (2008). Newborn oto-acoustic emission hearing screening tests: Preliminary evidence for a marker of susceptibility to SIDS. *Early Human Development, 84,* 225–259.

Ruggero, M. A., Kramek, B., & Rick, N. C. (1984). Spontaneous otoacoustic emissions in a dog. *Hearing Research, 13,* 293–296.

Ruggieri-Marone, M., & Schochat, E. (2007). Distortion product otoacoustic emissions in newborns treated by ototoxic drugs [Article in French]. *Review of Laryngology Otology Rhinology (Bordeaux), 128,* 41–46.

Ryan, S., & Piron, J. P. (1994). Functional maturation of the medial olivocochlear system in human neonates. *Acta Oto-Laryngologica, 144,* 485–489.

Sadri, M., Thornton, A. R., & Kennedy, C. R. (2007). Effects of maturation on parameters used for pass/fail criteria in neonatal hearing screening programmes using evoked otoacoustic emissions. *Audiology & Neurootology, 12,* 226–233.

Sahley, T. L., Nodar, R. H., & Musiek, F. E. (1999). *Efferent auditory system: Structure and function.* San Diego, CA: Singular Publishing Group.

Saitoh, Y., Sakoda, T., Hazama, M., Funakoshi, H., Ikeda, H., Shibano, A., . . . Kitano, H. (2006). Transient evoked otoacoustic emissions in newborn infants: Effects of ear asymmetry, gender, and age. *Journal of Otolaryngology, 35,* 133–138.

Sakashita, T., Kubo, T., Kyani, K., Ueno, K., Hikawa, C., Shibata, T., . . . Uyama, T. (2001). Changes in otoacoustic emission during the glycerol test in the ears of patients with Meniere's disease [Article in Japanese]. *Nippon Jibiinkoka Gakkai Kaiho, 104,* 682–693.

Sakashita, T., Minowa, Y., Hachikawa, K., Kubo, T., & Nakai, Y. (1991). Evoked otoacoustic emissions from ears with idiopathic sudden deafness. *Acta Otolaryngolica Supplement., 486,* 66–72.

Sakashita, T., Shibata, T., Yamane, H., & Hikawa, C. (2004). Changes in input/output function of distortion product otoacoustic emissions during the glycerol test in Ménière's disease. *Acta Otolaryngologica Supplement., 554,* 26–29.

Saleem, Y., Ramachandran, S., Ramamurthy, L., & Kay, N.J. (2007). Role of otoacoustic emission in children with middle-ear effusion and grommets. *Journal of Laryngology and Otology, 121,* 943–946.

Salvi, R. J., Ding, D., Wang, J., & Jiang, H. Y. (2000). A review of the effects of selective inner hair cell lesions on distortion product otoacoustic emissions, cochlear function and auditory evoked potentials. *Noise and Health, 2,* 9–26.

Salvinelli, F., D'Ascanio, L., & Casale, M. (2004). Staging rheumatoid arthritis: What about otoacoustic emissions? *Acta Otolaryngologica, 124,* 874–875.

Salvinelli, F., D'Ascanio, L., Casale, M., Vadacca, M., Rigon, A., & Afeltra, A. (2006). Auditory pathway in rheumatoid arthritis. A comparative study and surgical perspectives. *Acta Otolaryngologica, 126,* 32–36.

Salvinelli, F., Firrisi, L., Greco, F., Trivelli, M., & D'Ascanio, L. (2004). Preserved otoacoustic emissions in postparotitis profound unilateral hearing loss: A case report. *Annals of Otology, Rhinology and Laryngology, 113,* 887–890.

Sanches, S. G., & Carvallo, R. M. (2006). Contralateral suppression of transient evoked otoacoustic emissions in children with auditory processing disorder. *Audiology & Neurootology, 11,* 366–372.

Sano, M., Kaga, K., Kitazumi, E., & Kodama, K. (2005). Sensorineural hearing loss in patients with cerebral palsy after asphyxia and hyperbilirubinemia. *International Journal of Pediatric Otorhinolaryngology, 69,* 1211–1217.

Sano, M., Kitahara, N., & Kunikata, R. (2004). Progressive bilateral sensorineural hearing loss induced by an antithyroid drug. *ORL Journal of Otorhinolaryngology and Related Specialties, 66,* 281–285.

Santarelli, R., Cama, E., Scimemi, P., Dal Monte, E., Genovese, E., & Arslan, E. (2008). Audiological and electrocochleography findings in hearing-impaired children with connexin 26 mutations and otoacoustic emissions. *European Archives of Otorhinolaryngology, 265,* 43–51.

Santos, L., Morata, T. C., Jacob, L. C., Albizu, E., Marques, J. M., & Paini, M. (2007). Music exposure and audiological findings in Brazilian disc jockeys (DJs). *International Journal of Audiology, 46,* 223–231.

Saravanappa, N., Mepham, G. A., & Bowdler, D. A. (2005). Diagnostic tools in pseudohypacusis in children. *International Journal of Pediatric Otorhinolaryngology, 69,* 1235–1238.

Sardesai, M. G., Tan, A. K., & Fitzpatrick, M. (2003). Noise-induced hearing loss in snorers and their bed partners. *Journal of Otolaryngology, 32,* 141–145.

Satar, B., Kapkin, O., & Ozkaptan, Y. (2003). Evaluation of cochlear function in patients with normal hearing and tinnitus: a distortion product otoacoustic emission study (Article in Turkish]. *Kulak Burun Boğaz Ihtisas Dergisi, 10,* 177–182.

Schairer, K. S., Clukey, C., & Gould, H. J. (2000). Distortion product otoacoustic emissions to single and simultaneous tone pairs. *Journal of the American Academy of Audiololgy, 11,* 411–417.

Scheperle, R. A., Neely, S. T., Kopun, J. G., & Gorga, M. P. (2008). Influence of in situ, sound-level calibration on distortion-product otoacoustic emission variability. *Journal of the Acoustical Society of America, 124,* 288–300.

Schmuziger, N., Lodwig, A., & Probst, R. (2006). Influence of artifacts and pass/refer criteria on otoacoustic emission hearing screening. *International Journal of Audiology, 45,* 67–73.

Schmuziger, N., Patscheke, J., & Probst, R. (2006). Automated pure-tone threshold estimations from extrapolated distortion product otoacoustic emission (DPOAE) input/output functions. *Journal of the Acoustical Society of America, 119,* 1937–1939.

Schweinfurth, J. M., Cacace, A. T., & Parnes, S. M. (1997). Clinical applications of otoacoustic emissions in sudden hearing loss. *Laryngoscope, 107,* 1457–1463.

Seixas, N. S., Goldman, B., Sheppard, L., Neitzel, R., Norton, S., & Kujawa, S. G. (2005). Prospective noise induced changes to hearing among construction industry apprentices. *Occupational and Environmental Medicine, 62,* 309–317.

Seixas, N. S., Kujawa, S. G., Norton, S., Sheppard, L., Neitzel, R., & Slee, A. (2004). Predictors of hearing threshold levels and distortion product otoacoustic emissions among noise exposed young adults. *Occupational and Environmental Medicine, 61,* 899–907.

Serra, A., Cocuzza, S., Caruso, E., Mancuso, M., & La Mantia, I. (2003). Audiological range in Turner's syndrome. *International Journal of Pediatric Otorhinolaryngology, 67,* 841–845.

Sevik, O., Akdogan, O., Gocmen, E. S., Ozcan, K. M., Yazar, Z., & Dere, H. (2008). Auditory brainstem response and otoacoustic emissions in Duane retraction syndrome. *International Journal of Pediatric Otorhinolaryngology, 72,* 1167–1170.

Shahnaz, N. (2008). Transient evoked otoacoustic emissions (TEOAEs) in Caucasian and Chinese young adults. *International Journal of Audiology, 47,* 76–83.

She, W. D., Zhang, Q., Chen, F., Jiang, P., & Wang, J. (2004). Peri-uvulopalatopharyngoplasty otoacoustic emissions in patients with obstructive sleep apnea-hypopnea syndrome [Article in Chinese]. *Zhonghua Er Bi Yan Hou Ke Za Zhi, 39,* 48–51.

Shehata-Dieler, W., Völter, C., Hildmann, A., Hildmann, H., & Helms, J. (2007). Clinical and audiological findings in children with auditory neuropathy [Article in German]. *Laryngorhinootologie, 86,* 15–21.

Shera, C. A. (2003). Mammalian spontaneous otoacoustic emissions are amplitude-stabilized cochlear standing waves. *Journal of the Acoustical Society of America, 114,* 244–262.

Shera, C. A., & Guinan, J. J. Jr. (1999). Evoked otoacoustic emissions arise by two fundamentally difference mechanisms: A taxonomy for mammalian OAEs. *Journal of the Acoustical Society of America, 105,* 782–798.

Shera, C. A. & Zweig, G. (1992). Middle-ear phenomenology: The view from the three windows. *Journal of the Acoustical Society of America, 92,* 1356–1370.

Sheykholeslami, K., Schmerber, S., Habiby Kermany, M., & Kaga, K. (2005). Sacculo-collic pathway dysfunction accompanying auditory neuropathy. *Acta Otolaryngologica, 125,* 786–791.

Shi, G. Z., Gong, L. X., Nie, W. Y., Lin, Q., Xiang, L. L., Xu, X. H., & Qi, Y.S. (2005). Mutations of GJB2 gene in infants with non-syndromic hearing impairment [Article in Chinese]. *Zhonghua Yi Xue Za Zhi, 85,* 689–692

Shiomi, Y., Tsuji, J., Naito, Y., Fujiki, N., & Yamamoto, N. (1997). Characteristics of DPOAE audiogram in tinnitus patients. *Hearing Research, 108,* 83–88.

Shupak, A., Tal, D., Sharoni, Z., Oren, M., Ravid, A., & Pratt, H. (2007). Otoacoustic emissions in early noise-induced hearing loss. *Otology & Neurotology, 28,* 745–752.

Sideris, I., & Glattke, T. J. (2006). A comparison of two methods of hearing screening in preschool population. *Journal of Communication Disorders, 39,* 391–401.

Siegel, J. H. (1995). Cross-talk in otoacoustic emission probes. *Ear and Hearing, 16,* 150–158.

Siegel, J. H. (2005). The biophysical origin of Otoacoustic emissions. In A. L. Nuttal (Ed.), *Mechanics of hearing: Processes and models.* Singapore, Peoples Republic of China: World Scientific.

Siegel, J. H. (2007). Calibrating otoacoustic emission probes. In M. S. Robinette & T. J. Glattke (Eds.), *Otoacoustic emissions: Clinical applications.* New York, NY: Thieme.

Siegel, J. H., & Dreisbach, L. E. (1995). Optical placement of a probe tube in the occluded human ear canal. *Journal of the Association for Research in Otolaryngology, 18,* 119.

Sievert, U., Eggert, S., Goltz, S., & Pau, H.W. (2007). Effects of electromagnetic fields emitted by cellular phone on auditory and vestibular labyrinth [Article in German]. *Laryngorhinootologie, 86,* 264–270.

Silva, D. P., & Martins, R. H. (2009). Analysis of transient otoacoustic emissions and brainstem evoked auditory potentials in neonates with hyperbilirubinemia. *Brazilian Journal of Otorhinolaryngology, 75,* 381–386.

Sindhusake, D., Golding, M., Newall, P., Rubin, G., Jakobsen, K., & Mitchell, P. (2003). Risk factors for tinnitus in a population of older adults: The blue mountains hearing study. *Ear and Hearing, 24,* 501–507.

Sindhusake, D., Golding, M., Wigney, D., Newall, P., Jakobsen, K., & Mitchell, P. (2004). Factors predicting severity of tinnitus: A population-based assessment. *Journal of the American Academy of Audiology, 15,* 269–280.

Sininger, Y., & Starr, A. (2001). *Auditory neuropathy: A new perspective on hearing disorders.* New York, NY: Cengage.

Sisto, R., Chelotti, S., Moriconi, L., Pellegrini, S., Citroni, A., Monechi, V., . . . Moleti, A. (2007). Otoacoustic emission sensitivity to low levels of noise-induced hearing loss. *Journal of the Acoustical Society of America, 122,* 387–401.

Sliwinska-Kowalska, M., & Kotylo, P. (2001). Otoacoustic emissions in industrial hearing loss assessment. *Noise and Health, 3,* 75–84.

Sliwinska-Kowalska, M., & Kotylo, P. (2002). Occupational exposure to noise decreases otoacoustic emission efferent suppression. *International Journal of Audiology, 41,* 113–119.

Sliwinska-Kowalska, M., Kotylo, P., & Hendler, B. (1999). Comparing changes in transient-evoked otoacoustic emission and pure-tone audiometry following short exposure to industrial noise. *Noise and Health, 1,* 50–57.

Sindhusake, D., Golding, M., Newall, P., Rubin, G., Jakobsen, K., & Mitchell, P. (2003). Risk factors for tinnitus in a population of older adults: the Blue Mountains Hearing Study. *Ear and Hearing, 24,* 501–507.

Smurzynski, J., & Probst R. (1999). Intensity discrimination, temporal integration, and gap detection by normally hearing subjects with weak and strong otoacoustic emissions. *Audiology, 38,* 251–256.

Sockalingam, R., Kei, J., & Ho, C.D. (2007). Test-retest reliability of distortion-product otoacoustic emissions in children with normal hearing: A preliminary study. *International Journal of Audiology, 46,* 351–354.

Somers, T., Offeciers, F. E., & Schatteman, I. (2003). Results of 100 vestibular schwannoma operations. *Acta Otorhinolaryngologica Belgium, 57,* 155–156.

Sone, M., Katayama, N., Otake, N., Sato, E., Fujimoto, Y., Ito, M., & Nakashima, T. (2007). Characterizing the auditory changes in tumor metastasis to the bilateral internal auditory canals. *Journal of Clinical Neuroscience, 14,* 470–473.

Spankovich, C., & Lustig, L. R. (2007). Restoration of brain stem auditory-evoked potential in maple syrup urine disease. *Otology & Neurotology, 28,* 566–569.

Spektor, Z., Leonard, G., Kim, D. O., Jung, M. D., & Smurzynski, J. (1991). Otoacoustic emissions in normal and hearing impaired children and adults. *Laryngoscope, 101,* 965–976.

Stanton, S. G., Ryerson, E., Moore, S. L., Sullivan-Mahoney, M., & Couch, S. C. (2005). Hearing screening outcomes in infants of pregestational diabetic mothers. *American Journal of Audiology, 14,* 86–93.

Starr, A., Isaacson, B., Michalewski, H. J., Zeng, F. G., Kong, Y. Y., Beale, P., . . . Lesperance, M. M. (2004). A dominantly inherited progressive deafness affecting distal auditory nerve and hair cells. *Journal of the Association of Research in Otolaryngology, 5,* 411–426.

Starr, A., Picton, T. W., Sininger, Y., Hood, L. J., & Berlin, C. I. (1996). Auditory neuropathy. *Brain, 119,* 741–753.

Starr, A., Sininger, Y., Nguyen, T., Michalewski, H. J., Oba, S., & Abdala, C. (2001). Cochlear receptor (microphonic and summating potentials, otoacoustic emissions) and auditory pathway (auditory brainstem potentials) activity in auditory neuropathy. *Ear and Hearing, 22,* 91–99.

Starr, A., Sininger, Y. S., & Pratt, H. (2000). The varieties of auditory neuropathy. *Journal of Basic Clinical Physiology Pharmacology, 11,* 215–230.

Stavroulaki, P., Vossinakis, I.C., Dinopoulou, D., Doudounakis, S., Adamopoulos, G., & Apostolopoulos, N. (2002). Otoacoustic emissions for monitoring aminoglycoside-induced ototoxicity in children with cystic fibrosis. *Archives of Otolaryngology-Head and Neck Surgery, 128,* 150–155.

Stenklev, N. C., & Laukli, E. (2003). Transient evoked otoacoustic emissions in the elderly. *International Journal of Audiology, 42,* 132–139.

Stinson, M. R. (1985). The spatial distribution of sound pressure within scaled replicas of the human ear canal. *Journal of the Acoustical Society of America, 78,* 1596–1602.

Stinson, M. R., Shaw, E. A., & Lawton, B. W. (1982). Estimation of acoustical energy reflectance at the eardrum from measurements of pressure distribution in the human ear canal. *Journal of the Acoustical Society of America, 72,* 766–773.

Størmer, C. C., & Stenklev, N. C. (2007). Rock music and hearing disorders [Article in Norwegian]. *Tidsskr Nor Laegeforen, 29,* 874–877.

Stover, L., & Norton, S. J. (1993). The effects of aging on otoacoustic emissions. *Journal of the American Acoustical Society of America, 93,* 3308–3319.

Strenzke, N., Pauli-Magnus, D., Meyer, A., Brandt, A., Maier, H., & Moser, T. (2008). Update on physiology and pathophysiology of the inner ear: Pathomechanisms of sensorineural hearing loss [Article in German]. *HNO, 56,* 27–36.

Strickland, E. A., Burns, E. M., & Tubis, A. (1985). Incidence of spontaneous otoacoustic emissions in children and infants. *Journal of the Acoustical Society of America, 78,* 931–935.

Strumberg, D., Brügge, S., Korn, M. W., Koeppen, S., Ranft, J., Scheiber, G., . . . Scheulen, M. E. (2002). Evaluation of long-term toxicity in patients after cisplatin-based chemotherapy for non-seminomatous testicular cancer. *Annals of Oncology, 13,* 187–189.

Stuart, A., Jones, S. M., & Walker, L.J. (2006). Insights into elevated distortion product otoacoustic emissions in sickle cell disease: Comparisons of hydroxyurea-treated and non-treated young children. *Hearing Research, 212,* 83–89.

Stuart, A., & Mills, K. N. (2009). Late-onset unilateral auditory neuropathy/dysynchrony: A case study. *Journal of the American Academy of Audiology, 20,* 172–179.

Sułkowski, W. J., Kowalska, S., Matyja, W., Guzek, W., Wesołowski, W., Szymczak, W., & Kostrzewski, P. (2002). Effects of occupational exposure to a mixture of solvents on the inner ear: A field study. *International Journal of Occupational Medicine and Environmental Health, 15,* 247–256.

Sun, X. M. (2008a). Contralateral suppression of distortion product otoacoustic emissions and the middle-ear muscle reflex in human ears. *Hearing Research, 237,* 66–75.

Sun, X. M. (2008b). Distortion product otoacoustic emission fine structure is responsible for variability of distortion product otoacoustic emission contralateral suppression. *Journal of the Acoustical Society of America, 123,* 4310–4320.

Süslü, A. E., Polat, M., Köybaşi, S., Biçer, Y. O., Funda, Y. O., & Parlak, A. H. (2010). Inner ear involvement in Behçet's disease. *Auris Nasus Larynx, 37,* 286–290.

Swanepoel, D., & Hall, J. W., III. (2010). Football match spectator sound exposure and effect on hearing: A pretest-post-test study. *South African Medical Journal, 100,* 239–242.

Sztuka, A., Pośpiech, L., Gawron, W., & Dudek, K. (2006). DPOAE in tinnitus patients with cochlear hearing loss considering hyperacusis and misophonia [Article in Polish]. *Otolaryngology Polska, 60,* 765–772.

Sztuka, A., Pospiech, L., Gawron, W., & Dudek, K. (2010). DPOAE in estimation of the function of the cochlea in tinnitus patients with normal hearing. *Auris Nasus Larynx, 37,* 55–60.

Tadros, S. F., Frisina, S. T., Mapes, F., Kim, S., Frisina, D. R,. & Frisina, R. D. (2005). Loss of peripheral right-ear advantage in age-related hearing loss. *Audiology & Neurootology, 10,* 44–52.

Takeda, T., Kakigi, A., Takebayashi, S., Ohono, S., Nishioka, R., & Nakatani, H. (2010). Narrow-band evoked oto-acoustic emission from ears with normal pathologic conditions. *ORL Journal of Otorhinolaryngology and Related Specialties, 71,* 41–56.

Talmadge, C. L., Long, G. R., Tubis, A., & Dhar, S. (1999). Experimental confirmation of the two-source interference model for the fine structure of distortion product otoacoustic emissions. *Journal of the Acoustical Society of America, 105,* 275–292.

Talmadge, C. L., Tubis, A., Long, G. R., & Piskorski, P. (1998). Modeling otoacoustic emission and hearing threshold fine structures. *Journal of the Acoustical Society of America, 104,* 1517–1543.

Talmadge, C. L., Tubis, A., Long, G. R., & Tong, C. (2000). Modeling the combined effects of basilar membrane nonlinearity and roughness on stimulus frequency otoacoustic emission fine structure. *Journal of the Acoustical Society of America, 108,* 2911–2932.

Tamagawa, Y., Ishikawa, K., Ishikawa, K., Ishida, T., Kitamura, K., Makino, S., . . . Ichimura, K. (2002). Phenotype of DFNA11: A nonsyndromic hearing loss caused by a myosin VIIA mutation. *Laryngoscope, 112,* 292–297.

Tang, T. P., McPherson, B., Yuen, K. C., Wong, L. L., & Lee, J. S. (2004). Auditory neuropathy/auditory dys-synchrony in school children with hearing loss: Frequency of occurrence. *International Journal of Pediatric Otorhinolaryngol., 68,* 175–183.

Tanon-Anoh, M. J., Sanogo-Gone, D., & Kouassi, K. B. (2010). Newborn hearing screening in a developing country: results of a pilot study in Abidjan, Côte d'Ivoire. *International Journal of Pediatric Otorhinolaryngology, 74,* 188–191.

Tas, A., Yagiz, R., Tas, M., Esme, M., Uzun, C., & Karasalihoglu, A. R. (2007). Evaluation of hearing in children with autism by using TEOAE and ABR. *Autism, 11,* 73–79.

Tas, A., Yagiz, R., Uzun, C., Adali, M. K., Koten, M., Tas, M., & Karasalihoglu, A. R. (2004). Effect of middle ear effusion on distortion product otoacoustic emission. *International Journal of Pediatric Otorhinolaryngology, 68,* 437–440.

Tatli, M. M., Bulent Serbetcioglu, M., Duman, N., Kumral, A., Kirkim, G., Ogun, B., & Ozkan, H. (2007). Feasibility of neonatal hearing screening program with two-stage transient otoacoustic emissions in Turkey. *Pediatrics International, 49,* 161–166.

Taylor, C. L., & Brooks, R. P. (2000). Screening for hearing loss and middle ear disorders in children using TEOAEs. *American Journal of Audiology, 9,* 50–55.

Tekin, M., Akcayoz, D., & Incesulu, A. (2005). A novel missense mutation in a C2 domain of OTOF results in autosomal recessive auditory neuropathy. *American Journal of Medical Genetics, 138,* 6–10.

Telischi, F. F., Widick, M. P., Lonsbury-Martin, B. L., & McCoy, M. J. (1995). Monitoring cochlear function intraoperatively using distortion product otoacoustic emissions. *American Journal of Otology, 16,* 597–608.

Thabet, E. M. (2010). Transient evoked otoacoustic emissions in superior canal dehiscence syndrome. *European Archives of Otorhinolaryngology.* Epub ahead of publication 2010 June 26.

Tharpe, A. M., Bess, F. H., Sladen, D. P., Schissel, H., Couch, S., & Schery, T. (2006). Auditory characteristics of children with autism. *Ear and Hearing, 27,* 430–441.

Thornton, A. R. (1993). Click-evoked otoacoustic emissions: new techniques and applications, *British Journal of Audiology, 27,* 109–115.

Thornton, A. R., Marotta, N., & Kennedy, C.R. (2003). The order of testing effect in otoacoustic emissions and its consequences for sex and ear differences in neonates. *Hearing Research, 184,* 123–130.

Tilanus, S. C., Stenis, D. V., & Snik, A. F. M. (1995). Otoacoustic emission measurements in evaluation of the immediate effect of ventilation tube insertion in children. *Annals of Otology, Rhinology, and Laryngology, 104,* 297–300.

Tognola, G., Parazzini, M., de Jager, P., Brienesse, P., Ravazzani, P., & Grandori, F. (2005). Cochlear maturation and otoacoustic emissions in preterm infants: a time-frequency approach. *Hearing Research, 199,* 71–80.

Topolska, M. M., Hassmann-Poznańska, E., & Sołowiej, E. (2002). Assessment of hearing in children with infantile cerebral palsy. Comparison of psychophysical and electrophysical examination [Article in Polish]. *Otolaryngology Polska, 56,* 467–474

Toral-Martiñón, R., Poblano, A., Collado-Corona, M. A., & González, R. (2003). Effects of cisplatin on auditory function in children with cancer. Otoacoustic emission evaluation [Article in Spanish]. *Mexico Medical Gazette, 139,* 529–534.

Toral-Martiñón, R., Shkurovich-Bialik, P., Collado-Corona, M. A., Mora-Magaña, I., Goldgrub-Listopad. S., & Shkurovich-Zaslavsky, M. (2003). Distortion product otoacoustic emissions test is useful in children undergoing cisplatin treatment. *Archives of Medical Research, 34,* 205–208.

Torre III, P., Cruickshanks, K. J., Klein, B. E., Klein, R., & Nondahl, D. M. (2005). The association between cardiovascular disease and cochlear function in older adults. *Journal of the Speech, Language and Hearing Research, 48,* 473–481.

Torre III, P., Cruickshanks, K. J., Nondahl, D. M., & Wiley, T. L. (2003). Distortion product otoacoustic emission response characteristics in older adults. *Ear and Hearing, 24,* 20–29.

Torre III, P., Dreisbach, L. E., Kopke, R., Jackson, R., & Balough, B. (2007). Risk factors for distortion product otoacoustic emissions in young men with normal hearing. *American Academy of Audiology, 18,* 724, 749–759.

Torre III, P., & Howell, J. C. (2008). Noise levels during aerobics and the potential effects on distortion product otoacoustic emissions. *Journal of Communicative Disorders, 41,* 501–511.

Torrico, P., Gómez, C., López-Ríos, J., de Cáceres, M. C., Trinidad, G., & Serrano, M. (2004). Age influence in otoacoustic emissions for hearing loss screening in infants [Article in Spanish]. *Acta Otorrinolaringologica Espana, 55,* 153–159.

Triebig, G., Bruckner, T., & Seeber, A. (2008). Occupational styrene exposure and hearing loss: A cohort study with repeated measurements. *International Archives of Occupational and Environmental Health.* Epub ahead of print 2008 Sept. 2.

Trine, M. B., Hirsch, J. E., & Margolis, R. H. (1993). Effects of middle ear pressure on evoked otoacoustic emissions. *Ear and Hearing, 14,* 401–407.

Trnovec, T., Sovcíková, E., Pavlovcinová, G., Jakubíková, J., Jusko, T. A., Husták, M., . . . Wimmerová, S. (2010). Serum PCB concentrations and cochlear function in 12-year-old children. *Environmental Science and Technology, 44,* 2884–2889.

Truong, M.T., Winzelberg, J., & Chang, K.W. (2007). Recovery from cisplatin-induced ototoxicity: A case report and review. *International Journal of Pediatric Otorhinolaryngology, 71,* 1631–1638.

Truy, E., Ionescu, E., Lina-Granade, G., Butnaru, C., Thai-Van, H., Furminieux, V., & Collet, L. (2005). Auditory neuropathy: clinical presentation of seven cases and review of the literature [Article in French]. *Annals of Otolaryngology Chirugia Cervicofacial, 122,* 303–314.

Truy, E., Veuillet, E., Collet, L., & Morgon, A. (1993). Characteristics of transient otoacoustic emissions in patients with sudden idiopathic hearing loss. *British Journal of Audiology, 27,* 379–385.

Uchida, Y., Ando, F., Nakata, S., Ueda, H., Nakashima, T., Niino, N., & Shimokata, H. (2006). Distortion product otoacoustic emissions and tympanometric measurements in an adult population-based study. *Auris Nasus Larynx, 33*, 397–401.

Uchida, Y., Ando, .F, Shimokata, H., Sugiura, S., Ueda, H., & Nakashima, T. (2008). The effects of aging on distortion-product otoacoustic emissions in adults with normal hearing. *Ear and Hearing, 29*, 176–184.

Ugur, A. K., Kemaloglu, Y. K., Ugur, M. B., Gunduz, B., Saridogan, C., Yesilkaya, E., . . . Goksu, N. (2009). Otoacoustic emissions and effects of contralateral white noise stimulation on transient evoked otoacoustic emissions in diabetic children. *International Journal of Pediatric Otorhinolaryngology, 73*, 555–559.

Urban, I., Namysłowski, G., Morawski, K., & Wojtacha, M. (2004). Otoneurologic symptoms associated with Arnold-Chiari syndrome type I [Article in Polish]. *Otolaryngology Polska, 58*, 281–288.

Urbaniec, N., Namysłowski, G., Morawski, K., Urbaniec, P., Turecka, L., & Bazowska, G. (2004). Effects of intrauterine hypotrophy and perinatal hypoxia on cochlear function evaluated by click evoked otoacoustic emissions TEOAE) [Article in Polish]. *Otolaryngology Polska, 58*, 365–372.

Van Huffelen, W. M., Mateijsen, N. J., & Wit, H. P. (1998). Classification of patients with Ménière's disease using otoacoustic emissions. *Audiology & Neurotology, 3*, 419–430.

van Looij, M. A., Meijers-Heijboer, H., Beetz, R., Thakker, R. V., Christie, P. T., Feenstra, L. W., & van Zanten, B. G. (2006). Characteristics of hearing loss in HDR (hypoparathyroidism, sensorineural deafness, renal dysplasia) syndrome. *Audiology & Neurootology, 11*, 373–379.

Van Zyl, A., Swanepoel, D., & Hall, J. W. III. (2009). Effect of prolonged contralateral acoustic stimulation on transient evoked otoacoustic emissions. *Hearing Research, 254*, 77–81.

Vento, B. A., Durrant, J. D., Sabo, D. L. & Boston, J. R. (2004). Development of f2/f1 ratio functions in humans. *Journal of the Acoustical Society of America, 115*, 2138–2147.

Verde, P., Marciano, E., De Falco, R., Testa, R., Buonamassa, S., & Mariniello, G. (2003). Objective pulsatile tinnitus: Case report. *Acta Otorhinolaryngology Italia, 23*, 383–387.

Verhulst, S., Harte, J. M., & Dau, T. (2007). Temporal suppression of long-latency click-evoked otoacoustic emissions. *Conference Proceedings of the IEEE Engineering Medicine Biological Society*, 2007, pp. 1932–1936

Verhulst, S., Harte, J. M., & Dau, T. (2008). Temporal suppression and augmentation of click-evoked otoacoustic emissions. *Hearing Research, 246*, 23–35.

Viveiros, C. M., Pereira, L. D., & Kirsztajn, G. M. (2006). Auditory perception in Alport's syndrome. *Review of Brasilian Otorrinolaringology (Engl. ed.), 72*, 811–816.

Vlastarakos, P. V., Nikolopoulos, T. P., Tavoulari, E., Papacharalambous, G., & Korres, S. (2008). Auditory neuropathy: endocochlear lesion or temporal processing impairment? Implications for diagnosis and management. *International Journal of Pediatric Otorhinolaryngology, 72*, 1135–1150.

Voss, S. E., Adegoke, M. F., Horton, N. J., Sheth, K. N., Rosand, J., & Shera, C. A. (2010). Posture systematically alters ear-canal reflectance and DPOAE properties. *Hearing Research, 263*, 43–51.

Voss, S. E., & Allen, J. B. (1994). Measurement of acoustic impedance and reflectance in the human ear canal. *Journal of the Acoustical Society of America, 95*, 372–384.

Voss, S. E., Horton, N. J., Tabucchi, T. H., Folowosele, F. O., & Shera, C. A. (2006). Posture-induced changes in distortion-product otoacoustic emissions and the potential for noninvasive monitoring of changes in intracranial pressure. *Neurocritical Care, 4*, 251–257.

Wable, J., & Collet, L. (1994). Can synchronized otoacoustic emissions really be attributed to SOAEs?. *Hearing Research, 80*, 141–145.

Wagner, W., Frey, K., Heppelmann, G., Plontke, S. K., & Zenner, H. P. (2008). Speech-in-noise intelligibility does not correlate with efferent olivocochlear reflex in humans with normal hearing. *Acta Otolaryngologica, 128*, 53–60.

Wagner, W., Heppelmann, G., Kuehn, M., Tisch, M., Vonthein, R., & Zenner, H. P. (2005). Olivocochlear activity and temporary threshold shift-susceptibility in humans. *Laryngoscope, 115*, 2021–2028.

Wagner, W., Heppelmann, G., Vonthein, R., & Zenner, H.P. (2008). Test-retest repeatability of distortion product otoacoustic emissions. *Ear and Hearing, 29*, 378–391.

Wagner, W., Plinkert, P. K., Vonthein, R., & Plontke, S. K. (2008). Fine structure of distortion product otoacoustic emissions: Its dependence on age and hearing threshold and clinical implications. *European Archives of Otorhinolaryngology, 265*, 1165–1172.

Wagner, W., Staud, I., Frank, G., Dammann, F., Plontke, S., & Plinkert, P.K. (2003). Noise in magnetic resonance imaging: no risk for sensorineural function but increased amplitude variability of otoacoustic emissions. *Laryngoscope, 113*, 1216–1223.

Walker, L. J., Stuart, A., & Green, W. B. (2004). Outer and middle ear status and distortion product otoacoustic emissions in children with sickle cell disease. *American Journal of Audiology, 13,* 164–172.

Wang, Bu, X., Zhou, A., Xing, G., & Shi, Q. (2007). Auditory neuropathy in deaf school students [Article in Chinese]. *Lin Chung Er Bi Yan Hou Tou Jing Wai Ke Za Zhi, 21,* 457–459.

Wang, D. Y., Bu, X. K., Xing, G. Q., & Lu, L. (2003). Neurophysiological characteristics of infants and young children with auditory neuropathy [Article in Chinese]. *Zhonghua Yi Xue Za Zhi, 83,* 281–284.

Wang, H., Jiang, C., Zhou, F., Wu, P., Zhang, S., Liu, W., & Huang, Y. (2004). A clinical study about occupational noise-induced hearing loss measured and diagnosed with transient evoked otoacoustic emissions [Article in Chinese]. *Lin Chuang Er Bi Yan Hou Ke Za Zhi, 18,* 209–211.

Wang, H. T., Zhong, N. C. (2000). Effects of selective attention on distortion product otoacoustic emissions: depend on parameters [Article in Chinese]. *Lin Chuang Er Bi Yan Hou Ke Za Zhi, 14,* 14–16.

Wang, J., Duan, J., Li, Q,. Huang, X., Chen, H., Jin, J., Gong, S., & Kong, W. (2002). Audiological characteristics of auditory neuropathy [Article in Chinese]. *Zhonghua Er Bi Yan Hou Ke Za Zhi, 37,* 252–255.

Wang, J., Gao, L., Xue, F., Meng, M., Zha, D., & Deng, Y. (2002). Auditory neuropathy [Article in Chinese]. *Lin Chuang Er Bi Yan Hou Ke Za Zhi, 16,* 518–520.

Wang, J., Shi, L., Gao, L., Xie, J., & Han, L. (2007). Audiological characteristics of unilateral auditory neuropathy: 11 case study [Article in Chinese]. *Lin Chung Er Bi Yan Hou Tou Jing Wai Ke Za Zhi, 21,* 436–440.

Wang, W. Q., & Wu, L. W. (2000). Neurootologic study of acoustic neuroma [Article in Chinese]. *Lin Chuang Er Bi Yan Hou Ke Za Zhi, 14,* 149–151.

Wang, Y. (2003). The study of distortion products otoacoustic emissions in pseudo-anacousia [Article in Chinese]. *Fa Yi Xue Za Zhi, 19,* 22–23, 26.

Wang, Y. P., Wang, M. C., Lin, H. C., & Lee, K.S. (2006). Conversion deafness presenting as sudden hearing loss. *Journal of the Chinese Medical Association, 69,* 289–293.

Wecker, H., & Laubert, A. (2004). Reversible hearing loss in acute salicylate intoxication [Article in German]. *HNO, 52,* 347–351.

Whitehead, M. L., Lonsbury-Martin, B. L., & Martin, G. K. (1992a). Evidence for two discrete sources of 2f1-f2 distortion-product otoacoustic emission in rabbit. I: Differential dependence on stimulus parameters. *Journal of the Acoustical Society of America, 91,* 1587–1607.

Whitehead, M. L., Lonsbury-Martin, B. L., & Martin, G. K. (1992b). Evidence for two discrete sources of 2f1-f2 distortion-product otoacoustic emission in rabbit.

II: Differential physiologic vulnerability. *Journal of the Acoustical Society of America, 92,* 2662–2682.

Widziszowska, A., Namysłowski, G., Genge, A., Buczyńska, G., Hajduk, A., & Godula-Stuglik, U. (2005). Assessment of cochlea activity in a group of newborns with central nervous system impairment as an effect of perinatal asphyxia using click-evoked otoacoustic emissions CEOAEs) [Article in Polish]. *Pol Merkur Lekarski, 19,* 312–314.

Wimmer, E., Toleti, B., Berghaus, A., Baumann, U., & Nejedlo, I. (2010). Impedance audiometry in infants with a cleft palate: The standard 226-Hz probe tone has no predictive value for the middle ear condition. *International Journal of Pediatric Otorhinolaryngology, 74,* 586–590.

Wittekindt, A., Gaese, B. H., & Kössl, M. (2009). Influence of contralateral acoustic stimulation on the quadratic distortion product f2-f1 in humans. *Hearing Research, 247,* 27–33.

Viveiros, C. M., Pereira, L. D., & Kirsztajn, G. M. (2006). Auditory perception in Alport's syndrome. *Review of Brasilian Otorrinolaringology (Engl. ed.), 72,* 811–816.

Xiao, D. J., Zhang, Y. S., & Zhang, Y. H. (2006). Exploration of the auditory effects of diabetes mellitus using distortion product otoacoustic emissions [Article in Chinese]. *Zhonghua Er Bi Yan Hou Tou Jing Wai Ke Za Zhi, 41,* 924–927.

Xing, G., Bu, X., Yan, M., Lu, L., & Yang, S. (2000). Audiological findings and mitochondrial DNA mutation in a large family with matrilineal sensorineural hearing loss [Article in Chinese]. *Zhonghua Er Bi Yan Hou Ke Za Zhi, 35,* 98–101.

Xing, G., Cao, X., Tian, H., Chen, Z., Li, X., Wei, Q., & Bu, X. (2007). Clinical and genetic features in a Chinese pedigree with autosomal dominant auditory neuropathy. *ORL Journal of Otorhinolaryngology and Related Specialties, 69,* 131–136.

Xoinis, K., Weirather, Y., Mavoori, H., Shaha, S. H., & Iwamoto, L. M. (2007). Extremely low birth weight infants are at high risk for auditory neuropathy. *Journal of Perinatology, 27,* 718–723.

Xu, F. L., Xing, Q. J., & Cheng, X. Y. (2008). A comparison of auditory brainstem responses and otoacoustic emissions in hearing screening of high-risk neonates [Article in Chinese]. *Zhongguo Dang Dai Er Ke Za Zhi, 10,* 460–463.

Xu, J., Huang, W., Liu, G., Zhou, J., & Gao, B. (2005). Patterns of hearing disorders in normal otoacoustic emissions [Article in Chinese]. *Lin Chuang Er Bi Yan Hou Za Zhi, 19,* 1023–1025.

Xue, X., & Zhong, N. (2002). A study on the effects of contralateral acoustic suppression to transient evoked otoacoustic emissions after noise exposure

[Article in Chinese]. *Lin Chuang Er Bi Yan Hou Ke Za Zhi, 16,* 164–165, 168.

Yalçınkaya, F., Yılmaz, S. T., & Muluk, N. B. (2009). Transient evoked otoacoustic emissions and contralateral suppressions in children with auditory listening problems. *Auris Nasus Larynx.* Epub ahead of print 2009 May 2.

Yardley, M. P., Davies, C. M., & Stevens, J. C. (1998). Use of transient otoacoustic emissions to detect and monitor cochlear damage caused by platinum-containing drugs. *British Journal of Audiology, 32,* 305–316.

Yates, G. K., & Withnell, R. H. (1999). The role of intermodulation distortion in transient-evoked otoacoustic emissions. *Hearing Research, 136,* 49–64.

Yeo, S. W., Park, S. N., Park, Y. S., & Suh, B. D. (2002). Effect of middle-ear effusion on otoacoustic emissions. *Journal of Laryngology and Otology, 116,* 794–799.

Yeo, S. W., Park, S. N., Park, Y. S., & Suh, B. D. (2003). Prognostic value of otoacoustic emissions in children with middle ear effusion. *Otolaryngology-Head and Neck Surgery, 129,* 136–140.

Yildirim, C., Yağiz, R., Uzun, C., Taş, A., Bulut, E., & Karasalihoğlu, A. (2006). The protective effect of oral magnesium supplement on noise-induced hearing loss [Article in Turkish]. *Kulak Burun Boğaz Ihtisas Dergisi, 16,* 29–36.

Yilmaz, I., Erbek, S., Erbek, S., Ulusoy, O., & Calişaneller, T. (2007). Sudden hearing loss in a patient with a 3-mm acoustic tumor [Article in Turkish]. *Kulak Burun Boğaz Ihtisas Dergisi, 17,* 120–125.

Yilmaz, I., Yilmazer, C., Erkan, A. N., Aslan, S. G., & Ozluoglu, L. N. (2005). Intratympanic dexamethasone injection effects on transient-evoked otoacoustic emission. *American Journal of Otolaryngology, 26,* 113–117.

Yilmaz, M., Baysal, E., Gunduz, B., Aksu, A., Ensari, N., Meray, J., & Bayazit, Y. A. (2005). Assessment of the ear and otoacoustic emission findings in fibromyalgia syndrome. *Clinical and Experimental Rheumatology, 23,* 701–703.

Yilmaz, S., Karasalihoglu, A. R., Tas, A., Yagiz, R., & Tas, M. (2006). Otoacoustic emissions in young adults with a history of otitis media. *Laryngology and Otology, 120,* 103–107.

Yilmaz, S. T., Sennaroglu, G., Sennaroglu, L., & Köse, S. K. (2007). Effect of age on speech recognition in noise and on contralateral transient evoked otoacoustic emission suppression. *Journal of Laryngology and Otology, 121,* 1029–1034.

Yilmaz, Y. F., Aytas, F. I., Akdogan, O., Sari, K., Savas, Z. G., Titiz. A., . . . Unal, A. (2008). Sensorineural hearing loss after radiotherapy for head and neck tumors: A prospective study of the effect of radiation. *Otology & Neurotology, 29,* 461–463.

Yin, L., Bottrell, C., Clarke, N., Shacks, J., & Poulsen, M. K. (2009). Otoacoustic emissions: A valid, efficient first-line hearing screen for preschool children. *Journal of School Health, 79,* 147–152.

Yu, H., Ke, X. M., Yu, D. L., Li, Q., Liu, Y. H., & Ding, G. Y. (2000). Measuring and analyzing of otoacoustic emission tests of a family with genetic progressive sensorineural hearing loss [Article in Chinese]. *Lin Chuang Er Bi Yan Hou Ke Za Zhi, 14,* 157–159.

Zang, Z., & Jiang, Z. D. (2007). Distortion product otoacoustic emissions during the first year in term infants: A longitudinal study. *Brain Development, 6,* 346–351.

Zang, Z., Wilkinson, A. R., & Jiang, Z. D. (2008). Distortion product otoacoustic emissions at 6 months in term infants after perinatal hypoxia-ischaemia or with a low Apgar score. *European Journal of Pediatrics, 167,* 575–578.

Zaouche, S., Ionescu, E., Dubreuil, C., & Ferber-Viart, C. (2005). Pre- and intraoperative predictive factors of facial palsy in vestibular schwannoma surgery. *Acta Otolaryngologica., 125,* 363–369.

Zeng, X., Zhong, N., & Li, Q. (2006). The diagnostic value of TEOAE-glycerine test in patients with vertigo on the first attack [Article in Chinese]. *Lin Chuang Er Bi Yan Hou Ke Za Zhi, 20,* 16–18, 25.

Zhang, F., Boettcher, F. A., & Sun, X. M. (2007). Contralateral suppression of distortion product otoacoustic emissions: Effect of the primary frequency in Dpgrams. *International Journal of Audiology, 46,* 187–195.

Zhang, J., Zhou, H., Xu, Y., & Zhang, G. (2009). A study on relationship between distortion product otoacoustic emissions and therapeutic effects in tinnitus [Article in Chinese]. *Lin Chung Er Bi Yan Hou Tou Jing Wai Ke Za Zhi, 23,* 591–593.

Zhang, Q., Deng, Y., Xing, G., & Cheng, Z. (1999). Observation of distortion product otoacoustic emissions in recovery course of sudden deafness [Article in Chinese]. *Lin Chuang Er Bi Yan Hou Ke Za Zhi, 13,* 443–445.

Zhang, Q. R., Lei, Z. X., Shi, J. H., Wang, K. X., Huang, J., Wu, H. F., & Xiong, X. (2000). Detection of distortion-product otoacoustic emissions in well-drilling workers [Article in Chinese]. *Lin Chuang Er Bi Yan Hou Ke Za Zhi, 14,* 78–80.

Zhang, V. W., McPherson, B., & Zhang, Z. G. (2008). Tone burst-evoked otoacoustic emissions in neonates: normative data. *BMC Ear Nose and Throat Disorders, 8,* 3.

Zhang, Y., Qin, Y., Zhang, Y., & Pan, S. (2004). The effects of tympanotomy tube insertion on distortion product otoacoustic emissions [Article in Chinese]. *Lin Chuang Er Bi Yan Hou Ke Za Zhi, 18,* 408–410.

Zhang, Y., Zhang, X., Zhu, W., Zheng, X., & Deng, X. (2004). Distortion product of otoacoustic emissions as a sensitive indicator of hearing loss in pilots. *Aviation Space Environmental Medicine, 75,* 46–48.

Zhao, F., Wada, H., Koike, T., Ohyama, K., Kawase, T., & Stephens, D. (2003). Transient evoked otoacoustic emissions in patients with middle ear disorders. *International Journal of Audiology, 42,* 117–131.

Zhao, F., & Stephens, D. (2006). Distortion product otoacoutic emissions in patients with King-Kopetsky syndrome. *International Journal of Audiology, 45,* 34–39.

Zhong, N., Li, Q., & Guo, Y. (2001). Effects of selective attention and contralateral acoustic stimulation on latency of distortion product otoacoustic emissions [Article in Chinese]. *Lin Chuang Er Bi Yan Hou Ke Za Zhi, 15,* 437–438.

Zorwka, P. G. (1993). Otoacoustic emissions: A new method to diagnose hearing impairment in children. *European Journal of Pediatrics, 152,* 626–634.

Zou, J., Bi, A., Yang, W., Zhou, Y., Ding, W., & Hao, X. (2000). An observation on long-term influence of middle ear bacterial infection on inner ear function and systemic immune reaction [Article in Chinese]. *Zhonghua Er Bi Yan Hou Ke Za Zhi, 35,* 196–199.

Zwicker, E. (1990). On the frequency separation of simultaneously evoked otoacoustic emissions' consecutive extrema and its relation to cochlear traveling waves. *Journal of the Acoustical Society of America, 88,* 1639–1641.

Index